Lecture Notes in Networks and Systems

Volume 150

The series "Lecture Notes in Networks and Systems" publishes the latest developments in Networks and Systems—quickly, informally and with high quality. Original research reported in proceedings and post-proceedings represents the core of LNNS.

Volumes published in LNNS embrace all aspects and subfields of, as well as new challenges in, Networks and Systems.

The series contains proceedings and edited volumes in systems and networks, spanning the areas of Cyber-Physical Systems, Autonomous Systems, Sensor Networks, Control Systems, Energy Systems, Automotive Systems, Biological Systems, Vehicular Networking and Connected Vehicles, Aerospace Systems, Automation, Manufacturing, Smart Grids, Nonlinear Systems, Power Systems, Robotics, Social Systems, Economic Systems and other. Of particular value to both the contributors and the readership are the short publication timeframe and the world-wide distribution and exposure which enable both a wide and rapid dissemination of research output.

The series covers the theory, applications, and perspectives on the state of the art and future developments relevant to systems and networks, decision making, control, complex processes and related areas, as embedded in the fields of interdisciplinary and applied sciences, engineering, computer science, physics, economics, social, and life sciences, as well as the paradigms and methodologies behind them.

Indexed by SCOPUS, INSPEC, WTI Frankfurt eG, zbMATH, SCImago.

All books published in the series are submitted for consideration in Web of Science.

More information about this series at http://www.springer.com/series/15179

Shailesh Tiwari · Erma Suryani ·
Andrew Keong Ng · K. K. Mishra ·
Nitin Singh
Editors

Proceedings of International Conference on Big Data, Machine Learning and their Applications

ICBMA 2019

Springer

Editors
Shailesh Tiwari
Department of Computer Science and Engineering
ABES Engineering College
Ghaziabad, Uttar Pradesh, India

Andrew Keong Ng
Singapore Institute of Technology
Singapore, Singapore

Nitin Singh
Department of Electrical Engineering
Motilal Nehru National Institute of Technology Allahabad
Prayagraj, Uttar Pradesh, India

Erma Suryani
Department of Information Systems
Institut Teknologi Sepuluh Nopember
Surabaya, Indonesia

K. K. Mishra
Department of Computer Science and Engineering
Motilal Nehru National Institute of Technology Allahabad
Prayagraj, Uttar Pradesh, India

ISSN 2367-3370 ISSN 2367-3389 (electronic)
Lecture Notes in Networks and Systems
ISBN 978-981-15-8376-6 ISBN 978-981-15-8377-3 (eBook)
https://doi.org/10.1007/978-981-15-8377-3

This Springer imprint is published by the registered company Springer Nature Singapore Pte Ltd.
The registered company address is: 152 Beach Road, #21-01/04 Gateway East, Singapore 189721, Singapore

Preface

The ICBMA is a unique international conference that provides a forum for academics, researchers and practitioners from academia and industries to exchange ideas and share recent developments in the field of big data and machine learning and their applications. The conference will provide a platform for researchers to get recognition for their research innovations and provide wider publicity to the institute as well as industries. ICBMA-2020 focuses on both theory and applications in the broad areas of big data and machine learning and its application in various areas. The conference will provide the opportunities of collaboration with national and international organizations of repute to the research community. It is planned to have activities such as keynote addresses from academic and industry experts, technical sessions involving research paper presentation, tutorials and workshops.

Technological developments all over the world are dependent upon globalization of various research activities. Exchange of information, innovative ideas is necessary to accelerate the development of technology. Keeping this ideology in preference, the International Conference on Big Data, Machine Learning and Their Applications (ICBMA-2020) has been organized at Motilal Nehru National Institute of Technology Allahabad, Prayagraj, UP, India, during June 5–7, 2020.

This is the first time the International Conference on Big Data, Machine Learning and Their Applications has been organized with a foreseen objective of enhancing the research activities at a large scale. Technical Program Committee and Advisory Board of ICBMA include eminent academicians, researchers and practitioners from abroad as well as from all over the nation.

ICBMA-2020 received around 145 submissions. Each submission has been gone through the plagiarism check. On the basis of plagiarism report, each submission was rigorously reviewed by at least two reviewers with an average of 2.2 paper per reviewer. Even some submissions have more than two reviews. On the basis of these reviews, 34 high-quality papers were selected for publication in this proceedings volume, with an acceptance rate of 23.4%. A sincere effort has been made to make it an immense source of knowledge for all and includes 34 manuscripts.

The selected manuscripts have gone through a rigorous review process and are revised by authors after incorporating the suggestions of the reviewers.

We are thankful to the keynote speakers—Prof. Rajeev Tripathi, Director, Motilal Nehru National Institute of Technology Allahabad, UP, India, to enlighten the participants with his knowledge and insights. We are also thankful to delegates and the authors for their participation and their interest in ICBMA-2020 as a platform to share their ideas and innovation. We are also thankful to the Prof. JanuszKacprzyk, Series Editor, LNNS, Springer Nature, and Mr. Aninda Bose, Senior Editor, Springer Nature, for providing guidance and support. Also, we extend our heartfelt gratitude to the reviewers and Technical Program Committee Members for showing their concern and efforts in the review process. We are indeed thankful to everyone directly or indirectly associated with the conference organizing team leading it toward the success.

Allahabad, India

Shailesh Tiwari
Erma Suryani
Andrew Keong Ng
K. K. Mishra
Nitin Singh

About This Book

Big data, machine learning and their technological developments are increasingly growing areas that deal with the solutions to the engineering problems. This book provides significant contributions of the authors in a structured way so that prospective readers can understand how these techniques can be used in finding out the solutions of engineering problems efficiently.

Nowadays, the applications of big data and machine learning entered in a new era of technological innovations and development. Due to massive computational power of machine learning techniques, we can easily manage, analyze and use the data more effectively and efficiently.

The applications of big data and machine learning termed as 'intelligent automation' deal with integration of data with smart technologies to improve the productivity as well as the ability to cater the various problems. Keeping this ideology in preference, this book includes the insights that reflect the advances in these fields from upcoming researchers and leading academicians across the globe. It contains the high-quality peer-reviewed papers of 'The International Conference on Big Data, Machine Learning and Their Applications (ICBMA-2020)' held at Motilal Nehru National Institute of Technology Allahabad, Prayagraj, UP, India, during June 5–7, 2020. This book helps the perspective readers from computer science industry and academia to derive the immediate surroundings developments in the field of big data and machine learning and shape them into real-life applications.

Contents

Editors and Contributors

About the Editors

Prof. Shailesh Tiwari, Ph.D. currently works as a Director, ABES Engineering College, Ghaziabad, UP, India. He is an alumnus of Motilal Nehru National Institute of Technology Allahabad, India and having more than 20 years of experience in Teaching, Research and Administration. His primary areas of research are software testing, implementation of optimization algorithms and machine learning techniques in various engineering problems. He has published more than 90 publications in International Journals and in Proceedings of International Conferences of repute. He has edited special issues of several SCOPUS, SCIE and E-SCI-indexed journals. He has also edited several books published by Springer Nature. He has published 4 Indian patents as IPRs. He has organized several international conferences under the banner of IEEE, ACM and Springer. He is a Senior Member of IEEE, member of IEEE Computer Society, Fellow of Institution of Engineers (FIE). He is also a member of IEEE Computer Society- TCCC, TCSE, TCDE, TCSysE, IEEE Computer Society Test Technology Technical Council.

Dr. Erma Suryani received her B.Sc. and M.T. degrees from the Institut Teknologi Sepuluh Nopember (ITS), Indonesia, in 1994 and 2001, respectively. She obtained her Ph.D. degree from the National Taiwan University of Science and Technology (NTUST), Taiwan, in 2010. She has been working as a Professor at the Institut Teknologi Sepuluh Nopember (ITS), Indonesia, since 2005. Her current research interests are model-driven decision support systems, supply chain management, operation research, system dynamics with their applications in infrastructure, food and agriculture, sustainable transport, health and medicine, e-business, manufacturing and services.

Dr. Andrew Keong Ng is an Associate Professor with Singapore Institute of Technology. He is the Deputy Programme Director of the BEng (Honours) and MEngTech programmes in Sustainable Infrastructure Engineering (Land), which focus on railway engineering and total preventive maintenance. He is a Chartered Engineer with the UK Engineering Council and serves on the committees of various local and international professional engineering institutions, such as Institute of Electrical and Electronics Engineers (IEEE), Institution of Engineers Singapore (IES), Institution of Engineering and Technology (IET), and Institution of Railway Signal Engineers (IRSE). He is a Senior Member of IEEE and IES, as well as a Consultant and Advisor to startups and multinational corporations. In addition, Dr. Ng is a Principal Investigator of several grants amounting to more than SGD 500,000. He holds one international patent and has over 20 publications as both first and corresponding author. His research and development innovations have also garnered him several prestigious awards, such as Outstanding Researcher Award, Teaching Excellence Award, Amity Researcher Award, Young Investigator Award, and National Instruments Editor's Choice Engineering Impact Award. Furthermore, Dr. Ng has been frequently quoted and interviewed by news media on railway transportation issues and current affairs. He has also been a keynote and an invited speaker at various international conferences.

Dr. K. K. Mishra is currently working as an Assistant Professor, Department of Computer Science and Engineering, Motilal Nehru National Institute of Technology Allahabad, India. He has also been worked as Visiting Faculty in the Department of Mathematics and Computer Science, University of Missouri, St. Louis, USA. His primary area of research includes evolutionary algorithms, optimization techniques and design and analysis of algorithms. He has also published more than 50 publications in international journals and in proceedings of internal conferences of repute. He is serving as a programme committee member of several conferences and also edited Scopus and SCI-indexed journals. He is also member of reviewer board of Applied Intelligence Journal, Springer.

Dr. Nitin Singh is currently working as an Associate Professor in the Department of Electrical Engineering, MNNIT Allahabad, Prayagraj, India. He is an alumnus of Motilal Nehru National Institute of Technology Allahabad, India. His primary areas of research are power system planning, electricity markets, artificial intelligence, implementation of optimization algorithms and machine learning techniques for power system optimization. He has published more than 45 publications in international journals and in proceedings of international conferences of repute. He is serving as reviewer of several Scopus, SCI and E-SCI-indexed journals. He has organized several international conferences under the banner of IEEE and Springer. He is member of IEEE, member of IEEE Computational Intelligence Society, Power Energy Society, and Industrial Application society. He is currently serving as Secretary, IEEE Joint Chapter of Industrial Electronics/Power Electronics/ Control System at MNNIT Allahabad.

Contributors

Alka Agrawal Department of ECE, GLA University, Mathura, India

Ayush Agrawal Department of Computer Science & Engineering, Motilal Nehru National Institute of Technology, Prayagraj, Allahabad, India

Monika Agrawal Department of Electronics Communication and Engineering, GLA University, Mathura, India

Darpan Anand Chandigarh University, Gharuan, Mohali, India

Manan Arora ABES Engineering College, Ghaziabad, India

Sauvik Banerjee Department of Civil Engineering, Indian Institute of Technology Bombay, Mumbai, India

B. S. Charulatha Department of CSE, Rajalakshmi Engineering College, Chennai, Tamil Nadu, India

Shubham Chaudhary Data Science, Amity University, Noida, Uttar Pradesh, India

Ankit Chhillar Amity School of Engineering and Technology, Amity University, Noida, Uttar Pradesh, India

Arjun Choudhary Sardar Patel University of Police Security and Criminal Justice, Jodhpur, India;
Center for Cyber Security, Sardar Patel University of Police, Security and Criminal Justice, Jodhpur, India

Rupesh Kumar Dewang Department of Computer Science & Engineering, Motilal Nehru National Institute of Technology, Prayagraj, Allahabad, India

Nagaraj V. Dharwadkar Department of Computer Science and Engineering, Rajarambapu Institute of Technology, Uran Islampur, India

Gaurav Dubey ABES Engineering College, Ghaziabad, UP, India

Neetu Faujdar CSE Department, ASET, Amity Univesity, Noida, Uttar Pradesh, India

Garima Garg Department of CSE, Graphic Era Deemed to be University, Dehradun, India

Sudeshna Ghosh LNCTE, Bhopal, India

Vishal Goyal Department of Electronics Communication and Engineering, GLA University, Mathura, India

Anirban Guha Department of Mechanical Engineering, Indian Institute of Technology Bombay, Mumbai, India

Aurobindo Gupta Amity School of Engineering and Technology, Amity University Uttar Pradesh, Noida, India

Manish Gupta GLA University, Mathura, India

Rishi Raj Gupta Computer Engineering Department, NIT, Kurukshetra, Haryana, India

Shivam Gupta Department of CSE, Graphic Era Deemed to be University, Dehradun, India

Pawan Jaiswal CSE Department, ASET, Amity Univesity, Noida, Uttar Pradesh, India

R. C. Joshi Department of CSE, Graphic Era Deemed to be University, Dehradun, India

Misha Kakkar Computer Science and Engineering, Amity University, Noida, Uttar Pradesh, India

Megha Kamble LNCT, Bhopal, India

Ashish Khare Department of Electronics and Communication, University of Allahabad, Allahabad, India

Hardeep Kour Department of Computer Science and Engineering, Chandigarh University, Mohali, Punjab, India

Abhijeet Kumar Department of Mechanical Engineering, Indian Institute of Technology Bombay, Mumbai, India

Aashish Kumar Chandigarh University, Chandigarh, Punjab, India

Divesh Kumar GLA University, Mathura, India

Kuldeep Kumar ABES Engineering College, Ghaziabad, India

Nitin Kumar ABES Engineering College, Ghaziabad, India

Praveen Kumar Computer Science & Engineering, Amity University, Uttar Pradesh, India

Raj Kumar ABES Engineering College, Ghaziabad, India

Rajneesh Kumar Department of Computer Science and Engineering, Amity University, Noida, Uttar Pradesh, India

Utpal Kant Kumar Birla Institute of Technology, Patna, Bihar, India

Arati Kushwaha Department of Electronics and Communication, University of Allahabad, Allahabad, India

Prakhar Maheshwari Department of Electronics Communication and Engineering, GLA University, Mathura, India

Megha Malhotra CSE Department, ASET, Amity Univesity, Noida, Uttar Pradesh, India

Preeti Mishra Department of CSE, Graphic Era Deemed to be University, Dehradun, India

Ratnesh Mishra Birla Institute of Technology, Mesra, Patna, India

Smita Mondal Sardar Patel University of Police Security and Criminal Justice, Jodhpur, India

Akash Nigam CSE Department, ASET, Amity Univesity, Noida, Uttar Pradesh, India

Dasangam Venkat Nikhil Department of Computer Science & Engineering, Motilal Nehru National Institute of Technology, Prayagraj, Allahabad, India

Anagha Pakhare Department of Computer Science and Engineering, Rajarambapu Institute of Technology, Uran Islampur, India

Subhash Chandra Pandey Birla Institute of Technology, Patna, Bihar, India

Ashwin Perti Department of Information Terchnology, ABES Engineering College, Ghaziabad, India

Akash Punhani Department of Computer Science and Engineering, Amity University, Noida, Uttar Pradesh, India;
CSE Department, ASET, Amity Univesity, Noida, Uttar Pradesh, India

Shriyansh Rai ABES Engineering College, Ghaziabad, UP, India

Virender Ranga Computer Engineering Department, NIT, Kurukshetra, Haryana, India

Munish Sabharwal School of Computing Science and Engineering, Galgotias University, Greater Noida, Uttar Pradesh, India

Nausheen Sahar Birla Institute of Technology, Mesra, Patna, India

Paresh Chandra Sau GLA University, Mathura, India

Devasheesh Kumar Sharma Amity University, Noida, Uttar Pradesh, India

Vishnu Kumar Sharma Motilal Nehru National Institute of Technology, Allahabad, India

Ananta Narayan Shrestha Department of Computer Science & Engineering, Motilal Nehru National Institute of Technology, Prayagraj, Allahabad, India

S. Sidharth Department of CSE, Rajalakshmi Engineering College, Chennai, Tamil Nadu, India

Akash Singh ABES Engineering College, Ghaziabad, UP, India

Alka Singh Department of CSE, NIET Engineering College, Greater Noida, India

Anmol Singh LNCTE, Bhopal, India

Harivans Pratap Singh ABES Engineering College, Ghaziabad, India

Kush Kumar Singh ABES Engineering College, Ghaziabad, India

Pratyush Kumar Singh ABES Engineering College, Ghaziabad, India

Amit Sinha Department of Information Terchnology, ABES Engineering College, Ghaziabad, India

Sonia Chandigarh University, Gharuan, Mohali, India

Pranjal Srivastava Center for Cyber Security, Sardar Patel University of Police, Security and Criminal Justice, Jodhpur, India

Prabhat Kr. Srivastava Department of CSE, Dr. KNMIET, Ghaziabad, India

Yashdeep Srivastava ABES Engineering College, Ghaziabad, UP, India

Buvaneish Sundar Department of Computer Science & Engineering, Motilal Nehru National Institute of Technology, Prayagraj, Allahabad, India

Shakhzod Suvanov Faculty of Digital Technologies, Samarkand State University, Samarkand, Uzbekistan

Amit Tewari Motilal Nehru National Institute of Technology, Allahabad, India

Sanjeev Thakur Amity School of Engineering and Technology, Amity University, Noida, Uttar Pradesh, India

Akash Tiwari Department of Computer Science & Engineering, Motilal Nehru National Institute of Technology, Prayagraj, Allahabad, India

Ayush Tiwari ABES Engineering College, Ghaziabad, UP, India

Rajesh Tripathi Motilal Nehru National Institute of Technology, Allahabad, India

Munesh Chandra Trivedi Computer Science & Engineering, NIT Agartala, Agartala, India

Aryan Upadhyay ABES Engineering College, Ghaziabad, UP, India

Swati Vashisht Computer Science & Engineering, Amity University, Uttar Pradesh, India

Aman Verma ABES Engineering College, Ghaziabad, UP, India

Rajat Verma ABES Engineering College, Ghaziabad, India

Maneesh Vishwakarma ABES Engineering College, Ghaziabad, India

Sakshi Vyas Motilal Nehru National Institute of Technology, Allahabad, India

Neeraj Kumar Yadav ABES Engineering College, Ghaziabad, India

Analysis of Efficiency of Fractional Order Technique in a Controller for a Complex Nonlinear Control Process

Alka Agrawal

Abstract Proportional–Integral–Derivative (PID) has been the most popular choice as a controller of the industrialist for many years. However, with highly complex and nonlinear plants, the linear conventional PID is no-more effective to use. Thus, advancement in the computational technology leads to a new concept named as fractional order control which combined with PID to produce a high potential controller which can control the complex nonlinear plants more effectively. In this paper, the analysis of the fractional order PID in combination with fuzzy logic to control the movement of the end effector of a robotic manipulator has been done. The controller is compared with all other controllers with the integer order, the simulated results show that the fractional order provides extra flexibility to the controller design, and combined with the adaptive nature of the fuzzy logic, it performs in superior manner as compared to the other controllers even in the presence of unfavorable environment conditions like parameters uncertainties as well as disturbance and noise in the plant.

Keywords PID · Fuzzy logic · Fractional order · Optimization · SMO

1 Introduction

With the increasing complexity of the plants model day by day, a linear Proportional–Integral–Derivative (PID) controller is no longer able to control the functioning of the complex nonlinear plants accurately [1]. Thus, it is necessary to integrate some advanced intelligent techniques like neural networks (NN), fuzzy logic control (FLC) or meta-heuristic optimization algorithms to cope up the complexities and nonlinearities of the plants [2]. As the fuzzy logic has been introduced by L.A. Zadeh, it has been the most efficient tool used by the researchers in combination with PID applied in various fields like automation, robotics, surge tanks, etc. [3–6]. Since the

A. Agrawal (✉)
Department of ECE, GLA University, 17km Stone, NH2, Mathura 281406, India
e-mail: alka.agrawal@gla.ac.in

S. Tiwari et al. (eds.), *Proceedings of International Conference on Big Data, Machine Learning and their Applications*, Lecture Notes in Networks and Systems 150,
https://doi.org/10.1007/978-981-15-8377-3_1

fuzzy logic provides the capability of self-tuning and dealing with uncertainties, therefore it provides the system to work in adaptive manner to track the desired trajectory even in the case of parameters uncertainties as well as disturbance and noise accumulation [7–9].

In addition to this, with the advancement in the current technologies as well as the computational abilities, fractional order calculus has also been attracting the researchers these days as it provides additional flexibility to the system by changing the orders of the integral and the differential term [10, 11]. In 1999, Podlubny [12] applied the fractional order concept to the PID by replacing the integer order if the integral and the differential term with the fractional order and verified that the fractional order PID controller (FOPID) works better than the conventional PID. Since then, numerous different variants of PID have been published by utilizing the fractional order [13–15]. In 2012, the author combined the fuzzy logic with fractional order control to, and the simulation studies revealed that the FOFPID is much superior than the integer order fuzzy PID (IOFPID) [16].

In the present work, the analysis of the role of fractional order combined applied to the PID has been done by applying it to a highly complex nonlinear plant like two-link robotic manipulator system with a payload mass attached to the tip of the second link of the manipulator. For better efficacy of the fractional order, it has been combined with the fuzzy logic, and hence, the functioning of the four controllers named as PID, FOPID, IOFPID and the FOFPID was observed. In addition to this, a highly potential meta-heuristic optimization algorithm like Spider Monkey Optimization Algorithm has been used in this paper to provide best possible parameters of the controller by reducing the error upto zero. Since the two-link robotic manipulator is multi-input and the multi output system, it is very difficult to control its end-effecter's movement which gets affected by the environmental conditions like changing of parameters, disturbance at the controllers output and the noise accumulation at the sensor. Thus, after comparison of the controllers performance in these situation, it was found that the FOFPID outperforms all the controllers due to the ability of the fractional order to provide the flexibility in controller's design as well as dealing with the uncertainties in the plant.

The paper has been structured in five sections with the introduction part described in first section followed by the rest section as follows: The second section describes the dynamics and the parameters of the plant to be control. In third section, the description of the implementation of the controllers has been given with the description of the fractional order as well as the optimization technique. In fourth section, simulation of the results is given with the description of performance of the controllers in the presence of parameters uncertainties and disturbance or noise in the plant. At last, the fifth section concluded the whole results.

2 Model Description

To study the effectiveness of the fractional order controller, a nonlinear plant such as robotic manipulator system has been taken as a plant. Since the two-link robotic manipulator system is a multicoupled complex nonlinear system therefore, it is essential to devise a controller which can control the movement of its end effector in the desired track which cannot be fulfill by the conventional PID. Thus, in this section, dynamic model of the system has been discussed which will be further controlled by the designed controller.

Figure 1 shows the model of the two-link manipulator system in which the first link is attached to the hinge and another link is attached to the tip of the former. A motor is attached to the both joints which is actuated by the controller. At the tip of the second link, a payload mass has been attached.

The mathematical equation defining the dynamics of the manipulator is shown as below [17]:

$$\begin{bmatrix} S_{11} & S_{12} \\ S_{21} & S_{22} \end{bmatrix}\begin{bmatrix} \ddot{\theta}_1 \\ \ddot{\theta}_2 \end{bmatrix}+\begin{bmatrix} P_{11} \\ P_{21} \end{bmatrix}+\begin{bmatrix} f_{r1} \\ f_{r2} \end{bmatrix}+\begin{bmatrix} f_{n1p} \\ f_{n2p} \end{bmatrix}=\begin{bmatrix} \tau_1 \\ \tau_2 \end{bmatrix} \tag{1}$$

where θ_1 and θ_2 are the position of the links, τ_1 and τ_2 are the generated torques, f_{r1} and f_{r2}. represent dynamic friction.

The parameters values are taken from [17].

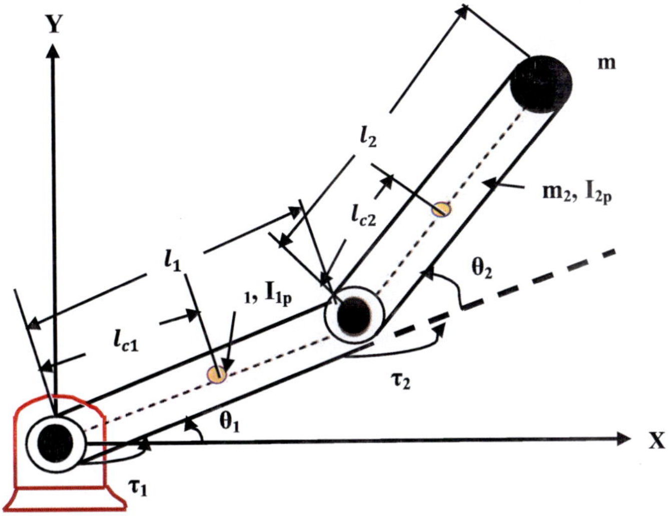

Fig. 1 Two-link manipulator system with payload

3 Controller Design for Two-Link Robotic Manipulator

Fig. 2 shows the block diagram of the closed loop control system of the two-link robotic manipulator system. Since both of the links are deflected by the angles θ_1 and θ_2 by their respective motors distinctly, therefore it requires two controllers producing outputs u_1 and u_2 to actuate the motors to produce the desired torques τ_1 and τ_2, respectively. The controllers are activated by the error signals e_1 and e_2 produced as a difference between the reference signal and the output of the link 1 and link 2, respectively.

3.1 *Design of the Fractional Order Fuzzy PID (FOFPID)/ Integer Order Fuzzy PID (IOFOID)*

Figure 3 shows the block diagram of FOFPID in which the fuzzy logic controller has the inputs as error and rate of change of error multiplied with the gains K_P and K_D,

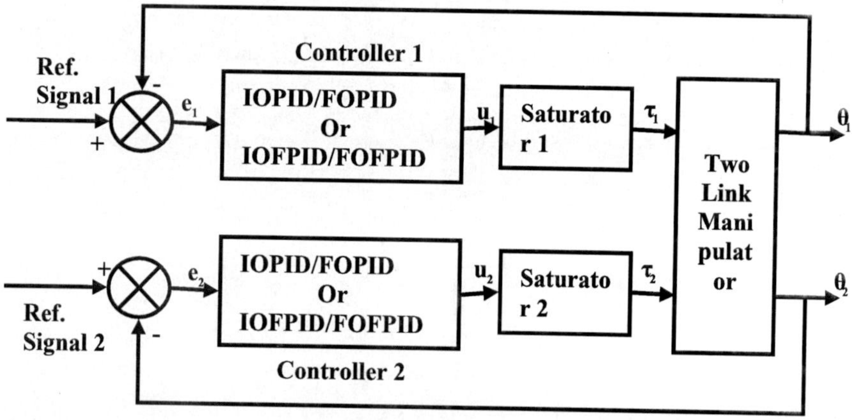

Fig. 2 Block diagram for the functioning of the closed loop system of the plant controlled by different controllers

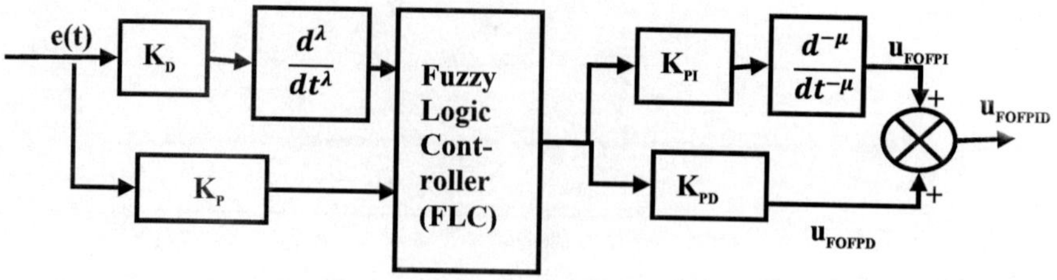

Fig. 3 Block diagram of implementation of fractional order fuzzy PID controller (FOFPID)

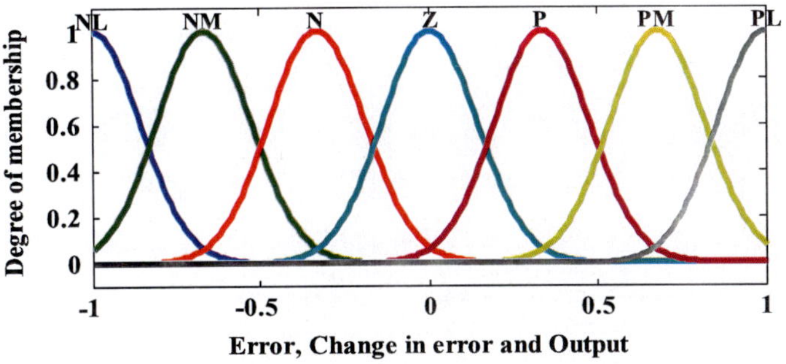

Fig. 4 Membership function for the inputs and output

Table 1 Knowledge-based rules for the FLC [18]

e de	NL	NM	N	Z	P	PM	PL
NL	NL	NL	NL	NM	N	N	Z
NM	NL	NM	NM	NM	N	Z	P
N	NL	NM	N	N	Z	P	PM
Z	NL	NM	N	Z	P	PM	PL
P	NM	N	Z	P	P	PM	PL
PM	N	Z	P	PM	PM	PM	PL
PL	Z	P	P	PM	PL	PL	PL

respectively. The order of the differential term is taken as λ to make it fractional. The fuzzy logic controller (FLC) works in three stages. In first stage, fuzzification of the data is done with the help of the membership functions as shown in Fig. 4.

After that in the second stage, the inference engine gives the controlled output as per the rule base as shown in Table 1. In the third stage, the fuzzified output is again converted into the crisp data with the help of center of gravity method. The output of the FLC goes in two ways. In first, it is passed through an integrator of order μ after multiplying with the gain K_{PI} and gives the output u_{FOFPI}, and in another way, it is multiplied with the gain K_{PD} to give an output u_{FOFPD} [18]. At the end, these two outputs are combined together to form FOFPID as given in Eq. (2).

$$u_{\mathrm{FOFPID}} = u_{\mathrm{FOFPI}} + u_{\mathrm{FOFPD}} \tag{2}$$

As per the architecture of the controller presented above and the setup for controlling the two-link manipulator system, since two controllers will be required for controlling each link distinctly, therefore the total parameters for the controllers will be twelve like K_{P1}, K_{D1}, K_{PI1}, K_{PD1}, λ_1 and μ_1 for the controller of link 1 and K_{P2}, K_{D2}, K_{PI2}, K_{P21}, λ_2 and μ_2 for the controller of link 2. The integer order fuzzy PID

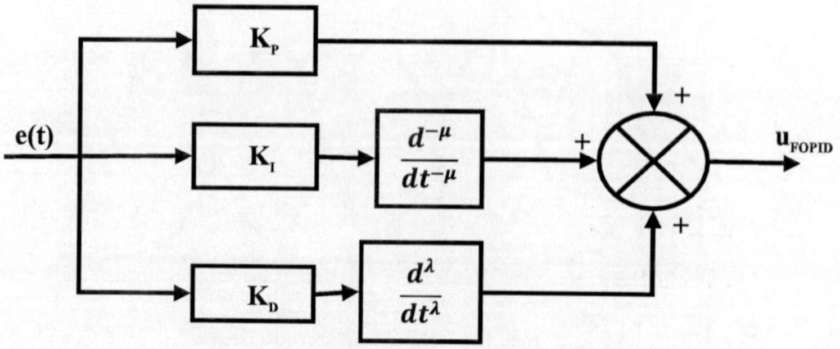

Fig. 5 Block diagram representation of fractional order PID (FOPID)

controller (IOFPID) has the same architecture as that of the FOFPID by taking the value of λ and μ as 1.

3.2 *Design of the Fractional Order PID (FOPID) / Integer Order PID (IOPID)*

Fig. 5 shows the block diagram of the FOPID in which the error signal $e(t)$ is applied to the proportional gain K_P, the integral gain K_I followed by the integration of the error with order as μ and the differential gain followed by the differentiation of the error having order as λ. Finally, the output of the controller will be sum of these three terms as shown in Eq. (3).

$$u_{\text{FOPID}} = K_P e(t) + K_I \frac{d^{-\mu} e(t)}{dt^{-\mu}} + K_D \frac{d^{\lambda} e(t)}{dt^{\lambda}} \tag{3}$$

For controlling the two-link manipulator system, the parameters of the controllers will be K_{P1}, K_{I1}, K_{D1}, λ_1 and μ_2 for link 1 while K_{P2}, K_{I2}, K_{D2}, λ_2 and μ_2 for link 2. The integer order PID (IOPID) has same configuration as FOPID where the values of λ and μ are taken as 1.

3.3 *Fractional Order Calculus*

In the literature, there are many methods available to implement the fractional order, but in the present work, Oustaloup's recursive approximation in frequency domain is used as it is best suitable for real hardware implementation [19].

3.4 Optimization Strategy for the Controllers

The controllers designed above have the gain parameters whose values will be determined by the highly potential meta-heuristic optimization algorithm such as Spider Monkey Optimization Algorithm (SMO). The optimization algorithm works in such a manner that they find out an optimum solution in order to minimize the cost function value. In the present study, the cost function is taken as the weighted sum of the time-averaged integral of the absolute error (ITAE) of both the links where the weights are taken as 1. The Eqs. (4) to (5) describe the cost function value (cost) as follows:

$$\text{Cost} = \int t.|e_1(t)|\mathrm{d}t + \int t.|e_2(t)|\mathrm{d}t \tag{4}$$

where,

$$e_1(t) = \text{ref.signal}_1(t) - \theta_1(t) \tag{5}$$

$$e_2(t) = \text{ref.signal}_2(t) - \theta_2(t) \tag{6}$$

SMO is inspired by the fission–fusion social structure (FFSS) organization of the spider monkeys, and it is proposed by the Jagdish et al. in 2014. It was proved by them [20] that the algorithm works superiorily than the other existing algorithms as it can remove the problem of local minima or premature convergence very efficiently.

4 Simulation Results

For simulation study, the closed loop control setup of the plant along with all the controllers has been designed on MATLAB/Simulink, and results are simulated using ODE solver having sampling rate of 0.001 s. The desired trajectories for the link 1 and link 2 are taken as described in Eqs. (7) and (8):

$$\text{ref.signal}_1(t) = -2\sin\left(\frac{\pi}{3}\right)t \tag{7}$$

$$\text{ref.signal}_2(t) = 2\sin\left(\frac{\pi}{2}\right)t \tag{8}$$

By applying the SMO algorithm for 100 iterations with the 40 population of spider monkeys in the range of 0.1–500, the obtained values of the parameters of the controllers are listed in Table 2. Fig. 6 shows convergence plot of the cost function of the controllers, and by the figure, it is clear that the FOFPID attains the minimum value of the cost function. The cost function value of the IOPID reaches 0.1182,

Table 2 Parameters of the controllers after optimization with SMO

Parameters	Controllers							
	IOPID		FOPID		IOFPID		FOFPID	
	Link 1	Link 2	Link 1	Link 2	Link 1	Link 2	Link 1	Link 2
K_P	500	500	499.99	499.96	500	53.81	470.04	296.54
K_I	36.52	500	499.98	442.43				
K_D	500	0.1	480.95	375.98	0.1	2.69	28.47	23.95
K_{PI}					499.1	500	54.94	496.5
K_{PD}					128.04	0.1	16.39	0.1
λ			0.653	0.1011			0.66	0.114
μ			0.1	0.2715			0.687	0.6
ITAE	0.0510	0.0671	0.0377	0.0320	0.00038	0.0037	0.0025	0.0004

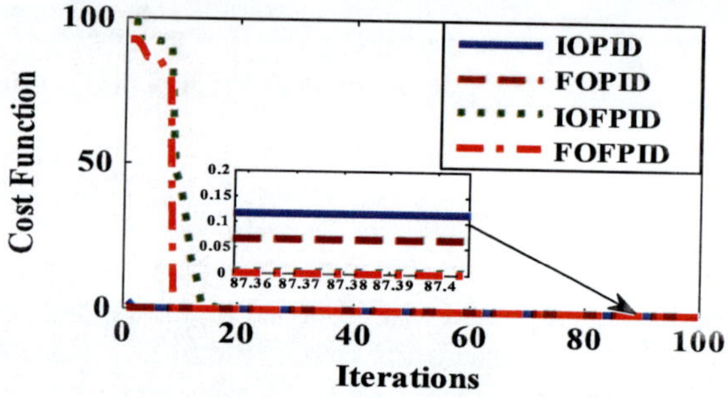

Fig. 6 Convergence plot of the cost function of the controllers during optimization

FOPID reaches upto 0.06977, IOFPID reaches to 0.0041, while the cost function of the FOFPID approaches about 0.00293. Thus, it shows that the FOFPID has the capability to track the desired signal with less error as compared to the others.

4.1 *Performance Testing of the Controllers Due to Disturbance at the Controllers Output.*

For testing the controllers against the disturbance d(t), a pulse signal given by Eq. (9) of amplitude 1 is applied at the controller's output.

After the simulation, it was found that the cost function of the IOPID becomes 0.1233, the FOPID cost function value becomes 0.0773, the cost function value of the IOFPID becomes 0.0051, while the cost function value of FOFPID becomes 0.0033. This shows

$$d(t) = u(t-4) - u(t-5) \tag{9}$$

that the FOFPID still acquires less value even in the presence of the disturbance signal. The variation in the cost function value of FOFPID is only about 14% with the actual value, while the variation in the cost function value of the others controller is about 24%.

4.2 Performance Testing for the Parameters Uncertainties

To test the performance of the controllers, the parameters of the plant like m_1, m_2, b_{1vp}, b_{2vp} and m_p have been changed by the 5%, and the simulated value of the cost function is listed in Table 3.

With the simulated values, it is clear that the FOFPID controller outperforms all the controller as the cost function value remains stable.

Table 3 Cost function values of the controllers with the variation in parameters

Change in parameters	Controllers	Cost function value
+5% in m_1	FOPID	0.0716
	IOFPID	0.0041
	FOFPID	0.0029
	IOPID	0.1184
+5% in m_2	FOPID	0.0767
	IOFPID	0.0051
	FOFPID	0.0029
	IOPID	0.1185
+5% in b_{1vp}	FOPID	0.0718
	IOFPID	0.0041
	FOFPID	0.0029
	IOPID	0.1208
+5% in b_{2vp}	FOPID	0.0765
	IOFPID	0.0040
	FOFPID	0.0029
	IOPID	0.1206
+5% in m_p	FOPID	0.0743
	IOFPID	0.0080
	FOFPID	0.0031
	IOPID	0.1185

5 Conclusion

The main aim of the paper was that to analyze the role of fractional order in a PID controller as well as fuzzy controller. In order to do this, four controllers have been designed along with the fractional order and its integer counterpart. After applying the controllers to a robotic manipulator,the simulation shows the efficacy of the fractional order as the fractional order controller tracks the desired signal more accurately as compared to its integer counterpart especially along with the fuzzy logic. Upon testing the controllers for disturbance rejection and robustness against the parametric variations, it is clear from the results that the fractional order fuzzy PID works in superior manner.

References

1. Xiao, B., & Yin, S. (2018). Exponential tracking control of robotic manipulators with uncertain dynamics and kinematics. *IEEE Transactions on Industrial Informatics, 15*(2), 689–698.
2. Sharma, R., Kumar, V., Gaur, P., & Mittal, A. P. (2016). An adaptive PID like controller using mix locally recurrent neural network for robotic manipulator with variable payload. *ISA Transactions, 62*(2016), 258–267.
3. Kumar, V., Nakra, B, & Mittal, A. P. (2011). A review of classical and fuzzy PID controller. *International Journal of Intelligent Control and System, 16*(3), 170–181.
4. Kumar, V., Nakra, B. C., & Mittal, A. P. (2012). Some investigations on fuzzy P + fuzzy I + fuzzy D controller for non-stationary process. *International Journal of Automation and Computing, 9*(5), 449–458.
5. Weiming, T., & Guanrong, C. (1994). A robust fuzzy PI controller for a flexible-joint robot arm with uncertainty. In *Proceedings of IEEE World Congress on Computational Intelligence on Fuzzy Systems* (Vol. 3, pp. 1554–1559).
6. Sooraksa, P., & Chen, G. (1998). Mathematical modeling and fuzzy control of a flexible-link robot arm. *Mathematical and Computer Modeling, 27*(6), 73–93.
7. Meng, J. E., & Ya, L. (2001). Hybrid fuzzy proportional-integral plus conventional derivative control of linear and nonlinear systems. *IEEE Transactions on Industrial Electronics, 48*(6), 1109–1117.
8. Li, W., Chang, X. G., Wahl, F. M., & Farrell, J. (2001, August). Tracking control of a manipulator under uncertainty by fuzzy P+ID controller. *Fuzzy Sets and Systems, 122*, 125–137.
9. Meza,J. L., Santibanez, V., Soto, R., & Llama, M. A. (2012). Fuzzy self-tuning PID semiglobal regulator for robot manipulators. *IEEE Transactions on Industrial Electronics, IE-59*(6), 2709–2717.
10. Goyal,V., Mishra, P., & Deolia, V. K. (2018). A robust fractional order parallel control structure for flow control using a pneumatic control valve with nonlinear and uncertain dynamics. *Arabian Journal for Science and Engineering, 44*(3), 2597–2611. https://doi.org/10.1007/s13369-018-3328-6.0123456789.
11. Goyal, V., Mishra, P., Shukla, A., Deolia, V. K., & Varshney, A. (2019). A fractional order parallel control structure tuned with meta-heuristic optimization algorithms for enhanced robustness. *Journal of Electrical Engineering, 70*(1), 16–24.
12. Podlubny, I. (1999). Fractional-order systems and PI∧λ D∧μ—controllers. *IEEE Transactions on Automatic Control, 44*(1), 208–214.
13. Kumar, V., Rana, K. P. S., & Mishra, P. (2016). Robust speed control of hybrid electric vehicle using fractional order fuzzy PD and PI controllers in cascade control loop. *Journal of the Franklin Institute, 353*(8), 1713–1741.

14. Kumar, J., Kumar, V., Rana, K. P. S. (2018). A fractional order fuzzy PD+I controller for three-link electrically driven rigid robotic manipulator system. *Intelligent Tools and Techniques for Signals, Machines and Automation, 35*(5), 5287–5299.
15. Kumar, J., Kumar, V., & Rana, K. P. S. (2018). Design of robust fractional order fuzzy sliding mode PID controller for two link Robotic manipulator system. *Intelligent Tools and Techniques for Signals, Machines and Automation, 35*(5), 5301–5315.
16. Das, S., Pan, I., Das, S., & Gupta, A. (2012). A novel fractional order fuzzy PID controller and its optimal time domain tuning based on integral performance indices. *Engineering Applications of Artificial Intelligence, 25*(2012), 430–442.
17. Sharma, R., Bhasin, S., Gaur, P., & Joshi, D. (2019). A switching based collaborative fractional order fuzzy logic controllers for robotic manipulator. *Applied Mathematical Modelling, 73*(2019), 228–246.
18. Kumar,J., Kumar, V., Rana, K. P. S. (2019). Fractional-order self-tuned fuzzy PID controller for three-link robotic manipulator system. *Neural Computing and Application.* https://doi.org/10.1007/s00521-019-04215-8(0123456789.
19. Oustaloup,A., Levron, F., Mathieu, B., & Nanot, F. M. (2000). Frequency-band complex non-integer differentiator: characterization and synthesis. *IEEE Transactions on Circuits and Systems-I: Fundamental Theory and Applications, 47*(1), 25–39.
20. Bansal, J. C., Sharma, H., Jadon, S. S., & Clerc, M. (2014). Spider Monkey optimization algorithm for numerical optimization. *Memetic Computing, 6*, 31–47.

NWP Models Combined with Ensemble Models for Hourly Estimation of Global Solar Irradiance

Megha Kamble, Sudeshna Ghosh, and Anmol Singh

Abstract The paper presents a 8-months-long dataset containing meteorological variables day of the year, sunrise and sunset time, maximum and minimum temperature, along with numerical weather forecasting model (NWP) dataset factors such as hourly solar irradiance, precipitation, humidity, wind direction, wind speed to train machine learning (ML) models and ensemble models for hourly global solar radiation prediction for City of lakes Bhopal, in the Central region of India. Bhopal has a very diverse climate. During the dry seasons of winter and summer, the sky is clear, and the sun shines brightly for most part of the day leading to solar radiation in a finite range every season. Thus, the climate of this city depicts the climate of the country and is suitable as solar radiation-based case study for further analysis of renewable energy source. The ensemble models were developed using Python scilearn kit code and trained over the dataset, and neural network model of machine learning is implemented through Keras library. The experimental setup identified two strong correlated factors wind direction and day of the year that affects the radiation the most, and results were obtained in the form of smaller values of root mean square error (RMSE), mean bias error(MBE) and r^2_score by varying kernel initializer and optimizer, to demonstrate good accuracy of the model. When compared with mathematical model for the given case study of the city, the experimental neural network model in the paper produced better results. Combination of NWP model and ensemble model is found to be fruitful for solar radiance prediction.

Keywords Global solar radiance · Ensemble model · Meteorological data · NWP model

---

M. Kamble (✉)
LNCT, Bhopal, India
e-mail: meghak@lnct.ac.in; megha_vi@yahoo.com

S. Ghosh · A. Singh
LNCTE, Bhopal, India

S. Tiwari et al. (eds.), *Proceedings of International Conference on Big Data, Machine Learning and their Applications*, Lecture Notes in Networks and Systems 150,
https://doi.org/10.1007/978-981-15-8377-3_2

1 Introduction

1.1 Background

Solar energy is the most useful, sustainable and commonly available renewable energy source in the world. Solar energy production system can be widely set up at specific region as an alternative to traditional energy systems. The major challenge in this venture is the constraints of unsteady and unpredictable nature of solar radiation resource, as due to its dependence on weather conditions. It is therefore important to design new effective system for prediction of solar radiation at any particular location.

Meteorological data such as humidity, pressure, wind speed, wind direction and solar radiance as well as geographical factors such as latitude, longitude have major role in prediction of solar energy output for effective implementation of solar energy setup at any location. However, availability of comprehensive basic solar radiation data are another challenge, whereas meteorological data is easily available on Internet resources. Some researchers have contributed methods of estimating solar radiance, solar energy and even preprocessed datasets for the same.

Numerous researchers [1–4] have predicted and estimated the solar radiance data for various locations in the world based on machine learning and deep learning algorithm. In the study [5], for development of solar technologies at Bhopal (Madhyapradesh), horizontal and tilted global solar radiation are estimated, and comprehensive database is presented. Inspired by [5], this paper presents experimental setup of combining NWP model with ensemble forecasting models for more accurate solar radiance estimation as compared to [5]. Bhopal occupies the central most region of India and has subtropical climate. It is one of the clean, environment friendly locations, and so, solar energy setup with effective prediction system is definitely beneficial contribution to environment.

With this motivation, this paper presents significant contribution in this research area, and highlights of the work are

- Identification of meteorological factors and dataset for potential location of central region of India, for solar radiance forecasting
- Identified wind direction and day of the year as, two strongly correlated factors of NWP model
- Dedicated neural network machine learning model and ensemble models with specified parameters for more accurate solar radiance prediction as in the available research on solar energy prediction, there are few ML models integrated with NWL model
- Neural network ML model is presented with extensive set of hyperparameters, kernel weight initializer and optimizer
- For validation of results, ML model results are compared with statistical mathematical model [5] implemented on same location of India.

The paper is organized as follows. This section covers introduction. Section 2 covers some significant contributions of recent researchers. Section 3 provides brief about ensemble learning model and its implementation on dataset, dataset description, data preprocessing. Section 4 presents results, discussions and comparative analysis followed by conclusion.

2 Literature Review

Literature [1–14] has paper proposed various methodologies for obtaining parameterized model to estimate generated power in PV generation systems, as they are weather independent and hard to predict site at the hourly level. Time series historical data are relatively rare, so in recent years, statistical approaches and NWP model-based exogenous datasets have been developed to help the research community to work on solar radiance prediction. Some significant work is summarized here in Table 1.

The authors in [13] presented a review on the application of neural networks for the estimation, forecasting, monitoring and classification of exogenous environmental variables, and the forecast of these variables allowed to explore renewable energy and

Table 1 Literature review

Reference No.	Method applied	Output	Advantages	Disadvantages
[2]	LSTM	RMSE	On time series historical radiation	Recurrent NN performance depends on activation function
[3]	Survey of neural network, regression	ROC Scatter plot	Comparative of different methods	ML regression model has weakness in prediction with time series
[4]	ANN, ARIMA	Prediction accuracy	Neural network reduces error	Computational burden
[7]	SVM Regression tree	Clustering performance Dense Sparse cluster	ML models on time series data	More key parameters, Historical data cannot help out weather dependence
[5]	Mathematical statistical equations	MAE MSE MAPE RMSE	Scientific equations based on longitude and latitude	Less feasible approach

water resources. Inspired by this work, this paper presented machine learning neural network model to explore renewable energy and solar energy prediction for particular location in India. One more important contribution for Indian environment is [14], in which the effects of wind speed and ambient temperature have been analyzed for location of Tripura, India, for the performance analysis of a monocrystalline silicon, solar photovoltaic module has been analyzed in a particular location called Tripura, India.

Existing literature provides worldwide instances of solar energy or radiance prediction for various regions of the world, China, Saudi Arbaia, USA, Uganda, whereas rare of such work is available for India. Most of the instances, as stated in Table 1, solar radiance prediction is done for time series analysis, historical data and various methods are studied for estimating solar radiation based on neural network, machine learning, AI, real-time series forecasting [1–4] with default parameters. Prediction with time series has additional computational burden due to nonlinearity of data variables in dataset that can be reduced by using NWP model-based dataset. The work contributed in this paper is different in the sense of hyperparameter tuning of ML neural network model, using kernel weight initializer and optimizer that has produced better results in the terms of r^2 score.

3 Proposed Model

Literature provides worldwide instances from various regions such as Uganda, Saudi Arabia where estimation of solar radiance is done. Most of the references followed non-supervised or supervised machine learning techniques on time series historical radiation data. The accuracy and robustness of estimation of solar radiance are improved by applying ensemble models on exogenous dataset(NWP model dataset) and cross-validation of results in this paper.

3.1 Prediction Framework

The popular technique from data analytics also overlapping with machine learning technique for prediction and forecasting is regression analysis. It is widely used to understand the relationship among dependent and independent data values so as to explore this association for prediction. The proposed work is using scikit-learn Python-based easy and well-organized tools for data mining and data analysis.

The primary machine learning algorithms for forecasting or prediction mechanism are regression, refer predicting a continuous-valued attributes associated with an object(data point). The limitations are variance, bias and outlier points.

To overcome these limitations, ensemble methods provide combination of several base estimators. The objective of ensemble model is to mingle the predictions of numerous base estimators built with a known learning algorithm in order to progress its performance. It is classified as boosting and averaging method.

The proposed model demonstrates random forest regress or under averaging method and AdaBoost regressor, under boosting category of ensemble machine learning.

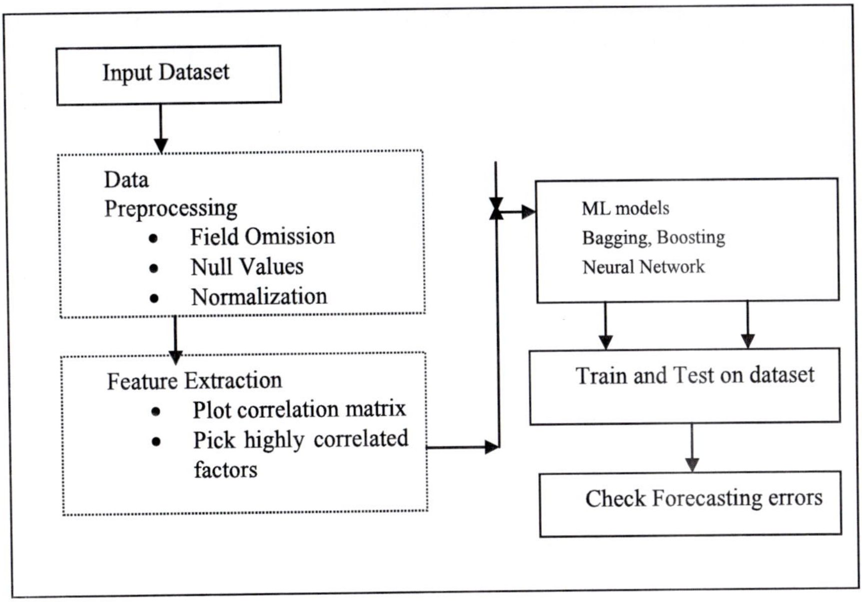

3.2 *Data Preprocessing and Attribute Correlation*

The dataset collected for the proposed framework implementation consists of following most common but significant factors representing climatic condition.

- Global Solar radiation: kW per h m^2 (radiant energy emitted by the sun, collected by solar energy system for electricity production)
- Temperature, maximum and minimum: °C
- Humidity, cloud: percentage, atmospheric pressure: MPa
- Wind direction: degrees, wind speed: km per h
- Gust: km per h, precipitation: mm
- Sunrise/sunset: Hawaii time (converted into daylength in seconds)
- Date, time: converted to day of year (unique no. from 1 to 365).

(Reference https://www.worldweatheronline.com/). The firm operates two high-tech weather data centers in the world along with mini setup at India to make available day to day weather and climatic situation.

Table 2 Correlation matrix of most significant independent factors with radiation

Pressure	Humidity	Precipitation	Gust	Cloud	Min. temp	Wind speed	Day of year
0.54	0.51	0.36	0.39	0.47	0.23	0.41	0.65

To make it compatible with regression model, the preprocessing is done for date, time and sunrise time and sunset time with the help of pandas Python library. Later on, day, time, sunrise, sunset columns are dropped and replaced by two columns day of year (unique integer) and length of day in seconds. The dataset consists of eight records for a day where three hourly record is available for all attributes. The correlation of the attributes is shown in Table 2.

3.3 *Neural Network Machine Learning Model*

A neural network is a computational system under machine learning that can help in predictions based on existing independent data. Neural network model is fruitful for nonlinear, complex data relationship, as we are finding NWP model-based dataset where radiation nonlinearly varies for number of weather facts. In this particular paper, neural network is built in Keras to solve regression problem with as much accuracy as possible. Keras is an API developed by Google to run high level neural network.

We are using the eleven (11) input variables (NWP dataset), along with seven hidden layers of 22 neurons, respectively, and finally using thelinear activation function to process the output. We are training our model over 1000 forward and backward passes, with the expectation that our loss will decrease with each epoch, meaning that our model is predicting the value of y more accurately as we continue to train the model. For superior training outcome of r^2 score, adaptive optimization methods Adam, RMSprop and NAdam have implemented in upper layer of gradient descent and neural network to close the generalization gap.

4 Results and Discussions

Dataset is split into training and testing in 80:20 ratio using Python scikit-learn functions to implement random forest regressor model with 100 decision trees and adaptive boosting regressor with bootstrap and random state of 100. Forecasting error and accuracy are evaluated for the following metrics [11, 12]:

- RMSE—Mean Square Error. The mean squared error(MSE) or mean squared deviation (MSD) of an estimator procedures the average of the squares of the errors—that is, the average squared difference between the estimated values and

Table 3 Accuracy of models comparative analysis

Prediction model	Accuracy	RMSE (ML) (kWh/m^2)	RMSE (Math model [5]) (kWh/m^2)
RF Regressor	0.93	0.24	1.61
AdaBoost Regressor	0.89	0.31	1.61

Table 4 Performance evaluation accuracy of models comparative analysis

Model	MBE (ML) (kWh/m^2)	MBE (Math.Model) [5] (kWh/m^2)	MAPE (%)	MAPE (Math Model [5]) (%)
RF regressor	0.70	0.5	2.96	−9.57
AdaBoost regressor	1.0	0.66	4.40	−9.57

the actual value. RMSE—Root mean square is the standard deviation of the residuals (prediction errors). Residuals are the measures of how far the data points are from the regression line.

- Accuracy: A test set is created for final evaluation, and cross-validation score gives accuracy of prediction on any set of possibilities of datasets.
- R^2_score: *R*-squared is a statistical measure of how close the data are to the fitted regression line. It is also known as the coefficient of determination. It explains to what extent the variance of one variable explains the variance of the second variable. In other words, how much variation of a dependent variable is explained by the independent variable(s) in a regression model.

Tables 3 and 4 demonstrate the performance of both ensemble models is better for all metrics for the given dataset as compared to traditional mathematical estimation. Mean absolute percentage error (MAPE) is scale-independent error and if the values are <10%, treated to be highly accurate prediction. Mean bias error (MBE) is representation of systematic error of forecast error, more reliable measure, and though higher values represent overforecast, it is reliable method for forecasting error. RMSE here is pretty low on datum of 0–8 global radiance value.

4.1 R^2_*Scores of Implemented Machine Learning Models*

Table 5 demonstrates the performance of five ML models trained for NWP dataset, most of them demonstrated competent r^2 score and neural network model is extended with hyperparameter tuning to find out suitable parameter values for neural network implementation.

These observations show that neural network ML model has demonstrated better score in comparison with other ML and ensemble models. Neural network ML model

Table 5 R^2 score of models comparative analysis

Prediction model	Hyperparameter	R^2_score
Bagging regressor		0.9085
Gradient boosting regressor		0.9311
Decision tree regressor		0.9683
K-nearest neighbor		0.74
Neural network	Layer = 8	
	Optimizer = adam	
	Activation = relu	
	Epochs = 1000	
	Batch_size = 32	
	Kernel initializer	
	Glorot uniform	0.9001
	He nomal	0.8145
	Lecun normal	0.929
	He uniform	0.952
	Glorot normal	0.9309
	Lecun uniform	0.9366
	Variance scaling	0.9349
	Orthogonal	0.9353
	Random normal	0.9397
	Random uniform	0.92511
	Truncated normal	0.911

and decision tree regressor when implemented with Keras provide several hyperparameter and thus stand at an upper edge than other classical ML and ensemble model techniques stated here.

4.2 *Significant Feature Extraction and Evaluation*

After observing the effect of several independent variables on radiation using decision tree regressor, it is observed that prediction of radiation is affected by the two independent variables, wind direction and day of the year. The predicted accuracy score r^2_score is 0.9722 with dependent variable wind direction and that of day of the year is 0.9528. The predicted accuracy score for independent variable solar irradiance, with the help of combined field of dependent variable wind direction and day of the year, is 0.9350. The significant features wind direction, day of the year, wind direction + day of the year are provided to neural network ML models.

Table 6 R^2 score of neural network model comparative analysis of hyperparameter tuning

Kernel Initializer	Optimizer	Optimizer specifications	R^2 score
glorot Uniform	Rmsprop	learning_rate = 0.001, rho = 0.9	0.9299
	Adam	learning_rate = 0.001, beta_1 = 0.9, beta_2 = 0.999)	0.8817
	Nadam	learning_rate = 0.002, beta_1 = 0.9, beta_2 = 0.999)	0.9374
Glorot Normal	Rmsprop	learning_rate = 0.001, rho = 0.9	0.9276
He normal	Nadam	learning_rate = 0.002, beta_1 = 0.9, beta_2 = 0.999	0.8918
Lecun normal	Rmsprop	learning_rate = 0.001, rho = 0.9	0.905
	adamax	learning_rate = 0.002, beta_1 = 0.9, beta_2 = 0.999	0.8822
	nadam	learning_rate = 0.002, beta_1 = 0.9, beta_2 = 0.999	0.912
Lecun uniform	Rmsprop	learning_rate = 0.001, rho = 0.9	0.8911
	nadam	learning_rate = 0.002, beta_1 = 0.9, beta_2 = 0.999	0.8958
Variance scaling Random normal	Rmsprop	learning_rate = 0.001, rho = 0.9	0.9221
Random normal	Adadelta	learning_rate = 1.0, rho = 0.95	0.9209
Truncated normal	nadam	learning_rate = 0.002, beta_1 = 0.9, beta_2 = 0.999)	0.9112

Neural network model with Layer = 8, Activation = relu, Epochs = 1000

With these significant features, neural network machine learning model is again implemented for 1000 epochs with 32 batch size to find out optimizer scores, presented in Table 6.

These results in Table 6 give complete picture of neural network implementation for solar energy prediction by varying number of layers, number of neurons in different hidden layers, different kernel initialize and several optimizers, with different learning rate for best possible results. The learning rate essentially controls how quickly an algorithm converges. If it is set too high, it may oscillate, and if it is set too low, it will take a very long time to converge. So, Table 6 demonstrates stable implementation with various optimizers with default learning rate value, predicting stable solar radiance from the give NWP dataset.

4.3 *Graphical Analysis*

Figures 1 and 2 show the scatter plot of solar radiance over day of year set from 1 to 250 number of days. Actual radiance values are shown in orange color, whereas predicted values are shown in blue color. Accuracy of estimation is reflected in the plot.

The regression in Figs. 1 and 2 illustrates that the relationship between the predicted output and the actual output is approximately same.

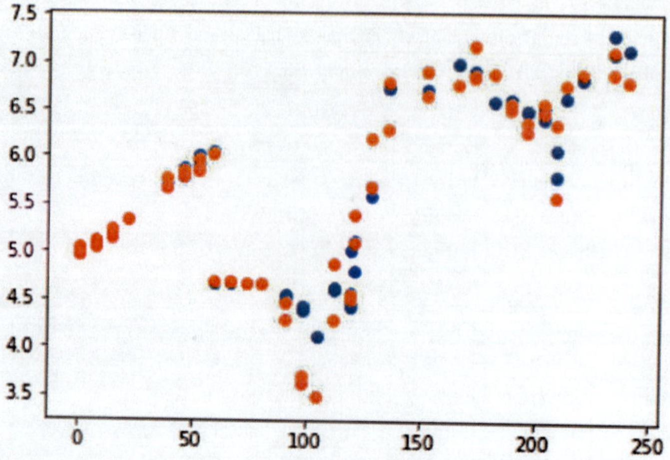

Fig. 1 Predicted versus observed hourly solar radiation with RF regressor

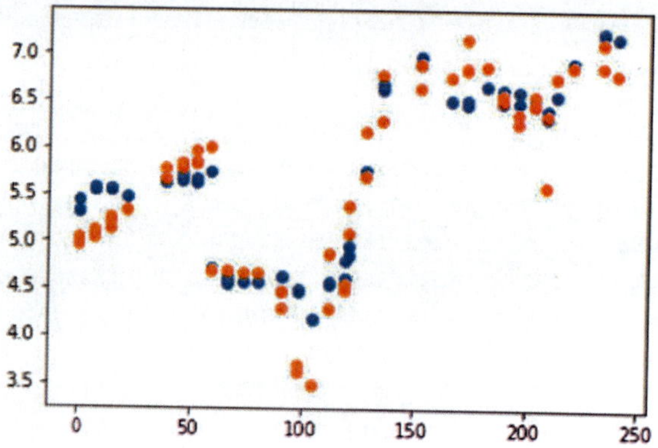

Fig. 2 Predicted versus observed hourly solar radiation with AdaBoost regressor

5 Conclusions

The two ensemble models random forest regressor and AdaBooster regressor successfully predicted the solar irradiation using NWP model dataset. Neural network machine learning model with extensive hyperparameter tuning provided the more accurate predictions with reduced forecasting errors comparable to mathematical estimation model. For hourly solar prediction of central region of India, the paper will be beneficial for renewable energy setup establishment. Future scope of the work is fine-tuning of ensemble model parameters for extended analysis of solar radiance forecasting and practical usage.

Acknowledgements This paper presents partial work for the research project funded under Collaborative Research Scheme (CRS), by TEQIP III RGPV(State Technical University of Madhya Pradesh, Bhopal, India) evaluated by NPIU.

References

1. Alzahrani, A., et al. (2014). Predicting solar irradiance using time series neural networks. *Procedia Computer Science, 36,* 623–628.
2. Qing, X., & Niu, Y. (2017). Hourly day-ahead solar irradiance prediction using weather forecasts by LSTM. *Energy, 148,* 461e468.
3. Voyant, C., Notton, G., Kalogirou, S., Nivet, M. L., Paoli, C., Motte, F., & Fouilloy, A. (2017). Machine learning methods for solar radiation forecasting: A review. *Renewable Energy, 105,* 569–582.
4. Mubiru, J. (2008). Predicting total solar irradiation values using artificial neural networks. *Renewable Energy, 33*(10), 2329–2332.
5. Shukla, K. N., et al. (2015). Estimation and validation of solar radiation incident on horizontal and tilted surface at Bhopal, Madhya Pradesh, India. *American-Eurasian Journal of Agriculture & Environment Sciences, 15*(1), 129–139.
6. Yacef, R., Benghanem, M., & Mellit, A. (2012). Prediction of daily global solar irradiation data using Bayesian neural network: A comparative study. *Renewable Energy, 48,* 146–154.
7. Tanaka, Y., & Takahashi, M. (2017). SVM-based Multi-point Prediction System of Solar Irradiance using DTW-based Cluster Analysis. *Transactions of the Institute of Systems, Control and Information Engineers, 30*(8), 305–313.
8. Govender, P., Brooks, M. J., & Matthews, A. P. (2018). Cluster analysis for classification and forecasting of solar irradiance in Durban, South Africa. *Journal of Energy in Southern Africa, 29*(2), 51–62.
9. Feng, C., Cui, M., Hodge, B. M., Lu, S., Hamann, H., &; Zhang, J. (2018). Unsupervised clustering-based short-term solar forecasting. *IEEE Transactions on Sustainable Energy*, 1–11. arXiv:1805.04193v1.
10. Seyedzadeh, S., et al. (2018). Machine learning for estimation of building energy consumption and performance: A review. *Visualization in Engineering 6*, 5. https://doi.org/10.1186/s40327-018-0064-7.
11. https://scikit-learn.org/stable/tutorial/index.html
12. Raschka, S. (2017). Python machine learning, 2nd edn, Epub.
13. Rodrigues, E., Gomes, A., Gaspar, A., & Antunes, C. (2018). Estimation of renewable energy and built environment-related variables using neural networks—A review. *Renewable and Sustainable Energy Reviews, 94,* 959–988. https://doi.org/10.1016/j.rser.2018.05.060
14. Bhattacharya, T., Chakraborty, A. K., & Pal, K. (2014) Effects of ambient temperature and Wind speed on performance of monocrystalline solar photovoltaic module in Tripura, India. *Hindawi Publishing Corporation Journal of Solar Energy, 2014,* Article ID 817078, 5 p. https://doi.org/10.1155/2014/817078.

Literature Review on Agricultural Internet of Things

Anagha Pakhare and Nagaraj V. Dharwadkar

Abstract This review paper focuses on the combinatorial implementation for agriculture and Internet of things (IoT). There are many prototype platforms that are built and demonstrated in order to construct the standard procedures. For the prediction and modelling environment of the field, deep learning and machine learning algorithms are being used. There are different units used for implementation independently like remote monitoring, image analysis, sensor network, irrigation system, cloud-mobile monitoring, drone surveillance, etc. It mainly focuses on the open challenges faced by the farmers and their solutions in the agricultural field. IoT plays a vibrant role in smart agriculture. Monitoring environmental factors is the major factor to improve the yield of the efficient crops, and the details are discussed.

Keywords Internet of things (IoT) · Sensors · Agriculture · Unmanned aerial vehicle · Smart grid · Remote monitoring

1 Introduction

Agricultural field has an indispensable role in human beings' lives as it began thousands of years ago [1–4]. From the middle age's revolution to organic and modern farming, lots of techniques were born. From the survey of USA—Food and Agriculture Organizations, with increasing population, so production of food worldwide should be increased by 70% of the current production within 2050 [5]. Agriculture has been the major source of food; also, it plays an important role in the country's economy growth [6]. Farmers still use some traditional methods of agriculture which results in low cultivation of crops. Besides, there are several new techniques that will be helpful for next generation farming. Many researchers mostly prefer the Internet of things (IoT) because of its prominent results. By implementing IoT in agriculture, farmers can raise the production with low cost by monitoring the efficiency of the

A. Pakhare (✉) · N. V. Dharwadkar
Department of Computer Science and Engineering, Rajarambapu Institute of Technology, Uran Islampur, India
e-mail: anagha2299@gmail.com

S. Tiwari et al. (eds.), *Proceedings of International Conference on Big Data, Machine Learning and their Applications*, Lecture Notes in Networks and Systems 150,
https://doi.org/10.1007/978-981-15-8377-3_3

soil, temperature, humidity, rainfall, fertilizers efficiency and theft detection in agriculture areas [7–9]. Because of the development of IoT, it is possible for farmers to implement it. But does anyone think about it why farmers are focusing only on IoT Technique? IoT consists of many independent techniques like UAV, monitoring and analysis, sensors, robotics. Moreover, technology such as IoT and automation has more efficiency due to some reasons such as wide range of connectivity with different devices, minimizing human labour, quick access, time efficient and efficient communication [10]. It demonstrates a high and intelligence, higher accuracy along with top notch identification. In addition, networks like 4G, wireless local area network (WLAN), wireless personal area network (WPAN), worldwide interoperability for microwave access (WiMAX), radio-frequency identification (RFID), Zigbee, near field communication (NFC) and bluetooth develop IoT so that smart system can be operated at remote places [11, 12]. Also, proper management of orphaned nodes reduces the cost of the irrigation systems for the field [13]. The productivity of crops is affected by different factors which includes (I) the use of proper insecticide and pesticides can help you to manage the attacks (II) as the crop grows they might be attacked by wild animals and birds (III) some unfortunate events such as unpredictable monsoon rainfalls, shortage of water and improper irrigation system that might damage the crop [14].

2 Literature Survey

Himadri Nath Sah et al. reported that the unmanned aerial vehicles or drones (UAVs) [3] are of humongous interest in the field of agriculture and crop cycle.

The reasons are as follows: (i) It will provide a drone-based outlook for improved harvest in less time. (ii) Efficient systematic monitoring of fields [15]. The production rate increases gradually with fewer consumption of energy from the upcoming technologies. Drones are focusing on the inquiry of soil and fields along with planting seeds and dusting nutrients. There are many applications of drones like hyper spectral, thermal-spectral or multispectral sensors which are embedded with drones and used to identify the dry land in the field so in future such problems do not occur. By scanning the crop, we can analyse the health using near-infrared and visible light. So, as per described in the above theories, it can be concluded that drones can be a more efficient belvedere for data collection in precision agriculture [16, 17].

Sridhar et al. reported that an agriculture irrigation system is formed using less complicated wireless circuits which can optimize the use of water and found a proper irrigation system for the crop fields [18, 19]. To detect the temperature and moisture of soil using the circuit, two sensors are used in the system to get very accurate real-time data. From different observations and experimental test results, it proves that the proposed method is one of the solutions for the field activities while considering parameters like time and maintenance cost [20]. Because of automation, generated data will be helpful for the farmers to make the right decision. Here, in this proposed model, two sensors are used to reduce the complexity of the hardware system, and for the calibration of the sensors, it focuses on the threshold voltage value. This value may vary with the crop and plantation (Fig. 1(c)).

Mr. Ankit Ranjan et al. reported that agriculture has been playing a primary role in India's economic development, but on the other hand, farmers are facing difficulties in agribusiness because of lack of technical knowledge. For that, this paper provides a solution in the form of IoT and automation techniques which will help in managing most of the agricultural work and also help the farmers to understand the strategy of the market and maintain their crops.

Shri pradha et al. reported that to detect the soil moisture content, the microcontroller is programmed with a soil moisture sensor and a pump motor. The tube inlet of a pump motor is connected with the microcontroller. The soil moisture sensor senses the moisture level in the soil and automatically switches the motor's ON/OFF system. IR sensors are used as a warning system against the intruders in the field, which provides an alert message for an unremitted entry. At that time, relevant data is stored in the flash memory of the microcontroller associated with the ESP8266 Wi-Fi module, which establishes a communication between the field and the server which posts all this on an open-source cloud [21]. This generated data can be accessed by farmers using mobile applications or some Website for analysis and monitoring (Fig. 1(b)).

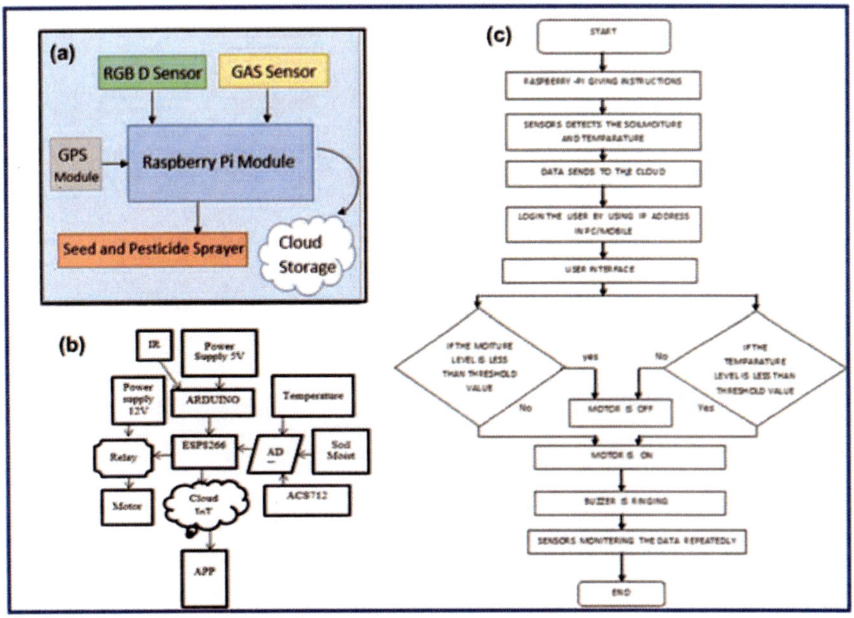

Fig. 1 Block diagram and flow chart of proposed methods

Prathibha et al. reported that the domain of "Internet of Things" consists of sensing devices which are used to gather statistics. The agricultural monitoring system is aided by IoT, and it serves a dependable and efficient system. Automation of farming will eventually replace manual labour, allowing farmers to harvest more yield resulting in more profit. As it is (IoT) inexpensive and uses less power, the resulting system is efficient and profitable. It provides the data regarding temperature and humidity in farms through multimedia messaging services (MMS) to the farmer [22]. This methodology can be used in greenhouse which host temperature-dependent plants. Adoption of these (IoT) technologies on the farm will improve the global farming yield (Fig. 2(a) and Table 1).

Authors presented smart farming concepts to enhance the crop yield using Internet of things (IoT) to overcome the limitation of traditional agricultural methods so that they can use it to fulfil the population's demand of food [2, 23]. The presented methodologies can achieve automation in the form of results which reduce farmer's effort and time in the field.

Above table summarizes the comparative study of existing methods which are used by authors to monitor crop fields, to automate irrigation systems, etc. It also contains results after applying proposed methods by authors, issues and limitations of methods that are faced.

The remaining part of the paper is planned as follows. The challenges and issues which are faced during automation in agricultural fields are briefly described in

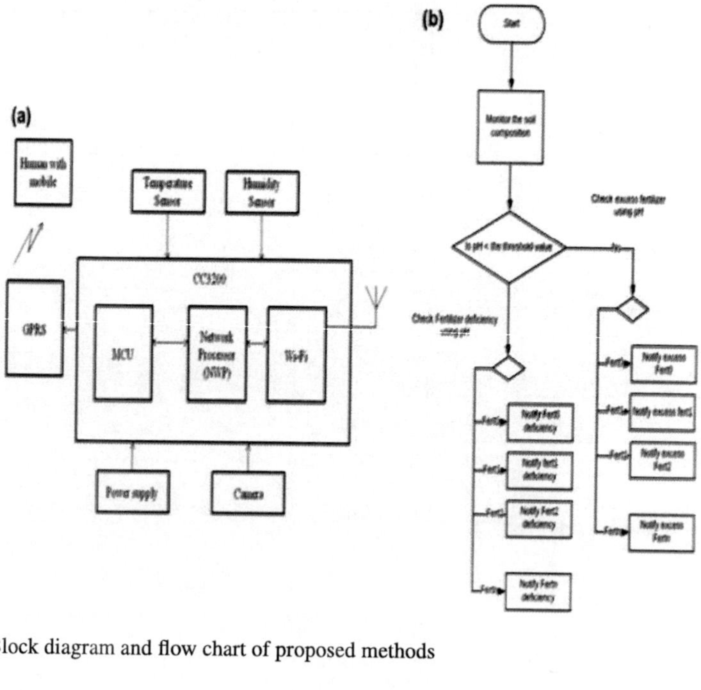

Fig. 2 Block diagram and flow chart of proposed methods

Table 1 Summary of existing methods

Author	Method	Result	Limitation
Himadri Nath Sah et al. [3]	Unmanned aerial vehicle integrated with machine learning algorithms and IoT concepts for a better quality of crops	Crop monitoring and scanning for crop health analysis, soil analysis, identification of dryness of land to improve irrigation system	External charging required by the drone Need more accurate classifiers for the classification of plants and yield
Rao and Sridhar [4]	A highly advanced and accurate monitoring and control agriculture automatic system with IoT technology such as IC, relay and Raspberry Pi 3 model	System-generated data is analysed by the farmer which is further useful in crop fields to run irrigation systems automatically. Using the water fertilizers efficiently while maximizing the yield of the crops and will also help in analysing the weather conditions of the field using the sensors	Calibration of the system sensors is required to minimize errors and the need of farmers, to analyse system-generated data
Ranjan et al. [1]	Sensors are used to detect moisture in the soil by calculating potential difference between soil and this value feed as an input to the microcontroller, if potential difference is less than threshold value, microcontroller is connected to the pump by a relay which is helpful in automation of irrigation system using IoT	Many farmers' problems are resolved by using IoT and automation techniques which manage agricultural work, focusing on market strategy rather than wasting time for crop maintenance and production	It is not a fully autonomous agricultural system; human intervention still exists
Shri Pradha et al. [2]	To make the best or most effective use of fields water usage and some different features that are focused on the security mechanisms of the field using cloud computing and IoT technologies	Soil moisture sensor detects the moisture level which is used to turn on/off the motor system for irrigation. The temperature sensor gives the approximate temperature in degree/Celsius, and the IR sensor detects unauthorized human or animal intervention in the field	Need to make system applications at the farmer level so that they can easily understand and access it

(continued)

Table 1 (continued)

Author	Method	Result	Limitation
Prathibha et al. [5]	Having the exact balance between the traditional methods along with the latest technologies such as IoT and wireless sensor, the result can be improved	Farmers are able to monitor the changes in the crop yield accurately using the wireless monitoring technique that gives the information such as temperature, humidity of the air in the field	The proposed method does not use renewable energy for power supply

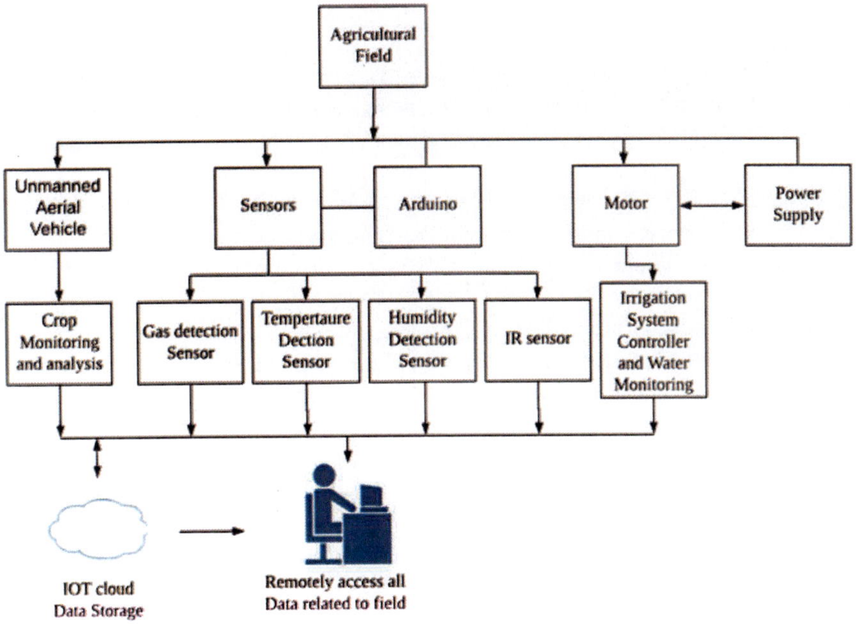

Fig. 3 Flow diagram of the system

Sect. 3. Section 4 reports the different approaches used to make a smart agricultural field. Section 5 epitomizes the conclusion.

3 Flow Diagram

Figure 3 represents the flow diagram of all methods explained in this paper. For monitoring and analysis of the field, all IoT devices, sensors, UAVs are used in the field. All data and information related to crops and field areas are collected in the IoT cloud, from which farmers can remotely access any data they want and take precautions to avoid conflict while making decisions [24]. This data is useful for the future life cycle of crops as well as to increase the production rate [25].

4 Challenges and Issues in Smart Agriculture

This section highlights some of the challenges, which are faced during automation in the agricultural industry. Agriculture is a vast area, which affects the production capacity of the ecosystem. Agriculture is a maintenance-intensive industry, requiring a hefty amount of supervision, so farmers cannot handle the large-scale farming manually [1].

Although it is named as an autonomous system, human intervention still exists. Instead of a human, it is needed to implement some robotic machines, which can monitor the whole system and generate data. Also, crops will be cultivated and harvested by using robotic machines will provide results as faster growth of crops in an organic manner. However, it is not cost-efficient for farmers. Smart automation using IoT is the major transformation of the agricultural domain from being manual and static to intelligent and dynamic which can help to reduce human supervision [2].

While monitoring the field, there are special tools and instruments used, which includes unmanned aerial vehicles (UAV), IoT-based sensors, ICs, arduinos, thermal cameras, etc. These instruments are widely used for applications such as monitoring the crop health, security and detecting environment change. However, these instruments require calibration, which means evaluating and adjusting the precision and accuracy of measurement equipment for the reduction of bias in instrument reading.

Use of hyperspectral imaging in UAVs, the only imaging format that enables the use of artificial intelligence for crop yield forecasting and other applications such as spraying pesticides, fertilizer using UAVs [26, 27]. But it is more challenging to implement hyperspectral imaging than multispectral techniques [3].

In addition, some technologies used during automation are complex to handle and understand for the farmers. So, there is a need to invent such types of technologies which can be easily made accessible to farmers based on their knowledge level.

The security issue is the main problem in IoT agriculture, there is a need to address the physical integrity of the field devices and end to end information security [28]. It is important during automation to deliver such kinds of solutions which can be integrated with existing infrastructure.

The proposed solution should be able to handle all the situations or environmental conditions so that it could be called as a complete solution for all problems [4].

Power management is the main issue while using IoT devices for automation purposes. There is a need to use energy-efficient and battery-free devices. For that, it is needed to reduce the power consumption of each electronic component. Using this technique, it will help to increase the lifespan of each module. Also, to reduce environmental impact, recycling strategies have to be taken into account. The IoT-based agricultural production system is designed to ensure the long-term desire of farmers so that their land remains productive in the future [29].

According to the above discussion, there are unique challenges faced by researchers when doing a study of IoT-based smart agriculture.

5 Approaches

According to the world population, the agricultural industry suffers from less production of crop yield. As a solution, some IoT-based techniques are being invented by researchers which will result in an increase in production at a low cost.

Researchers aim to combine modern techniques with traditional methods. In this section, various automation techniques are discussed.

Figure 2(b) shows the automation of soil care and fertilizer deficiency detection. There are two types of sensors used in this proposed method mainly to monitor the pH value of the soil along with sensors for nutrients check which monitor the composition of the soil which is further used to build an automatic irrigation system. The second set of sensors monitor the deficiency/excessiveness of a given fertilizer.

Thermal cameras and IR sensors are used to prevent animals as well as unauthorized people intervention in the field who can harm the field [30]. Also, thermal cameras are used to detect plant and fruit disease [31]. Figure 1(a) shows a block diagram of the proposed system, which includes a gas sensor and can be used to sense gases like ethylene, propane, methane. RGB-D sensor is a specific kind of depth sensor, which works in association with an RGB camera [32]. For the fruit ripeness detection, a search of ethylene gas in the field is mandatory [33]. Also, we can include SVM to predict the condition of the crop [34]. It is used to add the image, depth information, on a per-pixel basis. Adafruit AMG8833 IR thermal camera integrates with the Raspberry Pi module 3 B.

Figure 1(c) is the flow chart of the proposed method, and the data obtained from sensors is stored in the cloud which can be further monitored by the farmer using mobile or PC [35]. Farmers analyse system-generated data and provide water through an irrigation system which runs automatically. The situation in the automated system in the field can be known and checked by using mobile or PC through IP address. Also, they can analyse real-time data while considering past experiences of irrigation systems [36].

Figure 1(b) represents the block diagram of the methodology, which contains soil moisture sensor which measures volumetric content of water by using properties like resistance to the electrical interface, constant of dielectric and the neutron interaction. It has a pump motor which is attached to this sensor. It automatically turns off/on depending on the moisture level in the soil.

ESP8266 Wi-Fi module is used to store the significant information related to the field into the memory, and it includes IR and temperature sensors which are useful to give an alert message when unwanted activities or intervention is observed.

Figure 2(a) shows the block diagram of the proposed system model, which includes CC3200 single chip with an integrated microcontroller, network processor and Wi-Fi. Temperature infrared thermopile sensor-TMP007 is used to sense the temperature values in real time as well as for humidity detection. HDC1010 sensor is used which tracks the relative moisture of air within the farming field. The camera sensor is used to capture real-time images of the field, and these images are sent to the farmer through the GPRS system.

Sensors like soil moisture sensor, pH detector sensor, temperature and humidity detector sensors are useful in smart irrigation systems [37, 38]. Solar panels can be introduced in the sensors to charge the batteries of the nodes which improves the lifetime of the network [39]. It maintains pH and moisture level in the soil by remotely controlling the functions of motor pumps used in irrigation systems. Therefore, farmers can irrigate the land according to need. Unmanned aerial vehicles are used to capture hyperspectral images which collect accurate data. It is not only used to reduce time but also yield better cultivation based on analysed data. Crop management can be efficient due to the systematic monitoring of fields using UAV. It is also helpful in irrigation systems to check dry areas in the field and then irrigate that area.

6 Conclusion

This paper presents a review on automation and monitoring systems in an agricultural field using the Internet of things (IoT). Mainly focusing on smart agriculture, also many aspects are covered that include irrigation system, usage of drones and sensors, agribusiness and a huge impact on global production. Many applications can be connected using IoT. The better update is given to farming by minimizing farmer's hands-on work, where they can operate from one place. Drip irrigation becomes easy in farming for crop production, and analysis is done through IoT. Hence, IoT makes a good impact for considering it in the agricultural field.

7 Future Scope

Mobilizing IoT in agriculture is highly urged and will be very effective when considering long-term benefits such as automation and good production. When agriculture is articulated with IoT, the data acquired in the process will be more efficient in making huge progress towards enhancement in this field. Technical specifications can be improved in agricultural fields like greenhouses. Advanced drones can be of a huge benefit in taking forward for the crop analysis in the process of sowing seeds, spraying pesticides or fertilizers. Usage of robots can be increased in future with better-advanced methodologies like vegetable harvesting. Also, in future, farmers can introduce smart grid in the agricultural field to produce clean and renewable energy. Smart grid further promotes agricultural production and economic development using intelligent construction.

References

1. Puranik, V., Sharmila, Ranjan, A., & Kumari, A. (2019). Automation in Agriculture and IoT. In *2019 4th International Conference on Internet of Things: Smart Innovation and Usages (IoT-SIU)* (pp. 1–6), Ghaziabad, India.
2. Shri pradha, R., Suryaswetha, V. P., Senthil, K. M., Ajayan, J., Jayageetha, J., & Karhikeyan, A. (2019). Agricultural field monitoring using IoT. In: *2019 5th International Conference on Advanced Computing & Communication Systems (ICACCS)* (pp. 1–4), Coimbatore, India, 2019,.
3. Saha, K., et al. (2018). "IOT-based drone for improvement of crop quality in agricultural field. In *2018 IEEE 8th Annual Computing and Communication Workshop and Conference (CCWC)* (pp. 612–615), Las Vegas, NV.
4. Rao, R. N., & Sridhar, B. (2018). IoT based smart crop-field monitoring and automation irrigation system. In: *2018 2nd International Conference on Inventive Systems and Control (ICISC)* (pp. 478–483), Coimbatore.
5. Prathibha, S. R., Hongal, A., & Jyothi, M. P. (2017). IOT based monitoring system in smart agriculture. In: *2017 International Conference on Recent Advances in Electronics and Communication Technology (ICRAECT)* (pp. 81–84), Bangalore.
6. Sardare, M. D., & Admane, S. V. (2013). A review on plant without soil—hydronics, Vol. 2, Issue 3, pp. 299–304. Available: https://www.researchgate.net/publication/276320585_A_REVIEW_ON_PLANT_WITHOUT_SOIL_-_HYDROPONICS, India. ISSN: 2319-1163.
7. Aher, A., Kasar, J., Ahuja, P., & Jadhav, V. (2018). Smart agriculture using clustering and IoT. *International Research Journal of Engineering and Technology (IRJET), 05*(03). e-ISSN: 2395-0056. p-ISSN: 2395-0072.
8. Gayatri, M. K., Jayasakthi, J., & Mala, G. S. A. (2015). Providing smart agricultural solutions to farmers for better yielding using IoT. In *2015 IEEE Technological Innovation in ICT for Agriculture and Rural Development (TIAR)*. https://doi.org/10.1109/tiar.2015.
9. Gondchawar, N., & Kawitkar, R. S. (2016, June). IoT based smart agriculture. *International Journal of Advanced Research in Computer and Communication Engineering, 5*(6), ISSN (Online) 2278-1021 ISSN (Print) 2319 5940.
10. Mohanraj, I., Ashokumar, K., & Naren, J. (2016). Field monitoring and automation using IOT in agriculture domain. In: *6th International Conference On Advances In Computing & Communications, ICACC 2016* (Vol. 93, pp. 931–939), 6–8 September 2016, Cochin, India. Published: Elsevier; Verdouw, N., Wolfert, S., & Tekinerdogan, B. (2016). Internet of Things in agriculture, December, 2016. Available: https://www.researchgate.net/publication/312164156_Internet_of_Things_in_agriculture. Accessed December, 2016.
11. Lakshmisudha, K., Hegde, S., Kale, N., & Iyer, S. (2016, July). Smart precision based agriculture using sensors. *International Journal of Computer Applications, 146*(11), 36–38.
12. Mirabella, O., & Brischetto, M. (2011). A hybrid wired/wireless networking infrastructure for greenhouse management. *IEEE Transactions on Instrumentation and Measurement, 60*(2), 398–407. https://doi.org/10.1109/TIM.2010.2084250
13. Maheswararajah, S., Halgamuge, S., Dassanayake, K., & Chapman, D. (2011). Management of orphaned-nodes in wireless sensor networks for smart irrigation systems. *IEEE Transactions on Signal Processing, 59,* 4909–4922. https://doi.org/10.1109/TSP.2011.2160258
14. Muthunpandian, S., Vigneshwaran, S., Ranjitsabarinath, R. C., Manoj Kumar Reddy, Y. (2017, April). IOT based crop-field monitoring and irrigation automation (Vol. 4, Special Issue 19).
15. Patel, P. (2016). Agriculture drones are finally cleared for takeoff [News]. *IEEE Spectrum, 53*(11), 13–14. https://doi.org/10.1109/MSPEC.2016.7607013
16. Toal, R., Claggett, K., & Gosh, J. (2017). Smart and sustainable aquaponics. Santa Clara University- Scholar Comman-2017. Electrical Engineer Senior Thesis. Available: https://scholarcommons.scu.edu/elec_senior/36/.
17. Murugan, D., Garg, A., & Singh, D. (2017). Development of an adaptive approach for precision agriculture monitoring with drone and satellite data. *IEEE Journal of Selected Topics in Applied Earth Observations and Remote Sensing, 10*(12).

18. Gutiérrez, J., Villa-Medina, J. F., NietoGaribay, A., & Porta-Gándara, M. Á. (2017). Automated irrigation system using a wireless sensor network and GPRS module. *IEEE Transactions on Instrumentation and Measurement, 17*.
19. Sales, N., Remédios, O., & Arsenio, A. (2015).Wireless sensor and actuator system for smart irrigation on the cloud. In *2015 IEEE 2nd World Forum on Internet of Things (WF-IoT),* Milan (pp. 693–698). https://doi.org/10.1109/WF-IoT.2015.7389138.
20. Namala, K. K., AV, K. K. P., Math, A., Kumari, A., & Kulkarni, S. (2016). Smart irrigation with embedded system. In *2016 IEEE Bombay Section Symposium (IBSS)* (pp. 1–5), Baramati. https://doi.org/10.1109/IBSS.2016.7940199.
21. Pereira, R., Dupont, I., Carvalho, P., & Juca, S. (2017). IoT embedded Linux system BASED on Raspberry Pi applied to real-time cloud monitoring of a decentralized photovoltaic plant. *Measurement, 114*. https://doi.org/10.1016/j.measurement.2017.09.033.
22. Rajalakshmi, P., & Devi Mahalakshmi, S. (2016). IOT based crop-field monitoring and irrigation automation. In *2016 10th International Conference on Intelligent Systems and Control (ISCO)* (pp. 1–6). Coimbatore. https://doi.org/10.1109/ISCO.2016.7726900.
23. Aher, A., Kasar, J., Ahuja, P., & Jadhav, V. (2018). Smart agriculture using clustering and IOT. *International Research Journal of Engineering and Technology (IRJET), 05*(03). e-ISSN: 2395-0056. p-ISSN: 2395-0072. Sardare, M. D., & Admane, S. M. (2013). A review on plant without soil—hydronics, Vol. 2, Issue 3, pp. 299–304. Available: https://www.researchgate.net/publication/276320585_A_REVIEW_ON_PLANT_WITHOUT_SOIL_-_HYDROPONICS, India. ISSN: 2319–1163. Toal, R., Claggett, K., & Gosh, J. (2017). Smart and sustainable aquaponics. Santa Clara University-Scholar Comman-2017. Electrical Engineer Senior Thesis. Available: https://scholarcommons.scu.edu/elec_senior/36/.
24. Kulbacki, M., Segen, J., Knieć, W., Klempous, R., Kluwak, K., Nikodem, J., Kulbacka, J., & Serester, A. (2018). *Survey of drones for agriculture automation from planting to harvest*. https://doi.org/10.1109/INES.2018.8523943.
25. Maia, R. F., Netto, I., & Tran, A. L. H. (2017). Precision agriculture using remote monitoring systems in Brazil. In *IEEE Global Humanitarian Technology Conference (GHTC)*.
26. Mohd Kassim, M. R., Mat, I., & Harun, A. N. (2014). Wireless sensor network in precision agriculture application. In *2014 International Conference on Computer, Information and Telecommunication Systems (CITS), Jeju, 2014* (pp. 1–5). https://doi.org/10.1109/CITS.2014.6878963.
27. Yallappa, D., Veerangouda, M., Maski, D., Palled, V., & Bheemanna, M. (2017). Development and evaluation of drone-mounted sprayer for pesticide applications to crops. In *IEEE Global Humanitarian Technology Conference (GHTC), 2017*.
28. Baranwal, T., Nitika, & Pateriya, P. K. (2016). Development of IoT based smart security and monitoring devices for agriculture. In *2016 6th International Conference—Cloud System and Big Data Engineering (Confluence)* (pp. 597–602), Noida. https://doi.org/10.1109/CONFLUENCE.2016.7508189.
29. Lee, M., Hwang, J., & Yoe, H. (2013). Agricultural production system based on IoT. In *2013 IEEE 16th International Conference on Computational Science and Engineering,* Sydney, NSW (pp. 833–837). https://doi.org/10.1109/CSE.2013.126.
30. Yandun, F., Reina, G., Torres-Torriti, M., Kantor, G., & AuatCheein, F. (2017). A survey of ranging and imaging techniques for precision agriculture phenotyping. *IEEE Transactions on Mechatronics* MECH.
31. Jhuria, M., Kumar, A., & Borse, R. (2013). Image processing for smart farming: detection of disease and fruit grading. In *2013 IEEE 2nd International Conference on Image Information Processing, IEEE ICIIP 2013* (pp. 521–526). https://doi.org/10.1109/ICIIP.2013.6707647.
32. Tripicchio, P., Satler, M., Dabisias, G., Ruffaldi, E., & Avizzano, C. A. (2015). Towards smart farming and sustainable agriculture with drones. In *International Conference on Intelligent Environments (IE)*.
33. Pobkrut, T., Eamsaard, T., & Kerdcharoen, T., Sensor drone for aerial odor mapping for agriculture and security services.

34. Shi, L., Duan, Q., Ma, X., & Weng, M. (2012). The research of support vector machine in agricultural data classification. In D. Li & Y. Chen (Eds.), *Computer and Computing Technologies in Agriculture V. CCTA 2011. IFIP Advances in Information and Communication Technology* (Vol. 370). Berlin, Heidelberg: Springer.
35. Ghosh, S., Sayyed, S., Wani, K., Mhatre, M., & Hingoliwala, H. A. (2016). Smart irrigation: A smart drip irrigation system using cloud, android and data mining. In *2016 IEEE International Conference on Advances in Electronics, Communication and Computer Technology (ICAECCT)* (pp. 236–239), Pune. https://doi.org/10.1109/ICAECCT.2016.7942589.
36. Shekar, Y., Dagur, E., Mishra, S., Tom, R. J., Veeramanikandan, M., & Sankaranarayan, S. (2017). Intelligent IoT in automated irrigation. *International Journal of Applied Research, 12*(18). ISSN 0973-4562.
37. Wang, Q., Terzis, A., & Szalay, A. (2010). A novel soil measuring wireless sensor network. In *Conference Record—IEEE Instrumentation and Measurement Technology Conference* (pp. 412–415). https://doi.org/10.1109/IMTC.2010.5488224.
38. Chetan Dwarkani, M., Ganesh Ram, R., Jagannathan, S., & Priyatharshini, R. (2015). Smart farming system using sensors for agricultural task automation. In *IEEE International Conference on Technological Innovations in ICT for Agriculture and Rural Development (TIAR 2015).*
39. Heble, S., Kumar, A., Prasad, K. V. D., Samirana, S., Rajalakshmi, P., & Desai, U. B. (2017). A low power IoT network for smart agriculture. In *2018 IEEE 4th World Forum on Internet of Things (WF-IoT)* (pp. 609–614), Singapore. https://doi.org/10.1109/WF-IoT.2018.8355152.
40. Murrieta-Rico, F. N., Petranovskii, V., Raymond-Herrera, O., Sergiyenko, O., Lindner, L., Valdez-Salas, B., et al., Resolution improvement of accelerometers measurement for drones in agricultural applications.

Deep Learning Approach-Based Network Intrusion Detection System for Fog-Assisted IoT

Nausheen Sahar, Ratnesh Mishra, and Sidra Kalam

Abstract In an Internet of Things (IoT) network, miscellaneous devices exchange their resources as per their requirement via the Internet. The data aggregated from these IoT devices are stored in the cloud layer. Increment in the number of IoT devices and accessing real-time data leads to a huge latency issue. One solution is to use the fog layer which is an adjunct layer between the cloud layer and the end-user. It is very convenient for the user to access the data from the fog layer as it is at the brink of the network. Security in the fog layer is a major drawback. Easy accessing of resources from the fog layer makes the system more vulnerable to various attacks. In this research work, network intrusion detection system (NIDS) is developed based on the conception of deep learning. NIDS is a device implemented in the fog node for attack detection. This model proficiently detects any kind of malicious activity. UNSW-NB15 and NSL-KDD datasets are used to evaluate the performance of the proposed NIDS model. Accuracy of the model is also compared with these datasets. The result analysis shows that the model is working more efficiently for the NSL-KDD dataset. The accuracy of the proposed NIDS model for the UNSW-NB15 and NSL-KDD datasets is 91.20% and 95.40%, respectively.

Keywords IoT · Fog · IDS · NIDS · Deep learning · Anomaly detection

1 Introduction

IoT is nowadays very trendy as it grants access to the data from anywhere any time via the Internet. This makes a system highly interconnected with each other which increases the efficiency of the system. Data aggregated from IoT devices are stored at

N. Sahar (✉) · R. Mishra · S. Kalam
Birla Institute of Technology, Mesra, Patna campus, Patna 800014, India
e-mail: nausheen.jiet@gmail.com

R. Mishra
e-mail: r.mishra@bitmesra.ac.in

S. Kalam
e-mail: sidrakalam1@gmail.com

S. Tiwari et al. (eds.), *Proceedings of International Conference on Big Data, Machine Learning and their Applications*, Lecture Notes in Networks and Systems 150,
https://doi.org/10.1007/978-981-15-8377-3_4

the cloud layer which need not be accessed very frequently. Increment in the number of devices in the system increases the latency issue [1, 2]. Not only this, but it also makes the system an easy target for the attackers. Taking this into consideration, the fog layer is used to store the data which is frequently accessed by the end-user which is an intermediary layer between the cloud layer and the end-user. Fog layer includes several fog nodes which are the devices (routers, switches, etc.) that can be easily deployed anywhere in the network according to the requirement of the end-user. At the same time, it also increases the possibility of attacks. If it is not dealt for a longer period, the availability of the system to the end-user will get affected [3]. To get over these issues, several existing techniques can be used for the security of IoT networks. Traditional security measures cannot be applied to low-power and resource-constrained IoT device which is further solved by adding a fog layer that helps in enabling the computing services at the edge of the network. The security issue of the IoT generates the need for reliable IDS. IDS is a monitoring system that is used to detect malicious activities happening in the network. Various types of IDS are network-based intrusion system (NIDS), host-based intrusion detection system (HIDS) and hybrid-based intrusion detection system (HIDS) which helps in attack detection. In this research work, we have used the NIDS system which is based on a deep learning approach. Deep learning ensures more accurate results comparative to machine learning algorithms [4]. There are multiple applications in which deep learning concept is used in various ways. In the classification of brain tumors [5], autonomous concrete crack detection [6], prediction of the material defects [7], in HealthFog [8] and many more. It consists of multiple hidden layers that make the model more efficient to work on large-scale data. The dynamics of a network keep on changing with the addition of any new device. Considering this, the IDS must be flexible in such a way that it does not affect its functionality and hence can produce an effective result.

The main contributions of the proposed model are:

- Deep learning-based NIDS system is developed to detect malicious activities.
- Implementation and evaluation are done to check the model efficiency.

This paper continues as Sect. 2 gives a brief idea of the existing research works that are related with this proposed work, Sect. 3 explains about an overview of deep learning approach, Sect. 4 discusses the proposed work and then implementation detail is described in Sect. 5. In Sect. 6, result analysis and conclusion are discussed (Fig. 1).

1.1 *Author's Contribution*

The major contributions of the authors are listed below:

- A novel NIDS model is designed to classify the network traffic as attack and benign.

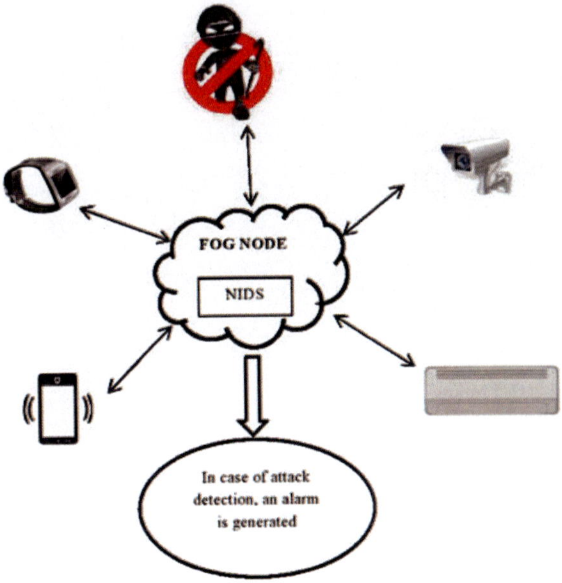

Fig. 1 Basic diagram of the proposed NIDS model

- Fog layer is deployed in an IoT environment to shift the processing and storage to the nearby fog nodes from sensor nodes to reduce the energy consumption.
- Deep neural network is deployed in the fog layer. Its working is encouraged from the way that human brain thinks and takes the decision. The hidden layers are used for feature extraction from the raw data
- Accuracy, precision, recall and F-score of NSL-KDD and UNSW-NB15 are compared.

2 Related Work

An IDS is a monitoring system that is used to detect malicious activities happening in the network. This helps the system to get rid of the attacks by implementing various technologies. Many research works are present regarding the IoT security using a deep neural network (DNN). Some of them are briefly discussed in this section.

Ad hoc networks are a self-organized network in the IoT environment. The probability of an attack is even more in this type of network [9]. Especially, denial-of-service (DoS) attack is very often used by the attacker. To deal with DoS, Fang Feng and Xin Liu have developed a model in which they have used the concept of the deep neural network [10]. They used DNN along with the convolutional neural network (CNN) and long short-term memory (LSTM) to detect Structured Query Language (SQL) and cross-site scripting (XSS) attack. To make an intruder-free network, firstly, we need to detect malicious activities. There is numerous research work that has developed an intrusion detection model based on the conceptualization

Table 1 Review of the datasets used and the attacks detected in the existing schemes

References	Type of dataset used	Attacks detected
Aloqaily et al. [16]	NS-3 and NSL-KDD	DoS, Probe, R2L, and U2R
Elsaeidy et al. [17]	The raw dataset generated from the smart water system	DDoS
Feng et al. [10]	KDD CUP'99	DoS, XSS, and SQL
Song et al. [18]	The real dataset is taken from a vehicle	DoS, message injection attack, and spoofing attack
Moustafa et al. [13]	NSL-KDD and UNSW-NB15	DoS, Normal, Probe, R2L and U2R Backdoor, Exploits, Fuzzer, Generic, Normal, Reconnaissance, Shellcode and Worms

of deep neural network [11, 12]. There are sundry types of deep learning systems that are the convolutional neural network, restricted Boltzmann machines, deep belief networks and deep auto-encoders to optimize the accuracy rate of anomaly detection. Muna AL-Hawawreh and NourMoustafa have used NIDS to detect the malicious activities [13]. Deep auto-encoder along with the deep feed-forward neural network is used to train the model. Denoising autoencoder [14] is another form of the autoencoder which is used for the development of the deep neural network. The main purpose is to design IDS especially for the detection of attacks in the cloud computing system because of its complex architecture. Deep learning models are also very efficiently working in the medical field [5–7, 15] etc. (Table 1).

3 Overview of Deep Learning

Deep learning is headway of machine learning that comprises multiple layers used to extricate high-level features from the training data. Each layer includes multiple neurons along with activation function to use in producing nonlinear outputs [19]. It extracts the features to formulate a model that gives the maximum accurate results. It is a type of feed-forward neural network in which the result of the output layer becomes the input of the input layer. Initially, all the inputs with its weighted sums are passed through a nonlinear activation function to produce an output. Then, the output of the previous layer with a bias value is computed by the activation function to get the weighted input of the next layer.

$$\overline{O}_n = \alpha\left[\omega_n I_{n-1}\right] \tag{1}$$

The calculated output is then compared with the target output in case of supervised learning. The loss function is applied to compute the loss between the predicted and the actual output.

Table 2 Key notations used in this paper

Key notation	Description
$\overline{O}_n$	The calculated output of the nth layer
ω_n	The weight associated with the nth layer
$I_{\mathrm{n}-1}$	The weight associated with the $(n-1)$th layer
£	Loss function
α	Activation function
$\Delta\omega_{ij}$	Weight update of the link connecting *i*th and *j*th neuron of the two neighboring layers
η	Learning rate parameter
$\frac{\delta\varepsilon}{\delta\omega_{ij}}$	Error gradient concerning the weight ω_{ij}

$$£ = \sum_{n=1}^{k} \left(\overline{O}_n - \tilde{O}_n\right)^2 \tag{2}$$

Now, to minimize the loss, gradient descent method is applied using weight and the activation function.

$$\Delta\omega_{ij} = \eta \frac{\delta\varepsilon}{\delta\omega_{ij}} \tag{3}$$

This is how the neural network learns in supervised learning by minimizing the error gradually in each step by adjusting the parameters (Table 2).

4 Proposed Work

In the proposed research work, the IoT system is protected against various attacks by classifying the attack using IDS. For this, a network-based intrusion detection system (NIDS) is developed using deep neural network implemented at the fog layer. This NIDS is signature based as the signature of each attack is stored in the NIDS knowledge database which helps in attack detection. This is a stand-alone device as it is easily deployed in the fog node irrespective of the heterogeneous network architecture. The NIDS is comprised of three phases, i.e., the monitoring phase, the anomaly detection phase and the response phase, as depicted in Fig. 3. The working of all the phases is discussed in the below sub-section.

4.1 The Monitoring Phase

In this phase, signature-based NIDS is deployed in the fog node to facilitate the smooth flow of network traffic. Mainly, it is responsible for monitoring the network connections, the incoming and the outgoing network traffic. It matches the attributes of the packet traffic with the stored set of signatures of the attacks in the NIDS knowledge database. Based on the training, NIDS is also responsible for the classification of network traffic as malicious or benign using the message details like IP of source, IP of destination, transmission rate, reception rate, transmission mode, duration, etc.

4.2 Anomaly Detection Phase

This phase includes the cleaning process of data, transformation of data into machine codes, feature extraction and feature selection to make the data simpler to work on. After the completion of the preprocessing of data, the concept of deep learning is used to train the proposed NIDS model which is preferable than the traditional machine learning algorithm. Deep learning helps in learning useful features by adding more numbers of hidden layers if required to achieve high accuracy. The proposed neural network is comprised of one input layer, two hidden layers, and one output layer which classify the attacker and the legitimate users. The input layer consists of 1024 neurons and hidden layer consists of 768 and 512 neurons, respectively. ReLU (Rectified linear unit) activation function is used in the hidden layer to provide the filtered output to the next hidden layer. Compared to nonlinear activation functions like sigmoid and tangent, ReLU is found to be more efficient for training large-scale data in terms of time and cost. Mathematically, it is defined in Eq. (4).

$$F(y) = \max(0,\ y) \tag{4}$$

ReLU function transforms the input (y) as-

Case 1. If $y \leq 0$, then $y = 0$

Case 2. If $y > 0$, then y will remain unchanged

The softmax function is also used as an activation function for the known number of labeled classes. We have used this in the second hidden layer of the proposed neural network architecture for the classification of the attacks. The mathematical representation of the softmax function (σ) is shown in Eq. (5).

$$\sigma(\acute{Z})_i = e\frac{\acute{Z}_i}{\sum_{j=1}^{m} e^{m_j}} \quad \text{for i} = 1, \ldots, m \tag{5}$$

$\acute{Z} = \left(\acute{Z}_1, \ldots, \acute{Z}_m\right)$ where m is the real numbers and Z is the input vector.

Fig. 2 Deep neural network for our proposed NIDS model

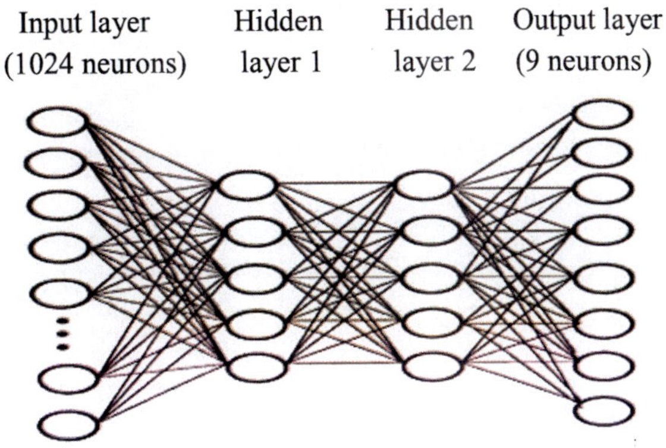

The output layer contains nine neurons as the attacks found in the UNSW-NB15 dataset are nine (Fig. 2).

After the training of the model, the softmax function helps in classifying the network traffic as benign or attack.

4.3 Response Generation Phase

This phase is responsible for decision making and response generation. As soon as any baleful activity is detected in the network, an alarm is generated as a security alert for all the devices in the IoT system. Immediately, it sends a notification to the security administrator to take an immediate action to avert the system to suffer from any major loss (Fig. 3).

5 Implementation and Evaluation

5.1 Datasets

A dataset is very important in testing, examining and evaluating the working of the proposed model. Many research works has used the NSL-KDD dataset which is popular in many of the IDS models for the detection of any kind of malicious activity. NSL-KDD is an upgraded form of KDD CUP 99 which consists of many redundant records. NSL-KDD dataset is mainly used for comparing the various IDS models to prove its efficiency. The reason behind this is the availability of public datasets for network-based intrusion detection system (NIDS). Data preprocessing is done by Python programming in which raw data is cleaned to remove redundancy and transform it in an appropriate format so that it can be used. NSL-KDD and UNSW-NB15 datasets are used for better clarity of the performance of our model.

Fig. 3 NIDS framework

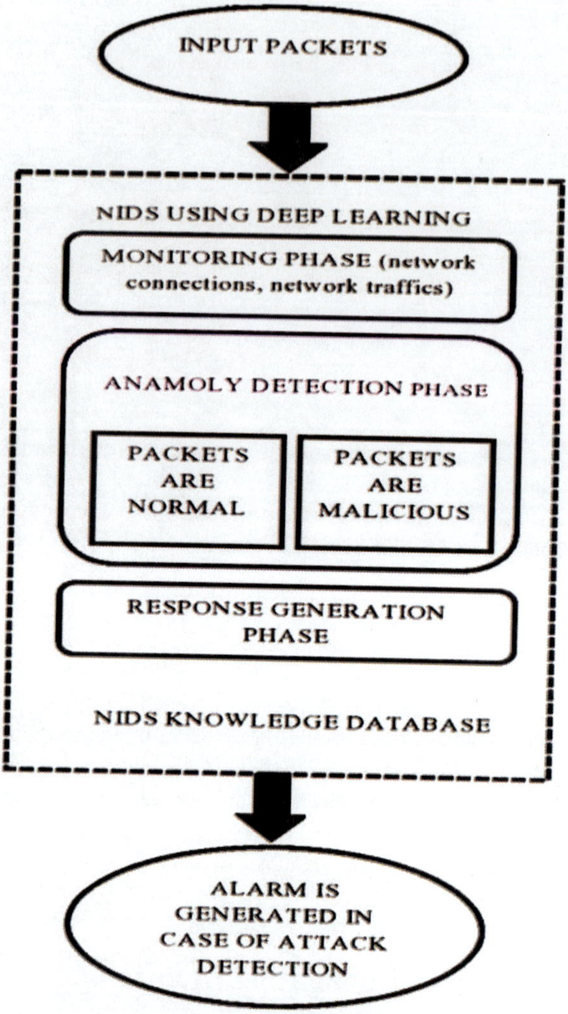

UNSW-NB15 is a current dataset that is mostly used now a day as it reflects the actual behavior of today's network traffic. Attacks found in this dataset are DoS, Backdoors, Analysis, Exploits, Generic, Reconnaissance, Shellcode, Worms and Fuzzers [20]. In the NSL-KDD dataset, overall records after the preprocessing are 148,518 in which 22,544 is the test dataset and 125,974 is the training dataset. Five types of attacks are found in this dataset that is DoS, Probe, User to the root (U2R), Remote to Local (R2L) and normal. In the UNSW-NB15 dataset, the total number of raw records was 2,800,004. After preprocessing, it is 259,675 in which 175,342 is used for testing and 82,333 is used for the training of data.

5.2 Evaluation Metrics

The performance of the proposed NIDS model is valuated on NSL-KDD and UNSW-NB15 to make a clear vision of the efficiency of the model. Accuracy, recall, precision and F-score are determined to check the model performance. The main evaluation metrics are calculated in the following way.

- Accuracy is the overall correctly classified observations among the total observations in the test dataset. It is calculated as

$$\text{Accuracy} = \frac{\text{TP} + \text{TN}}{\text{TP} + \text{TN} + \text{FP} + \text{FN}}$$

- Precision is the percentage of relevant results among the total retrieved instances. It is calculated as

$$\text{Precision} = \frac{\text{TP}}{\text{TP} + \text{FP}}$$

- The recall is the percentage of total relevant results that are correctly classified by the model. It is calculated as

$$\text{Recall} = \frac{\text{TP}}{\text{TP} + \text{FN}}$$

- F-score is the measure of a test's accuracy. It is calculated as

$$\text{F-Score} = 2 * \frac{\text{Precision} * \text{Recall}}{\text{Precision} + \text{Recall}}$$

5.3 Result Analysis

UNSW-NB15 and NSL-KDD datasets are used for the evaluation of the proposed NIDS model. The accuracy of the model is 81.2% and 95.4% respectively. The proposed scheme worked more efficiently with the NSL-KDD dataset. The comparison of the performance metrics of both the dataset is discussed in Table 4 (Figs. 4, 5, 6, 7 and 8).

Table 4 Comparison of NIDS model using NSL-KDD and UNSW-NB15 dataset

Dataset	Accuracy (%)	Precsion (%)	Recall (%)	F-score (%)
NSL-KDD	95.4	96.2	95.4	90.2
UNSW-NB15	91.2	90.4	91.2	86.9
UNSW-NB15	91.2	90.4	91.2	86.9

Fig. 4 Accuracy of UNSW-NB15 and NSL-KDD datasets

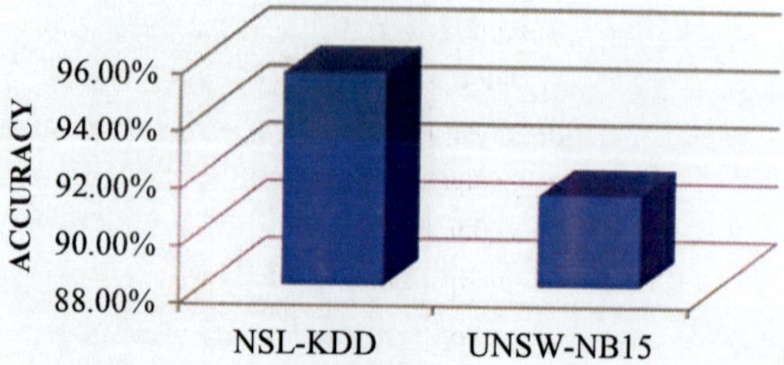

Fig. 5 The precision of UNSW-NB15 and NSL-KDD datasets

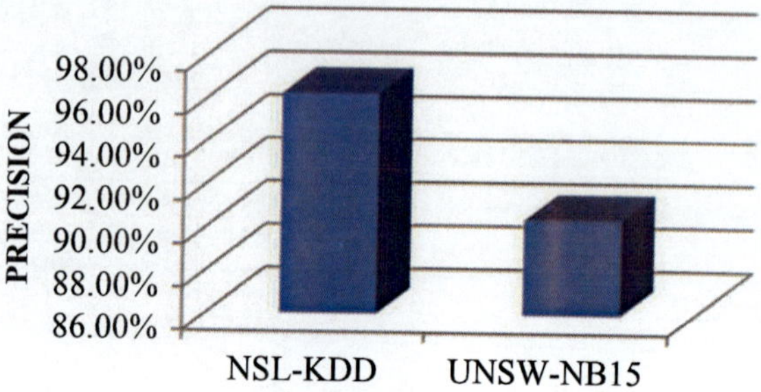

Fig. 6 Recall of UNSW-NB15 and NSL-KDD datasets

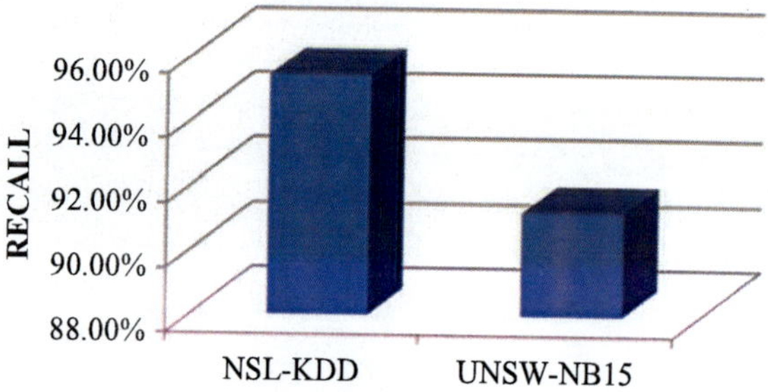

Fig. 7 F-score of UNSW-NB15 and NSL-KDD datasets

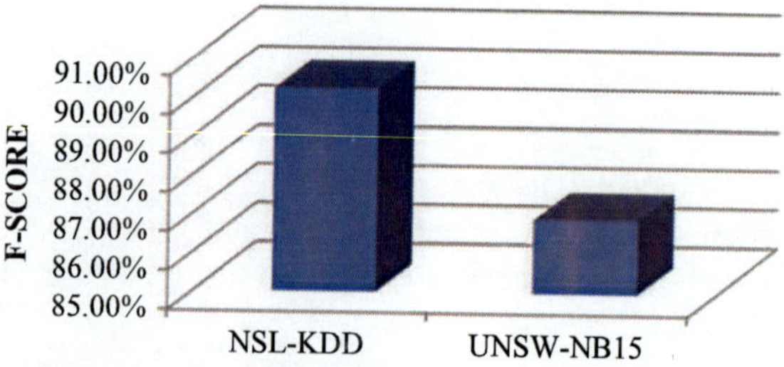

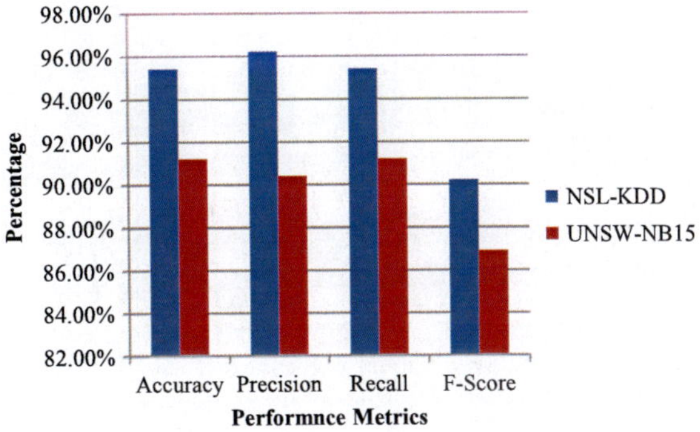

Fig. 8 Performance metrics of NSL-KDD and UNSW-NB15 datasets

6 Conclusion

In this paper, the proposed NIDS model is based on deep learning. This model is deployed in the fog node for attack detection. Preprocessing of datasets and evaluation of accuracy, precision, recall and F-score is done by Python programming in PyCharm. Implementation of deep neural network is done using keras which is an open-source neural network library in python. Two types of datasets are used in this model for performance measure that is NSL-KDD and UNSW-NB15. After the implementation and evaluation of the model on both the datasets, we came to know that the model is more effectively working on the NSL-KDD dataset comparative to the UNSW-NB15 dataset. The accuracy measure of the model for the NSL-KDD dataset and UNSW-NB15 datasets is 95.40% and 91.20%, respectively. In the future, we can extend our proposed NIDS model for the detection of new attacks.

References

1. Kaur, H., & Sood, S. K. (2019). Fog-assisted IoT-enabled scalable network infrastructure for wildfire surveillance. *Journal of Network and Computer Applications, 144,* 171–183.
2. Mutlag, A. A., Ghani, M. K. A., Arunkumar, N. A., Mohammed, M. A., & Mohd, O. (2019). Enabling technologies for fog computing in healthcare IoT systems. *Future Generation Computer Systems, 90,* 62–78.
3. Thamilarasu, G., & Chawla, S. (2019). Towards deep-learning-driven intrusion detection for the Internet of Things. *Sensors, 19*(9), 1977.
4. Diro, A. A., & Chilamkurti, N. (2018). Distributed attack detection scheme using deep learning approach for Internet of Things. *Future Generation Computer Systems, 82,* 761–768.
5. Mohsen, H., El-Dahshan, E. S. A., El-Horbaty, E. S. M., & Salem, A. B. M. (2018). Classification using deep learning neural networks for brain tumors. *Future Computing and Informatics Journal, 3*(1), 68–71.
6. Dung, C. V. (2019). Autonomous concrete crack detection using deep fully convolutional neural network. *Automation in Construction, 99,* 52–58.
7. Feng, S., Zhou, H., & Dong, H. (2019). Using deep neural network with the small dataset to predict material defects. *Materials & Design, 162,* 300–310.

8. Tuli, S., Basumatary, N., Gill, S. S., Kahani, M., Arya, R. C., Wander, G. S., & Buyya, R. (2020). HealthFog: An ensemble deep learning based Smart Healthcare System for Automatic Diagnosis of Heart Diseases in integrated IoT and fog computing environments. *Future Generation Computer Systems, 104,* 187–200.
9. Zhang, P., Wang, S., Guo, K., & Wang, J. (2018). A secure data collection scheme based on compressive sensing in wireless sensor networks. *Ad Hoc Networks, 70,* 73–84.
10. Feng, F., Liu, X., Yong, B., Zhou, R., & Zhou, Q. (2019). Anomaly detection in ad-hoc networks based on deep learning model: A plug and play device. *Ad Hoc Networks, 84,* 82–89.
11. Karatas, G., Demir, O., & Sahingoz, O. K. (2018, December). Deep learning in intrusion detection systems. In *2018 International Congress on Big Data, Deep Learning and Fighting Cyber Terrorism (IBIGDELFT)* (pp. 113–116). IEEE.
12. Vijayakumar, R., Alazab, M., Soman, K. P., Poornachandran, P., Al-Nemrat, A., & Venkatraman, S. (2019). Deep learning approach for the intelligent intrusion detection system. *IEEE Access, 7,* 41525–41550.
13. Muna, A. H., Moustafa, N., & Sitnikova, E. (2018). Identification of malicious activities in the industrial internet of things based on deep learning models. *Journal of Information Security and Applications, 41,* 1–11.
14. Abusitta, A., Bellaiche, M., Dagenais, M., & Halabi, T. (2019). A deep learning approach for the proactive multi-cloud cooperative intrusion detection system. *Future Generation Computer Systems, 98,* 308–318.
15. Kong, Y., Gao, J., Xu, Y., Pan, Y., Wang, J., & Liu, J. (2019). Classification of autism spectrum disorder by combining brain connectivity and deep neural network classifier. *Neurocomputing, 324,* 63–68.
16. Aloqaily, M., Otoum, S., & Al Ridhawi, I. (2019). & Jararweh, YAnintrusion detection system for connected vehicles in smart cities. *Ad Hoc Networks, 90,* 101842.
17. Elsaeidy, A., Munasinghe, K. S., Sharma, D., & Jamalipour, A. (2019). Intrusion detection in smart cities using Restricted Boltzmann Machines. *Journal of Network and Computer Applications, 135,* 76–83.
18. Song, H. M., Woo, J., & Kim, H. K. (2020). In-vehicle network intrusion detection using deep convolutional neural network. *Vehicular Communications, 21,* 100198.
19. Amanullah, M. A., Habeeb, R. A. A., Nasaruddin, F. H., Gani, A., Ahmed, E., Nainar, A. S. M., et al. (2020). Deep learning and big data technologies for IoT security. *Computer Communications.*
20. Choudhary, S., & Kesswani, N. (2020). Analysis of KDD-Cup'99, NSL-KDD and UNSW-NB15 Datasets using Deep Learning in IoT. *Procedia Computer Science, 167,* 1561–1573.

Improving Prediction Accuracy for Debonding Quantification in Stiffened Plate by Meta-Learning Model

Abhijeet Kumar, Anirban Guha, and Sauvik Banerjee

Abstract Improving prediction accuracy has been a major challenge for damage diagnosis system in the field of structural health monitoring. To tackle this issue, several machine learning algorithms have been used. This study presents effectiveness in improving prediction accuracy of meta-learning model over a range of individual machine learning algorithms in damage diagnosis. The learning algorithm chosen in this paper is support vector machine, random forest, vote method, gradient boosting regression and stacked regression as meta-model. The learning algorithms are employed for debonding quantification in metallic stiffened plate. The algorithms trained and tested on numerically simulated first mode shape vibration data. To check robustness of algorithms, artificial noise is added in numerically simulated data. The result showed that the prediction accuracy of the meta-model as stacked regression is better than the individual model.

Keywords Structural health monitoring · Debonding prediction · Stacked regression · Accuracy improvement

1 Introduction

Improving prediction accuracy of the machine learning techniques for damage assessment in structural health monitoring (SHM) is interested topic for researchers. Ever since the machine learning technique introduced in SHM by [1] in 1992, numerous studies have been done to boost the prediction accuracy for damage detection and damage characterization in mechanical, civil and aerospace structures [2]. There are numerous machine learning algorithms such as artificial neural network (ANN), support vector machine (SVM), decision tree (DT) and Bayesian analysis which

A. Kumar (✉) · A. Guha
Department of Mechanical Engineering, Indian Institute of Technology Bombay, Mumbai 400076, India
e-mail: abhijeetshandilya19@gmail.com

S. Banerjee
Department of Civil Engineering, Indian Institute of Technology Bombay, Mumbai 400076, India

S. Tiwari et al. (eds.), *Proceedings of International Conference on Big Data, Machine Learning and their Applications*, Lecture Notes in Networks and Systems 150,
https://doi.org/10.1007/978-981-15-8377-3_5

have been successfully implemented for damage assessment in structures [3]. Despite successfully implementation of machine learning algorithms in SHM field, there still need robust algorithms which could be able to show robust in performance with noisy sensor data collected from structure under analysis. For same type of data, different classifiers perform differently. One of the major issues with individual classifier is that it may give better classification accuracy for one data type, but fail to gives similar result with another type of data. The issue with classifier can be resolved by combining number of classifier. There are several methods to combine these classifiers such as minimum, maximum, median, sum, product of posterior probability and majority of vote [4–7]. These methods are simple and easy to implement even after training of individual classifier. Identifying and combining best-performing individual classifier so that the performance of combined classifier is better than individual classifier is termed as an ensemble classifier [8, 9].

The ensemble learning is the method by which a group of weak learner classifier combined to form a strong learner classifier model. The sole purpose of ensemble learning is to improve the performance of model. A state of the art on ensemble algorithms, methodology and combining rule has given in [10]. The most popularly used ensemble models are bagging [11], boosting, Adaboost [12] and stacked generalization [13], mixture of expert. The stacked generalization ensemble model introduced by Wolpert in the early 1990s [13]. In stacked generalization also commonly termed as stacking, the information gathers by lower-level classifier transformed to higher-level classifier also known as *meta-learner* to boost the prediction capability of model. The idea of stacked generalization further extended to regression problem by Breiman [14]. In stacked generalization ensemble classifier, the misclassification rate is minimized by careful selection of *level-1* (base) classifier [15]. The selection of best among *level-1* classifier is done by further training of predicted decision of base classifier to form *meta*-classifier model [16, 17]. These features of stacked generalization classifier attract the research of different fields, and there, it has been implemented in multiple field for better prediction accuracy [18–22].

This paper presents stacked regression meta-learner model for debonding length prediction in stiffened plate structure. The effectiveness of meta-learner model as compared to base learning is measured by adding noise level in raw dataset. The algorithms that are used include single learner SVM, as well as ensemble learner random forest (RF), vote method, gradient boosting regression (GBR) as base learner and SVM as meta-level leaner in stacked regression ensemble model. The training data is generated by finite element simulation of stiffened plate structure. Standard first mode vibration displacement data use as feature vector.

2 Theoretical Background

2.1 *Support Vector Machine (SVM)*

SVM is defined by a separating hyperplane. In soft margin SVM classifier, the hyperplane is obtained by minimizing cost function

$$\frac{1}{2}w^T w + C\sum_{i=1}^{N}\xi_i \tag{1}$$

Subjected to:

$$y_i\left(w^T\varphi(x_i) + b\right) \geq 1 - \xi_i \tag{1a}$$

$$\xi_i \geq 0,\ i = 1, \ldots, N \tag{1b}$$

where C is penalty factor, w weight vector perpendicular to the hyperplane, b is constant, x input data point, y is class label. φ is kernel function. The types kernel function uses are radial basis function (RBF), polynomial, linear and sigmoid.

2.2 *Random Forest (RF)*

RF is an ensemble learning model of number of decision tree classifier. The number of subsample dataset is formed by bootstrapping and random sub-feature of original input dataset.

The algorithm for RF classifier and regression [23].

1. For $n = 1$ to N:
   a. Bootstrap of sample D_i of size N (D_i is independent identically distributed (i.i.d))
   b. Develop a forest T_n to bootstrapped data, by recursively repeating the following steps for each terminal node of the tree, until minimum node size is reached.
      i. Select m features at random from p features ($p \gg m$).
      ii. Among m features, calculate the best split.
      iii. Using best split, split the node into daughter node.
2. Ensemble the output of trees $\{T_n\}_1^N$

For prediction at a new point x:

$$\text{Regression: } \hat{f}_{rf}^{N}(x) = \frac{1}{N}\sum_{n=1}^{N} T_n(x)$$

$$\text{Classification: } \hat{C}_{rf}^{N}(x) = \text{majority vote } \left\{\hat{C}_b(x)\right\}_1^N$$

2.3 *Gradient Boosting Regression (GBR)*

GBM is a combination of boosting and gradient descent. In GBM, the loss of weak learners is minimized and forms strong learner by combining all weak learners by assigning weight to each learner.

The residual of each weak model F is given by:

$$\begin{aligned} h(x_1) &= y_1 - f(x_1) \\ h(x_2) &= y_2 - f(x_2) \\ &\vdots \\ h(x_n) &= y_n - f(x_n) \end{aligned} \tag{2a}$$

where x_i is input and y_i is output variables, and there is n number of weak learner. Now, fit a regression to data

$$\{(x_1, y_1 - f(x_1)), (x_2, (y_2 - f(x_2)), \ldots, (x_n, y_n - f(x_n))\} \tag{2b}$$

by using gradient decent to minimization loss function h. If the new model $F + h$ is still not satisfactory, we can add another regression and repeat the entire process.

2.4 *Vote Model*

In vote model, the calculated posterior probability of each base classifier for each class is combined by either average of probability or multiplication of probability.

$$\hat{y} = \arg\max \sum_{j=1}^{m} w_i p_{ij} \tag{3}$$

where p predicted posterior probability for a classifier and w_i is weight assigned to j classifier.

2.5 *Stacked Model*

Stacked generalization is performed its tasks in two stages. In first stage, the layer-1 learner (also called base learner) is trained on the level-0 training dataset and then output prediction of layer-1 which created the level-1 training dataset use to train layer-2 learner known as meta-learner.

The learning models $M_1, M_2, \ldots, M_m$ train on dataset $X(x^* = (x_i, y_i))$. In first stage, base-level learner $f_1(x^*), f_2(x^*), \ldots, f_m(x^*)$ is created. To create training dataset for training for *Meta-learner*, a cross-validation (leave-one-out) procedure is applied. The cross-validation is applying to each base-level learner on original dataset:$\forall_i = 1, \ldots, n : \forall_j = 1, \ldots, m : f_j^i(x^*) = M_j(X - x^*)$. Then, use trained learner to generate predictions for $x^* : \hat{y}_i^j = f_j^i(x_i)$. This generates training dataset for *layer-2(meta-level)* learner of the form $\left\{\left(\hat{y}_i^1, \hat{y}_i^2, \ldots, \hat{y}_i^n\right), y_i\right\}$ which consist of both base-level prediction and correct target [17].

3 Finite Element Modelling

3.1 *Stiffened Plate Modelling*

An aluminium plate of length 530 mm, width 330 mm and thickness 1.5 mm and an inverted T-shape stiffener of the same material of length 530 mm, flange dimension 2 mm × 26 mm and Web dimension 26 mm × 2 mm was modelled in finite element packaged ANSYS 16.2. The stiffener was situated in the middle of the plate along the length of plate. Material properties of aluminium for both plate structure and stiffener for modelling are: Young's modulus $E = 70$ Gpa, density $(\rho) = 2700$ kg/m^3 and Poisson's ratio $(\nu) = 0.33$ (Fig. 1).

The boundary condition for the present study was fixed-free-fixed-free with fixed boundary conditions applied at the stiffener end. An eight-nodded quadrilateral shell element SOLSH190 (suitable for modelling of shell structure with a range of thickness) with three degrees of freedom in translation direction at each node is chosen for modelling of both plate and stiffener. After deciding element type, a convergence analysis is done to select the size of element, and 1 mm × 1 mm element size is finally selected for meshing of entire structure (Table 1).

3.2 *Modelling of Debonding and Contact*

Damages in the stiffened plate are introduced as debonding between plate and stiffener. Debondings are created in the middle of the stiffener. There is a total of 71 debonding are created with a size of 10 mm to 150 mm with a step length of 2 mm.

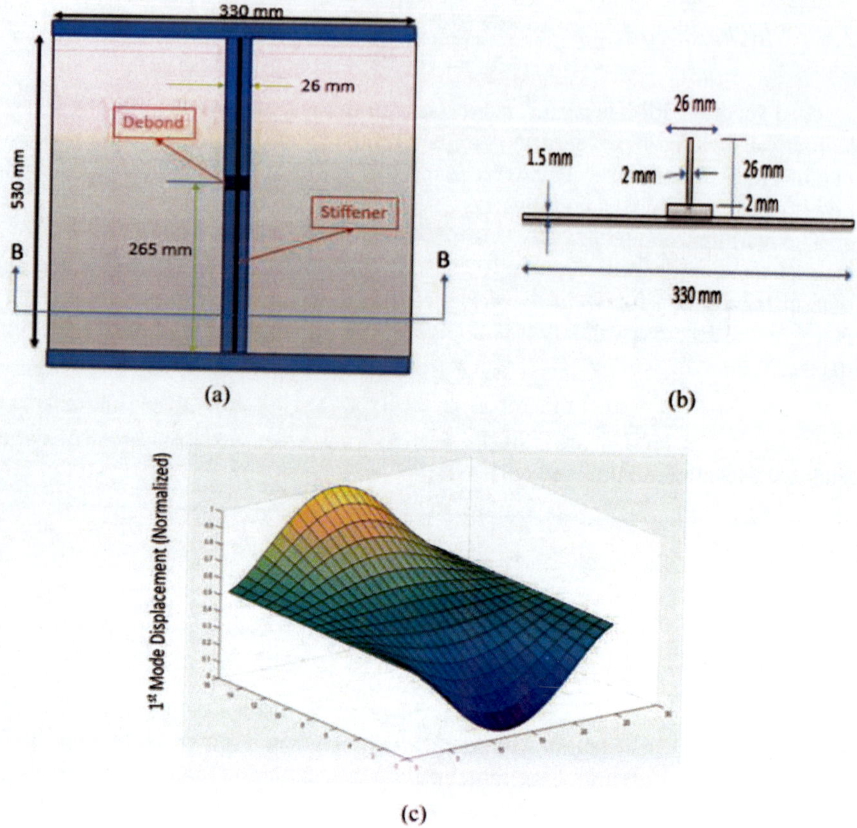

Fig. 1 **a** Geometry of stiffened plate. **b** Section B-B view. **c** First mode shape

Table 1 First three modal frequencies of undamaged stiffened plate

Frequency	Mode 1	Mode 2	Mode 3
FEM	49.953 Hz	70.009 Hz	110.26 Hz
Experimental	49.375 Hz	68.22 Hz	105.9 Hz

To model the contact between plate and stiffener in ANSYS© with parameter which are given in Table 2:

4 Methodology Framework

The procedure of implementations is explained in detail for the stiffened plate structure in this section. The procedure starts with collection 1st mode shape data from

Table 2 Contact modelling detail

Parameter	Type
Contact type	Flexible surface to surface
Target element	TARGE170
Contact element	CONTA174
Contact algorithm	Augmented Lagrange Method
Location of contact detection	Gauss point
Contact selection	Asymmetric
Gap/closure	No adjustment
Behaviour of contact surface	No Separation
Geometry	3-D

FE simulation result in data acquisition process. Then, there are 114 mode shape displacement data of different node points which have taken as feature vector for training and testing of learner model situated each at gap of 20 nodes starting from one of the corners of stiffened plate. Because of symmetry, only one side of mode shape data has taken (Fig. 2).

The prediction of debonding length in the stiffened plate by use of support SVM, RF, vote method, GBR and stacked regression model has been discussed. For the *SVM* algorithm, the hyperparameter of SVM ($\boldsymbol{C}$, $\boldsymbol{\gamma}$, $\boldsymbol{\epsilon}$) has been tuned. By using trial and error method, the tuned value of parameters is $C = 10000$, $\gamma = 0.03$, $\in = 0.0001$. In the present study, radial basis function (RBF) used as kernel function. Similarly, RF algorithm has been applied to predict debonding length. For *vote* combiner model, SVM and RF learning algorithms were used as two independent base learner models. The posterior probability distribution of the base learner output was used as input to vote combiner model. An average of posterior probability distribution combining rule is used to predict the final output of the model. In GBR model, RF is chosen as the base learner algorithm, and 10 number of iteration has performed for final prediction. The stacked regression model has implemented as the meta-model in the present study. The framework for the stacking model is shown in Fig. 3. SVM, RF,

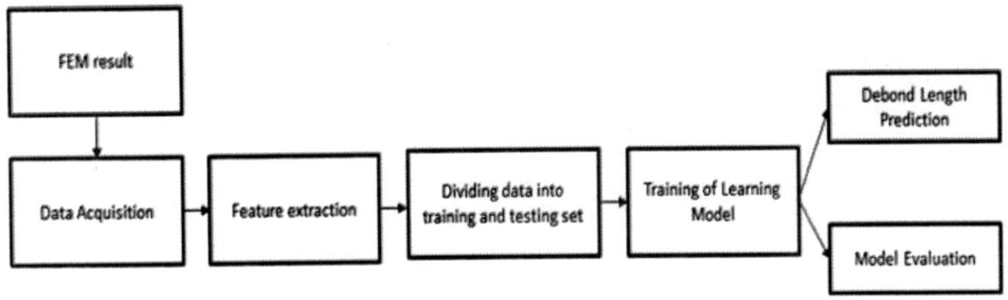

Fig. 2 Methodology framework

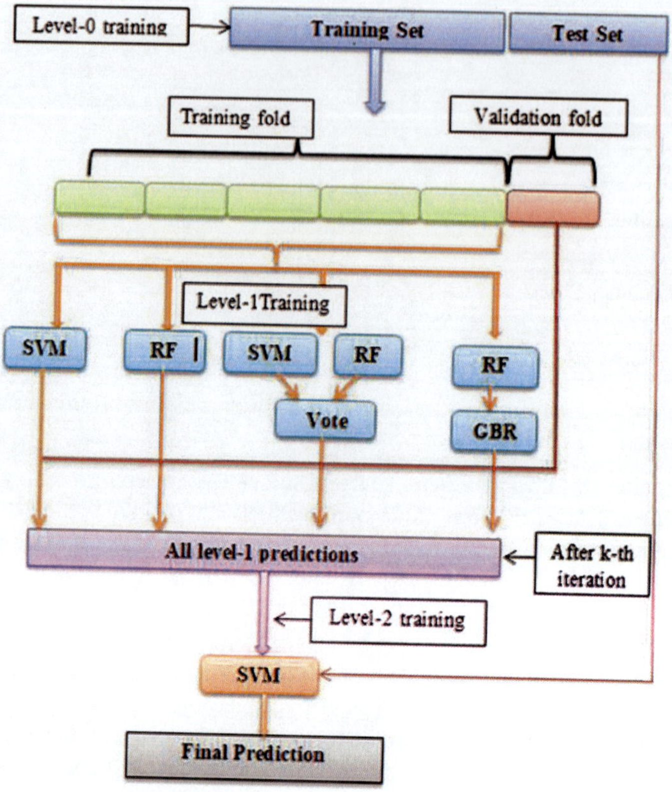

Fig. 3 Stacked regression framework

vote model and GBR algorithms are employed as the base regression model at layer-1 in stacking model. For meta-learner, many deferent algorithms such as Gaussian processes, linear regression, RF and SVM have been tried. The SVM algorithm is chosen as meta-learner for stacking model. All machine learning model/algorithm implemented in Waikato Environment for Knowledge Analysis (**weka**) [24].

5 Result and Discussions

5.1 Debonding Length Prediction

To predict the length of debonding, each instant took at a time as a test set and remaining all instances areas the training set (71-th fold cross-validation). To make simulation data more realistic, 90 dB, 80 dB, 70 dB of noise severity are added to simulation data. A MATLAB® function, *awgn(A, snr)*, is used to add noise where A is simulation data and *snr* is signal-to-noise ratio.

The results of the prediction of debonding length using different learning are shown in Fig. 4. From the literature, it has been observed that SVM is the most widely used machine learning algorithm in the field SHM. So, in the present study, we started with SVM regression algorithm. SVM algorithm produced satisfactory results in predicting length of debonding when the minimum length is not less than 25 mm at lower noise level (up to 90 dB) with reasonable accuracy. However, as noise

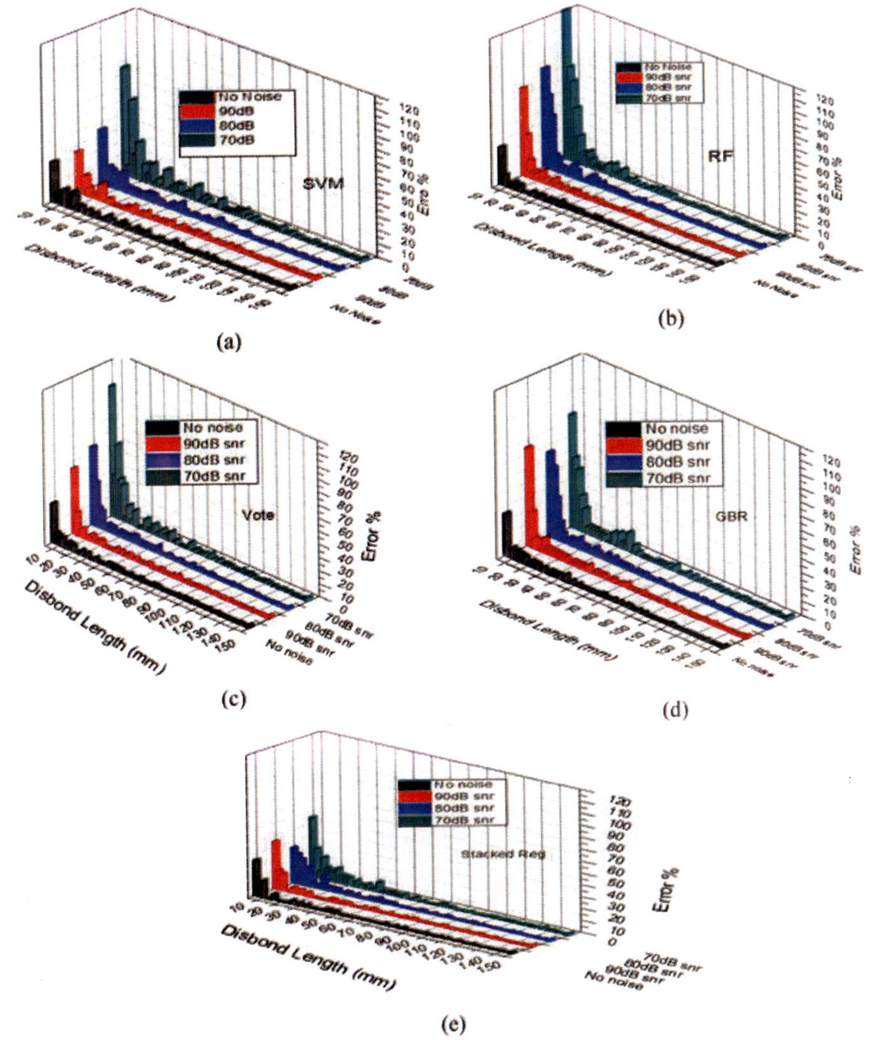

Fig. 4 Debonding length prediction error by **a** SVM, **b** RF, **c** GBR, **d** vote model, **e** stacked regression at different noise level

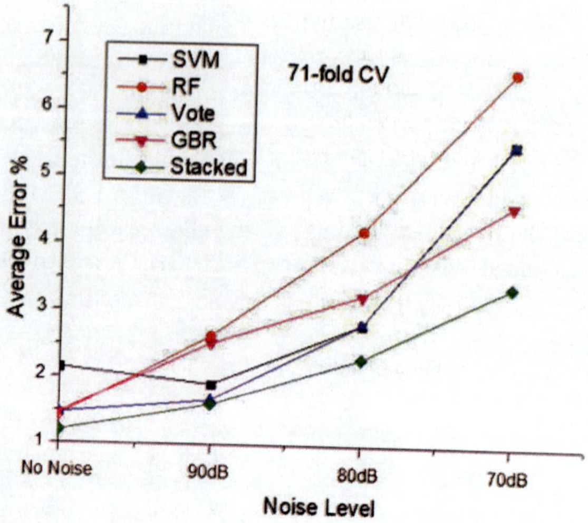

Fig. 5 Average error versus noise level

level increases, SVM starts giving false prediction for debonding length of dimension less than 90 mm at noise level 70 dB. RF has better prediction accuracy than SVM for noiseless data. However, as noise level increases, the prediction accuracy of SVM is better as compared to RF. Maximum error of prediction given by SVM is 80% as compared to 120% by RF at 10 mm debonding length. One thing also has noticed that performance of SVM at 90 dB noise level is better than at noiseless data. The vote combining model combines prediction result of RF and SVM, and GBR performed on RF as base learner. Vote model tries to fill the gap between the error produced by RF and SVM for debonding length as the difference between prediction error given by SVM and RF is widen, that is if prediction error by SVM is 30% and RF is 70%, then error of vote model is lying somewhere in between these two. However, when the prediction error given by both RF and SVM is similar, then prediction accuracy of vote model is better than of both (Fig. 5).

5.2 *Performance Evaluation of Learning Models*

The average prediction error of employed models trained on different training size of total dataset data at various noise level data has presented. The noise level is starting from noiseless data to 70 dB SNR data in step of 10 dB. The number of training decreasing from 70 to 35 in step of 1 and rest of dataset is used as a test set. As it has seen in the previous section with noiseless data, here also the prediction error of RF is slightly less than that of SVM. However, as noise level increases, SVM performed better than RF. It is also observed that the number of training data decreases, prediction error of RF is steeply increasing as compared to SVM. The trend of prediction error of GBR model with decreasing training size at different

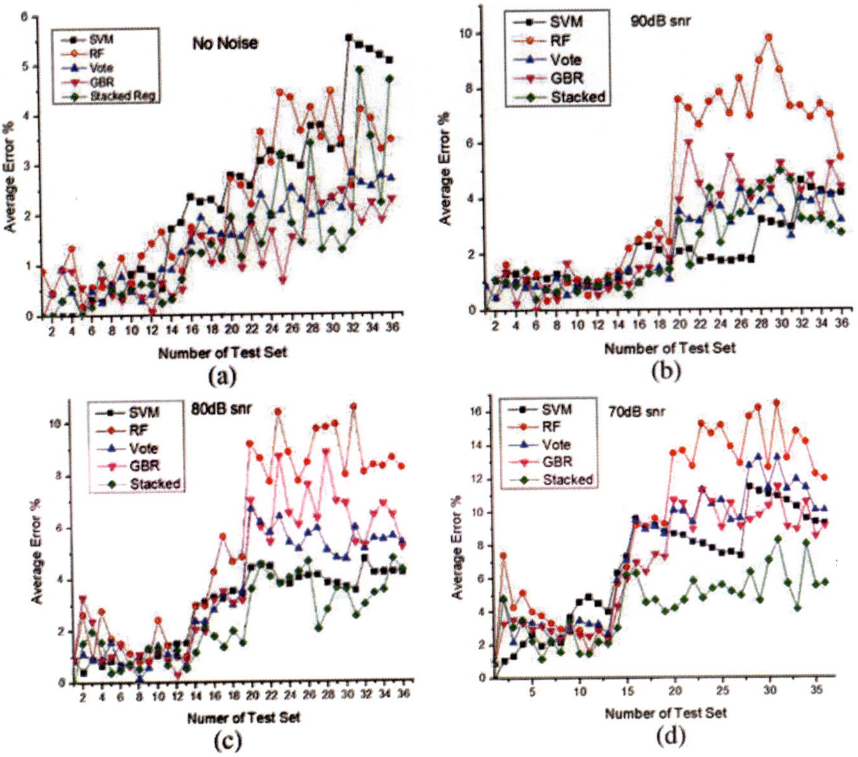

Fig. 6 Performance of algorithms with a different training sets

noise levels is just followed as of RF with better prediction accuracy. The overall prediction error of stacked regression model is lower than SVM and RF but higher than vote and GBR with noiseless data. As noise level increases, prediction error of stacked regression is lower than all employed learning models (Figs. 6 and 7).

6 Conclusion

In this paper, we analysed the effectiveness in improving prediction accuracy of debonding length of stacked regression meta-learning model over base learning model at various noise levels in input data. We also analysed the prediction capability of individual base model at these noise levels. The SVM gives good prediction accuracy with noiseless or lower noise level data. However, as noise level increases, prediction accuracy of SVM is deteriorating considerably. Similar behaviour for RF is observed with higher prediction error. However, when RF is used as base learner in GBR algorithm, prediction error of the model has gone down considerably. When difference between prediction error of SVM and RF is narrowed, then error produced

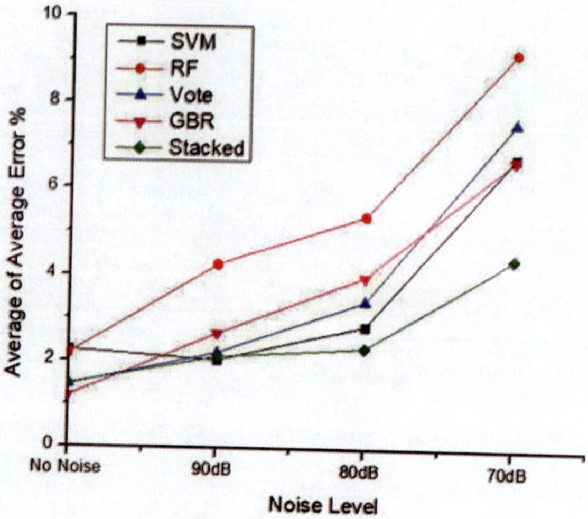

Fig. 7 Average trends

by vote model is lower than of both algorithms. However, as errors differences are widened, the error produced by vote lies somewhere in between errors produced by these two algorithms. The stacked regression model as it works as heterogeneous ensemble learning model comes out as strong ensemble learning model. This model gives better prediction accuracy with noisy data.

References

1. Kudva, J. N., Munir, N., & Tan, P. W. (1992). Damage detection in smart structures using neural networks and finite-element analyses. *Smart Materials and Structures, 1*(2), 108–112.
2. Farrar, C. R., Worden, K., & Wiley, J. (2012). *Structural Health monitoring: A machine learning perspective*. Wiley.
3. Razak, H. A. (2017). Recent developments in damage identification of structures using data mining (pp. 2373–2401).
4. Fairhurst, M. C., & Rahman, A. F. R. (4997, February). Generalised approach to the recognition of structurally similar handwritten characters using multiple expert classifiers. *IEE Proceedings-Vision, Image and Signal Processing, 144*(1), 15–22.
5. Dietterich, T. G. (1995). Solving multiclass learning problems via error-correcting output codes (Vol. 2).
6. Ho, T. K., Hull, J. J., & Srihari, S. N. (1994, January). Decision combination in multiple classifier systems. *IEEE Transactions on Pattern Analysis and Machine Intelligence, 16*(1), 66–75.
7. Jordan, M. I., & Jacobs, R. A. (1993). Hierarchical mixtures of experts and the EM algorithm. In *Proceedings of 1993 International Conference on Neural Networks (IJCNN-93-Nagoya, Japan)* (vol. 2, pp. 1339–1344).
8. Kittler, J. (1998). Combining classifiers: A theoretical framework. *Pattern Analysis and Applications, 1*(1), 18–27.
9. Jacobs, R. A. (1995). Methods for combining experts' probability assessments. *Neural Computation, 7*(5), 867–888.
10. Idave, R., & Dussault, J. P. (2014). Systematic ensemble learning for regression (pp. 1–38).

11. Breiman, L. (1994). Bagging predictors. UCR Statistics Department - University of California (No. 2, p. 19).
12. Freund, Y., & Schapire, R. E. (1996). Experiments with a new boosting algorithm. In *Proceedings of the 13th International Conference on Machine Learning* (pp. 148–156).
13. Wolpert, D. H. (1992). Stacked generalization. *Neural Networks, 5*(2), 241–259.
14. Breiman, L. (1996). Stacked regressions. *Machine Learning, 24*(1), 49–64.
15. Kuncheva, L. I. (2002). A theoretical study on six classifier fusion strategies. *IEEE Transactions on Pattern Analysis and Machine Intelligence, 24*(2), 281–286.
16. Woods, K., Kegelmeyer, W. P., & Bowyer, K. (1997). Combination of multiple classifiers using local accuracy estimates. *IEEE Transactions on Pattern Analysis and Machine Intelligence, 19*(4), 405–410.
17. Džeroski, S., & Ženko, B. (2004). Is combining classifiers with stacking better than selecting the best one? *Machine Learning, 54*(3), 255–273.
18. Zhai, B., & Chen, J. (2018). Development of a stacked ensemble model for forecasting and analyzing daily average PM 2.5 concentrations in Beijing, China. *Science of the Total Environment, 635,* 644–658.
19. Sesmero, M. P., Ledezma, A. I., & Sanchis, A. (2015). Generating ensembles of heterogeneous classifiers using stacked Generalization. *Wiley Interdisciplinary Reviews: Data Mining and Knowledge Discovery, 5*(1), 21–34.
20. Ozay, M., & Yarman-Vural, F. T. (2016). Hierarchical distance learning by stacking nearest neighbor classifiers. *Information Fusion, 29,* 14–31.
21. Cao, C., & Wang, Z. (2018). IMCStacking: Cost-sensitive stacking learning with feature inverse mapping for imbalanced problems. *Knowledge-Based Systems, 150,* 27–37.
22. Naimi, A. I., & Balzer, L. B. (2018). Stacked generalization: An introduction to super learning. *European Journal of Epidemiology, 33*(5), 459–464.
23. Hastie, T., Tibshirani, R., & Friedman, J. (2009). Springer Series in Statistics The elements of Statistical learning. *The Mathematical Intelligencer, 27*(2), 83–85.
24. Frank, E., Hall, M. A., & Witten, I. H. (2016). WEKA Workbench. Online appendix for data Mining: Practical machine learning tools and techniques [Online]. Available: www.cs.waikato.ac.nz/~ml/weka.

The Need of Advanced Assisting Devices for Blind People

Manan Arora, Harivans Pratap Singh, Nitin Kumar, and Maneesh Vishwakarma

Abstract Image processing has been advancing for the last many years and been upgraded to provide a better solution to the issues by collecting data from the images and processing them. We aim to compare the object detection algorithms to develop a critical solution that can be a game-changer for blind people across the world. All the image processing algorithms have some advantages and some disadvantages. Our aim in this paper is to find the algorithm that will best suite our solution for providing a thought-breaking solution for assisting blind people for obstacles overcoming across the world. Since we are fulfilling one of the five senses of our body, we want a reliable and trustworthy solution.

Keywords Object detection · YOLO · Contours · Edge tracing · Tensorflow · Mean shift · Camshift

1 Introduction

The technological advancements have been writing new definitions for the world for many years now. Every domain has now been linked up with technological advancement to pace up the development. Finding economical and better solutions for each problem has been one of the major requirements in today's world. The advancement in technology includes the replacement of man labor with machines and robots. We are also using technology to solve many social problems also. But one problem that

M. Arora · H. P. Singh (✉) · N. Kumar · M. Vishwakarma
ABES Engineering College, Ghaziabad, India
e-mail: harivans.singh@abes.ac.in

M. Arora
e-mail: manan.16bcs1034@abes.ac.in

N. Kumar
e-mail: nitin.16bcs1065@abes.ac.in

M. Vishwakarma
e-mail: maneesh.16bcs1153@abes.ac.in

S. Tiwari et al. (eds.), *Proceedings of International Conference on Big Data, Machine Learning and their Applications*, Lecture Notes in Networks and Systems 150,
https://doi.org/10.1007/978-981-15-8377-3_6

Fig. 1 An assisting dog helping a blind person to find the route

still needs much emphasis is a smart solution for providing visual assistance for visually impaired people or blind people. The technology currently available can provide a better medium to start working over this issue to create a safe and stable world for these people who often need support and help to perform their daily chores.

There are several devices to help blind people with their way. These devices include smart canes, smart glasses, assistive animals, and many more. But all of these solutions either include the cost to be paid or care to be given. Smart canes have a small range to that of the length of the stick. It will sense the obstacle and the sensor by creating the vibration to alert the person about the obstacle ahead. The smart glasses are a solution for the people who are suffering from either visual impairment or acute blindness. Assistive dogs are also used for providing support to blind people where assistive dogs guide the blind people and help them to guard themselves against the obstacles on the road as shown in Fig. 1.

In this journey to develop a more reliable and trustworthy solution, we are working to find more of a software solution for blind people than a hardware solution. We are trying to develop a solution that could easily run on a mobile phone thus reducing cost factors and also developing a trustworthy solution for blind people. For this solution, we are using image processing as a tool to develop a solution. Analyzing images of the live video of the objects which are present in front of the person through a mobile phone camera. Recognizing obstacles that can disrupt the walking of a blind person and can affect him/her in his/her way.

2 Background Details and Related Work

Yakimovsky [1] presented a computer solution for the problem of the automatic location of objects. This solution was developed to solve the problem in digital images. The use of planar fits and dynamic use of region features can yield much better results.

Shrivakshan et al. [2] worked on a case study that dealt with the observations of Shark Fist Classification. The observations were made through image processing with various filters which constitute mainly gradient-based edge detection operators, Laplacian, and Canny-based edge detectors. The software was implemented using

MATLAB. The evaluation of the result shows that Canny, Sobel, LoG, Robert's, Prewitt show better performance but Canny is quite expensive. In this comparison, Laplacian does show better results for some features(fins).

Schlecht et al. [3] presented a non-parametric representation of contours, curvatures, and junctions. This representation enables there accurate localization. They combined contour and appearance information into a general and voting-based detection algorithm. They also demonstrated that this approach reveals the most relevant contours and junctions supporting each object hypothesis. The result of this algorithm showed that contours and appearances are complimentary. Also, the combination of contour and appearance information has improved the performance of baseline voting algorithms.

Geethapriya et al. [4] proposed using the YOLO algorithm for object detection using a single neural network. The algorithm is simple to build and can be trained on a complete image. YOLO accesses to complete image in predicting boundaries with some false positives in background. This is the fastest and most efficient algorithm as compared to others.

Warade et al. [5] presented an approach of object detection in a video sequence. Their approach was for tracking moving objects using edge detection and feature matching. The feature extraction method would be applied on edge difference images and followed by feature matching technique. This method is advantageous as it also accounts for the edges in the images and can be operated on binary images directly.

Parekh et al. [6] presented a paper showing study of various phases of object tracking systems which includes object detection at the primary stage. The different methods mentioned for object detection are frame difference, background subtraction, and optical flow. Based on the study, they concluded that background subtraction is the simplest method for providing complete information of objects to be detected as compared to the other two mentioned methods.

Zhao et al. [7] presented a paper containing a detailed review of deep learning-based object detectionframework. These handle different sub-problems which can be low-resolution, clutter, and occlusion and with several degrees of modifications on R-CNN. They proposed several promising directions for a thorough understanding of the object detection landscape.

Kumar et al. [8] presented a paper for object detection and recognition. They compared the Easynet model with various other models. The model generalizes the domains and can be trained on full images very easily. The model can also be implemented on hardware. They used the background subtraction approach for object detection.

Tiwari et al. [9] reviewed different object detection, recognition, and tracking techniques. They noted some methods which would give accuracy but this includes high computational complexity.

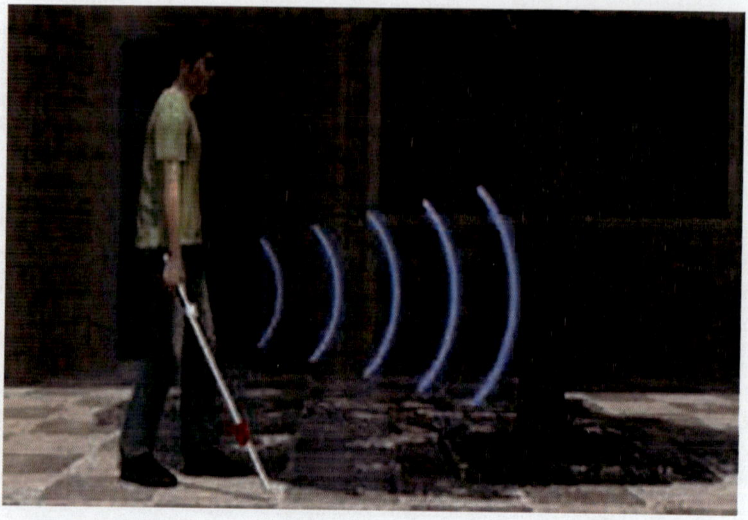

Fig. 2 Concept of smart canes to help blind people

3 Problem Statement

The blind people are dependent on canes for their travel. They can only detect obstacles with the help of canes and then can decide their path on which they can dodge the obstacles as you can see in Fig. 2. We want to complement canes with technology where a blind person can use his/her mobile phone for obstacle detection and thus can move more freely on the roads without the support of any person.

4 Detailed Study

Several image processing algorithms are used for object detection in the real world. These algorithms focus on different features in an image to detect an object. These include edge-based object detection methods, contour-based object detection methods, mean shift-based object detection methods, frame change object detection methods, and many more. These algorithms can be used to develop a smart aiding device for blind people so that they can move freely without their canes.

There are several object detection algorithms in image processing such as

1. YOLO [10]
2. Edge detection algorithm
3. Contour-based object detection algorithm

Fig. 3 How object detection algorithms is used for helping blind people

4. Mean shift object detection algorithm
5. Camshift object detection algorithm
6. Tensorflow-based object detection algorithm
7. Haar Cascade-based object detection (Table 1).

5 Proposed Solution

We propose a technical solution to complement canes. In our proposed solution, we are using a mobile phone for the solution to the problem of obstacle detection for blind people. Our solution includes using image processing object detection algorithms for detecting obstacles and then alarming the blind person about the obstacle on his/her way and providing a safer route to travel from one place to another. We aim to develop a handy solution so that it can be used by every individual and at the same time be highly affordable.

The expected output of this solution can be easily seen from Fig. 3.

6 Conclusions

All the image processing object detection algorithms are best suited to cater to their corresponding activities and lead to the output. YOLO scans complete image as a whole and detects for objects in the image. Edge detection algorithms search for all edges in the image and mark along the edges to detect objects. The other algorithms use frame difference in videos or real-time input to find the difference in the frame and thus recognizing objects.

Contour-based algorithm checks for movement based on frame difference. To meet the requirement, we would be comparing the working of all the algorithms on a common data set to calculate the efficiency and accuracy to make a better product to provide a safe solution to blind people. In the case of TensorFlow models, the algorithm uses a convolutionary neural network to perform object detection and also can be easily developed over android due to the presence of TensorFlow lite which converts the TensorFlow models into the lighter base so that it can be used to develop mobile application. According to the comparison and application of each algorithm, the TensorFlow algorithm can be used easily to develop a more trivial solution for blind people.

7 Future Works

We aim to create an assisting solution for blind people. Through this comparison, we will make ensure to use the best algorithm to develop the solution for helping blind people around the globe. We aim to complement assisting sticks with the advancement of technology and use image processing and some other technologies to develop a better solution for blind people so that they can move more independently on the roads without the fear of obstacles that they could not know about through their sticks and develop a more trusting relationship between human beings and technology. This solution will help people come more closer to technology and could also result in other breakthroughs in the field of science and technology.

Table 1 Comparison of various object detection image processing algorithms

Algorithm	Description
Contour-based object detection	This algorithm is based on the frame difference in two frames This algorithm finds the difference between the consecutive frames in a video. With the help of differences in objects in the frames, the detection takes place on moving things. This algorithm can detect even small changes in the frame when the algorithm is running on a fixed device
Edge-based object detection	This algorithm is based on detecting edges in the frame. This algorithm can work on a single frame. This algorithm analyzes all the sharp edges present in the frame and detects them
Mean shift-based object detection	This algorithm detected those objects whose appearance is defined as a histogram. This algorithm calculates the mean shift of pixel value. This algorithm creates a confidence map in the new image based on the color histogram of the object in the previous frame

(continued)

Table 1 (continued)

Algorithm	Description
Cam shift-based object detection	CAM stands for Continuously Adaptive Mean Shift. It is based on a mean shift object detection algorithm
YOLO	This stands for 'You Only Look Once.' Meanwhile, other algorithms analyze the frame multiple times, this algorithm analyzes the frame only once and divide the frame into 13 * 13 grid lines. After this, it checks for the difference between adjacent blocks and creates multiple boundaries. Based on the probabilistic score of each box, the objects are detected and marked with boxes in this algorithm
Tensorflow-based object detection	This algorithm works on TensorFlow-based object detection models. In this algorithm, the pre-trained models scan the frame and detect all the objects that can exist in the image. The model can be trained according to the requirements of the projects on different datasets This algorithm can also run on mobile phone applications just by simply converting the TensorFlow model to the TensorFlow Lite model and running the classifier by providing input
Haar Cascade object detection	This is one of the oldest object detection algorithms developed. This algorithm was based on the concepts proposed by Paul Viola and Michael Jones in their paper published in 2001 [11]. In this paper, they proposed various features. Haar Cascade algorithm is mainly used for facial recognition and detection. This works on the pixel difference among the pixels of the frame

References

1. Yakimovsky, Y. (1976). Boundary and object detection in real world images. *Journal of the ACM (JACM), 23*(4), 599–618.
2. Shrivakshan, G. T., & Chandrasekar, C. (2012) A comparison of various edge detection techniques used in image processing. *International Journal of Computer Science Issues (IJCSI), 9*(5), 269.
3. Schlecht, J., & Ommer, B. (2011). Contour-based object detection. *BMVC*. 2011.
4. Geethapriya, S., Duraimurugan, N., & Chokkalingam, S. P. Real time object detection with Yolo.
5. Warade, P. P., Kale, M. S., & Thakare, V. M. (2015). Edge detection for moving object tracking. *International Journal, 5*(4).
6. Parekh, H. S., Thakore, D. G., & Jaliya, U. K. (2014). A survey on object detection and tracking methods. *International Journal of Innovative Research in Computer and Communication Engineering, 2*(2), 2970–2978.
7. Zhao, Z.-Q., et al., (2019). Object detection with deep learning: A review." *IEEE transactions on neural networks and learning systems, 30*(11), 3212–3232.
8. Kumar, S., Balyan, A., & Chawla, M. (2017). Object detection and recognition in images. *International Journal of Engineering Development and Research (IJEDR), 5*(4), 1029–1034.
9. Tiwari, M., & Singhai, R. (2017). A review of detection and tracking of object from image and video sequences. *The International Journal of Computational Intelligence, 13*(5), 745–765.

10. Redmon, J., et al., (2016). You only look once: Unified, real- time object detection." *Proceedings of the IEEE conference on computer vision and pattern recognition.* 2016.
11. Viola, P., & Jones, M. (2001). Rapid object detection using a boosted cascade of simple features. In *Proceedings of the 2001 IEEE computer society conference on computer vision and pattern recognition. CVPR 2001.* Vol. 1. IEEE, 2001.

A Noval Fuzzy-Based Artificial Bee Colony Algorithm for Medical Records Classification

Subhash Chandra Pandey and Utpal Kant Kumar

Abstract Data mining is the core step of a broader perspective known as the knowledge discovery. Many nature-inspired algorithms have been designed for this pursuit. Many of them are functioning with higher classification accuracy. The artificial bee colony (ABC) is one of them. However, it has been visualized that subsequent performance enhancement can be rendered in this algorithm by assimilating the fuzzy concept with exiting state-of-the-art ABC algorithm. Therefore, attempt has been made to construct a fuzzy-based ABC algorithm. The performance of proposed Fuzzy-ABC is evaluated with Latent Dirichlet Allocation (LDA) algorithm for the classification of medical records. Three frequently used datasets are taken for this purpose. Experimental analysis reveals the fact that the proposed Fuzzy-ABC renders a superior performance over the LDA technique considered in the experimentation. Further, the proposed algorithm is also validated using fivefold cross-validation.

Keywords Data mining · Medical record classification · Fuzzy artificial bee colony algorithm · LDA

1 Introduction

Natural computing is the stream of computation inspired by the different computing paradigms manifested in nature and bio-inspired computing is considered as sub-domain of natural computing. Extensive literature survey reveals the fact that there is plethora of techniques in the domain of bio-inspired and evolutionary computation. This review provides a brief and general overview of natural computing and incorporates some state-of-art-algorithms. There are many algorithms in literature which mimics the flocking behavior of birds, bees, fishes, etc. such as particle swarm

S. C. Pandey · U. K. Kumar (✉)
Birla Institute of Technology, Patna, Bihar, India
e-mail: utpalk29@gmail.com

S. C. Pandey
e-mail: s.pandey@bitmesra.ac.in

S. Tiwari et al. (eds.), *Proceedings of International Conference on Big Data, Machine Learning and their Applications*, Lecture Notes in Networks and Systems 150,
https://doi.org/10.1007/978-981-15-8377-3_7

optimization (PSO), bee colony optimization (BCO), bee swarm optimization (BSO) [1].

Artificial bee colony (ABC) is a paradigm of nature-inspired algorithm and it mimics the functionality of swarm intelligence and the foraging of the bees. This technique was proposed in [2]. Indeed, ABC is very simple and easy to implement. Perhaps, because of these features, this algorithm has widely been implemented by different researchers in recent past. Moreover, different modified versions of this algorithm have also been proposed each of which entails its own merits and demerits. The working of ABC algorithm is lucidly elaborated in [3].

Further, the concept of fuzzy set theory and fuzzy logic has frequently been used by different researcher in different scenario as the fuzzy set theory and fuzzy logic are indispensable tool to work out the knowledge required for different purposes while the datasets are peppered with uncertainties created owing to fuzziness. Indeed, fuzzy set theory entails membership grades and its value can lie anywhere from 0 to 1. Mathematically, it is defined as: $\mu_A(x){:}X \rightarrow [0, 1]$ where μ is the membership grade for different elements $x \in A$, and A is the fuzzy set. In addition, the concept of dynamic programming is an important phenomenon to find the optimal solution. In this paradigm, the original problem is divided in discrete sub-problem and the outcome of these sub-problems is properly stored for future used and thus avoids the repeated computation of same task which in turn reduces the execution time.

In this paper, attempt has been made to modify the existing ABC algorithm so that it can effectively be used in the case of fuzzy scenario. This paper preludes a novel 'fuzzy-based ABC (FABC)' algorithm and it also assimilates the characteristics of dynamic programming. The proposed FABC algorithm can be implemented for many tasks such as classification and clustering.

Rest of the paper is organized as follows. Section 2 renders the brief literature review pertaining to work already have been done in the domain of bio-inspired computing. Further, the proposed FABC algorithm and its implementation details are discussed in Sect. 3. Section 4 displays the experimental work and analysis performed. Finally, the paper is concluded in Sect. 5.

2 Literature Review

Extensive literature survey revealed the fact that there are many nature-inspired techniques used for the task of classification and clustering. Indeed, the domain of nature-inspired algorithm for the pursuit of classification and clustering tasks entails plethora of algorithm and perhaps tens of hundreds of new algorithms are also developed every years. Thus, it is not possible to mention all of them. However, few state-of-the-art algorithm developed will briefly be mentioned in this section. Genetic algorithm is one of the techniques, which have been used for classification task [4]. Genetic algorithm is from the domain of evolutionary algorithm (EA) and is considered as an optimization technique. GA is useful when no mathematical analysis is available and is most suitable for solving the constraint optimization problems [5].

Particle swarm optimization was evolved in 1995 by Eberhart and Kennedy and is also used for solving the optimization problems. In PSO, each element of the swarm keeps its direction changing to optimize its position in the swarm [6]. Ant colony algorithm is also an optimization-based algorithm and it is a meta-heuristic technique developed by Dorigo et al. [7]. This algorithm has been used by different researchers for classification and clustering tasks [8, 9]. The virtual bee (VB) algorithm is also an optimization algorithm which modeled the hunting behavior of honey bees for searching the optimal solution. This algorithm is very much similar to the Genetic algorithm. There are various applications in which the paradigm of VB algorithm is effectively used. The artificial bee colony algorithm is also elaborated in [10]. In [11, 12], an innovative co-variance-based ABC technique is used. Moreover, different application domains of ABC algorithm are discussed in [13].

3 Proposed FABC

Different steps of proposed fuzzy-based artificial bees colony algorithm are shown in Fig. 1. The proposed FABC algorithm can effectively be used for different categories of datasets for the purposes of classification and clustering tasks. This characteristic enables FABC to be implemented in the different domains [14].

Proposed FABC
1. *Create initial population* χ_i *randomly, i=1, 2, 3,…., NS [Initialization]*
2. *Find the membership function* μ_i $\forall$ *individuals (solutions)* $\in \chi_i$.
3. *Keep the best individual* $\chi best$ *in* χ_i *[This is the memorization of solution]*
4. *Initial cycle setting=1*
5. *Reiterate*
6. *Create new individual* V_i *from old individual* χ_i, *using Mamdani approach*
7. *Evaluate the membership function* μ_i *for all individual*
8. *Keep the best individual between threshold limit [Dynamic programming approach]*
9. *Calculate the membership grade* μ_i $\forall$ *individual* χ_i *[using fuzzy theory]*
10. *Generate the new individual from* V_i *on the basis of its* μ_i *[Onlookers bees]*
11. *Find the membership function* μ_i $\forall$ *new individual* $\in$ *population*
12. *Incorporate the best individual between current individual. [Dynamic programming approach]*
13. *Determine the abandoned individual if exists, (i.e. those lying beyond the threshold value of* μ_i*) and randomly replace it with new individual* χ_i *[Scout bees]*
14. *Keep the best individual* ϕ_{best} *obtained so far in the individual*
15. *cycle setting=cycle setting+1*
16. *Until cycle setting<= maximum cycle number*

Fig. 1 The proposed FABC algorithm

Table 1 Datasets descriptions

Datasets	No. of documents	No. of term tokens	No. of unique terms	Average term per document	Details
M	1537	244,931	14,511	96.3	Papers pertaining to therapeutic domain
N	1617	298,449	111,059	125.8	Pertaining to nursing
O	21,092	198,798	15,668	96.1	Papers pertaining to therapeutic domain

4 Experimental Analysis

This section will render different experimentation details including the comparison of LDA and proposed FABC algorithm. LDA is a popular unsupervised topic model and it assembles the words with alike semantic [15]. Two main outcomes of this technique are the probability of each topic for each document and the probability of each word for each topic. Perhaps, this technique is the most effective method among supervised and unsupervised approaches [16].

4.1 Datasets

In this experimentation, three datasets pertaining to health sector have been taken. These datasets are described in Table 1 and also enumerated below.

- M-Dataset. This dataset is the labeled medical data.
- N-Dataset. This is an unlabeled corpus of 1617 numbers of documents.
- O-Dataset. This dataset incorporates different attributes of bacterial infections and mycoses as well as the virus infected diseases.

4.2 Document Classification

In our experimentation setup, the preliminary evaluation measure is performed for the purpose of document classification. We have taken two labeled datasets for this purpose. These are M and O-Datasets. Both datasets have two classes. It is pertinent to mention that in the document classification problem, we allocate the given document to a class, and subsequently, features are extracted from the textual data. Further, for the validation purposes, we observe the performance of FABC and LDA

Table 2 Confusion matrix

		Predicted	
		Negative	Positive
Actual	Negative	TN	FP
	Positive	FN	TP

and performed the 5-fold cross-validation. In this experimentation, we implemented different numbers of topics as input attributes of documents. These numbers are shown in Tables 3, 4, 5 and 6 for random forest (RF) technique owing to its high performance in classification tasks [17]. The output of RF displayed as a confusion matrix defined in Table 2. Further, Table 3 shows the M-Dataset classification for 50 and 100 topics and Table 4 represents the M-Dataset classification for 150 and 200 topics. Likewise, details for 50 and 100 topics in case of O-dataset are given in Table 5 and for 50 and 200 topics are shown in Table 6.

4.3 Document Clustering

The N-Dataset is used for document clustering purposes and we evaluated different numbers of themes and clusters and also used Calinski-Harabasz (CH) index for this pursuit. We also used K-means with 500 numbers of iterations. We observe the efficacy of FABC with LDA by considering the variations of clusters from two to eight. The numbers of topics considered are fifty, hundred, one hundred fifty, and two hundred, respectively. The Calinski-Harabasz (CH) index displays the superiority of FABC in comparison to the LDA in varying range of attributions.

5 Conclusions

The gigantic amount of therapeutic transcript data renders precious source of information which can subsequently be used for future references. However, extracting the useful information from this huge data requires something more than simple keyword search. Fuzzy paradigm of machine learning is considerably used for the sake of medical image processing. However, the use of fuzzy paradigm in text processing is considerably ignored by the researchers. The current state-of-the-art techniques entail mainly the linear algebra and statistical distributions for the pursuit of topic modeling. As a matter of fact, these techniques do not present satisfactory outcome in many situations. In contrast, the proposed FABC incorporates the fuzzy aspects and thus produces more accurate results. Indeed, FABC showed a superior performance

Table 3 M-Dataset classification for fifty and hundred topics

Technique	Percentage accuracy	*F*-measurement	MCC	Receiver operating characteristic (AUC)	Percentage accuracy	*F*-measurement	MCC	Receiver operating characteristic (AUC)
	For 50 Topics				For 100 Topics			
LDA	78.36	0.78	0.561	0.895	77.36	0.79	0.661	0.995
FABC	96.49	0.965	0.929	0.986	97.49	0.865	0.829	0.976

Table 4 M-Dataset classification for one hundred fifty and two hundred topics

Technique	Percentage accuracy	*F*-measurement	MCC	Receiver operating characteristic (AUC)	Percentage accuracy	*F*-measurement	MCC	Receiver operating characteristic (AUC)	
	For 150 Topics				For 200 Topics				
LDA	76.19	0.78	0.636	0.897	81.45	0.822	0.646	0.894	200
FABC	95.90	0.959	0.917	0.995	97.076	0.971	0.941	0.992	200

Table 5 O-Dataset classification for fifty and hundred topics

Technique	Percentage accuracy	F-measurement	MCC	Receiver operating characteristic (AUC)	Percentage accuracy	F-measurement	MCC	Receiver operating characteristic (AUC)
	For 50 Topics				For 100 Topics			
LDA	74.38	0.71	0.262	0.76	73.46	0.675	0.143	0.638
FABC	76.21	0.761	0.311	0.787	77.26	0.753	0.363	0.722

Table 6 O-Dataset classification for one hundred fifty and two hundred topics

Technique	Percentage accuracy	*F*-measurement	MCC	Receiver operating characteristic (AUC)	Percentage accuracy	*F*-measurement	MCC	Receiver operating characteristic (AUC)
	For 150 Topics				For 200 Topics			
LDA	71.80	0.592	0.145	0.542	70.08	0.657	0.069	0.65
FABC	75.87	0.698	0.334	0.734	74.32	0.711	0.243	0.71

in comparison to LDA. Moreover, FABC can perform efficiently in both discrete and continuous spectrum of datasets. Over and above, proposed FABC can also be applied on social media datasets and it can be considered as the future work of this research.

References

1. Pandey, S. C. (2018). New facet of honey bees dancing language for mining the induction rules. *Sādhanā, 43*(3), 35.
2. Harfouchi, F., & Habbi, H. (2018, April). A novel artificial bee colony learning system for data classification. In *International Conference on Computer Science and its Applications* (pp. 322–331). Cham: Springer.
3. Jayanth, J., Koliwad, S., & Kumar, T. A. (2015). Classification of remote sensed data using artificial bee colony algorithm. *The Egyptian Journal of Remote Sensing and Space Science, 18*(1), 119–126.
4. Shukran, M. A. M., Chung, Y. Y., Yeh, W. C., Wahid, N., & Zaidi, A. M. A. (2011). Artificial bee colony based data mining algorithms for classification tasks. *Modern Applied Science, 5*(4), 217.
5. Ali, R. S., Aldair, A. A., & Almousawi, A. K. (2014). Design an optimal PID controller using artificial bee colony and genetic algorithm for autonomous mobile robot. *International Journal of Computer Applications, 100*(16), 8–16.
6. Wang, D., Tan, D., & Liu, L. (2018). Particle swarm optimization algorithm: An overview. *Soft Computing, 22*(2), 387–408.
7. Dorigo, M. (1992). *Optimization, learning and natural algorithms*. Ph.D. Thesis, Politecnico di Milano.
8. Kefayat, M., Ara, A. L., & Niaki, S. N. (2015). A hybrid of ant colony optimization and artificial bee colony algorithm for probabilistic optimal placement and sizing of distributed energy resources. *Energy Conversion and Management, 92,* 149–161.
9. Shunmugapriya, P., & Kanmani, S. (2017). A hybrid algorithm using ant and bee colony optimization for feature selection and classification (AC-ABC Hybrid). *Swarm and Evolutionary Computation, 36,* 27–36.
10. Khan, L., Ullah, I., Saeed, T., & Lo, K. L. (2010). Virtual bee's algorithm based design of damping control system for TCSC. *Australian Journal of Basic and Applied Sciences, 4*(1), 1–18.
11. Kumar, D., & Mishra, K. K. (2017). Portfolio optimization using novel co-variance guided artificial bee colony algorithm. *Swarm and evolutionary computation, 33,* 119–130.
12. Kumar, D., & Mishra, K. K. (2018). Co-variance guided artificial bee colony. *Applied Soft Computing, 70,* 86–107.
13. Kumar, D., Mishra, K. K. (2017). Artificial bee colony as a frontier in evolutionary optimization: A survey. In: Bhatia, S., Mishra, K., Tiwari, S., & Singh, V. (Eds.), *Advances in Computer and Computational Sciences. Advances in Intelligent Systems and Computing*, (Vol. 553). Singapore: Springer.
14. Keller, J., Krisnapuram, R., & Pal, N. R. (2005) *Fuzzy models and algorithms for pattern recognition and image processing* (Vol. 4). Springer Science & Business Media.
15. Arnold, C., & Speier, W. (2012). A topic model of clinical reports. In *Proceedings of the 35th international ACM SIGIR conference on Research and development in information retrieval* (pp. 1031–1032). ACM.

16. Helgason, C. M., Malik, D., Cheng, S.-C., Jobe, T. H., & Mordeson, J. N. (2001). Statistical versus fuzzy measures of variable interaction in patients with stroke. *Neuroepidemiology, 20*(2), 77–84.
17. Qi, Y., Bar-Joseph, Z., & Klein-Seetharaman, J. (2006). Evaluation of different biological data and computational classification methods for use in protein interaction prediction. *Proteins: Structure, Function, and Bioinformatics, 63*(3):490–500.

A Comparative Study: Glaucoma Detection Using Deep Neural Networks

Paresh Chandra Sau, Manish Gupta, and Divesh Kumar

Abstract Glaucoma is one of the most reported eye diseases which lead to complete blindness, if not treated early. Glaucoma normally associated with intraocular pressure (IOP) which gradually damages the vision field of eye. Glaucoma is an alarming threat to global health problem as it causes irreversible blindness. Evidence of glaucoma indicates that the pathogenesis of glaucoma depends on many interacting mechanism of body. Glaucoma has two main types such as open-angle and close-angle. Angle term means contacting length of iris and cornea; if the length is large, then associated disease termed as open-angle, and for shorter length, it terms as close-angle glaucoma. Glaucoma effects not only vision it also associated with hearing problem of the patients (Greco et al. in The American Journal of Medicine, 2016). In this paper, a complete review of glaucoma detection techniques based on deep learning system, CAD system and other technique is represented.

Keywords Intraocular pressure · Convolution neural network · Computer added diagnosis system · Optic disc · Optic cup

1 Introduction

Glaucoma is extreme reason of blindness beyond the word [2]. Age of the person is biggest factor for glaucoma. After the age of 35–40, glaucoma is easily effect the person eye. Any accident may cause glaucoma in younger age. Glaucoma is cause of permanent blindness. If it is not detected early, then it causes permanent blindness. Large volume of intraocular pressure (IOP) is reason of glaucoma disease, [3] reduce

P. C. Sau (✉) · M. Gupta · D. Kumar
GLA University, Mathura, India
e-mail: praesh.chandra@gla.ac.in

M. Gupta
e-mail: manish.gupta@gla.ac.in

D. Kumar
e-mail: divesh.kumar@gla.ac.in

S. Tiwari et al. (eds.), *Proceedings of International Conference on Big Data, Machine Learning and their Applications*, Lecture Notes in Networks and Systems 150,
https://doi.org/10.1007/978-981-15-8377-3_8

the IOP inside eyes to stop the harm to eye. Glaucoma has two main types open-angle glaucoma and close-angle glaucoma; close-angle glaucoma is uncommon as compared to open glaucoma. Open glaucoma is asymptomatic because it grows slowly, and close angle is very pain full and needs immediate treatment [4]. Retinal nerve fibre layer (RNFL) thickness and ocular field argument are necessary for the find glaucoma [5]. Glaucoma mainly affects the ganglion cell complex (GCC), and these GCC are collection of three inner most layers such as RNFL, Ganglion cell layer and plexifrom layer [6]. Cause of this type disease is large amount of IOP.

A heterogeneity of probably recognized by mechanical and vessel framework that has been used for pathological proses of detected glaucoma. At least 67 million people in world are affected by permanent optic nerve damage and vision loss [7]. Ocular area monitoring is most significant methods for the glaucomatous patients [8]. For timely detection, glaucoma has no fast method, but new technology has efficiency to examine glaucoma timely [9]. Caucasian and African people have primarily open angle, and close-angle glaucoma is most common in Asian [10]. Deep learning is most advanced technology for image processing. With the deep learning, identifying the glaucoma is going to very easy task. This task is done by automatically calculation cup-to-disc ratio (CDR) and evaluated the texture features. Segmentation-based technique is one of techniques to detect glaucoma through image features that surround optic disc (OD) region. Hence, deep features were automatically segmented and classified by modern deep learning classifiers [11].

2 Detection of Glaucoma Techniques

In the literature, number of algorithm has been reported for detection of glaucoma. They can be broadly classified in four distinct classes, namely nerve fibre layer defect (NFL) detection and texture analyses of nerve fibre layer, neuro retinal optic cup detection (NROC), computer added diagnostics system and glaucoma detection using deep learning methods. Figure 1 shows the different glaucoma detection classification, and following paragraphs described the detailed discussion on these methods:

2.1 NFL Defect Detection and Texture Analysis of NFL

For early detection of glaucoma, authors used NFL and texture analyses. The most usual nerve fiber layer (NFL) affects in quality of fundus image because image background reflection is mellow, while contrast is less. Glaucoma is diffuse atrophy or thinness of the NFL. NFL detection is not easy because quality of fundus image. Image reflection background in fundus image is affable and has less contrast. Optical coherence tomography (OTC) enables detection of NFL fundus image [12, 13]. OTC is mostly used for optic nerve misalignment like glaucoma, diabetic retinopathy outer

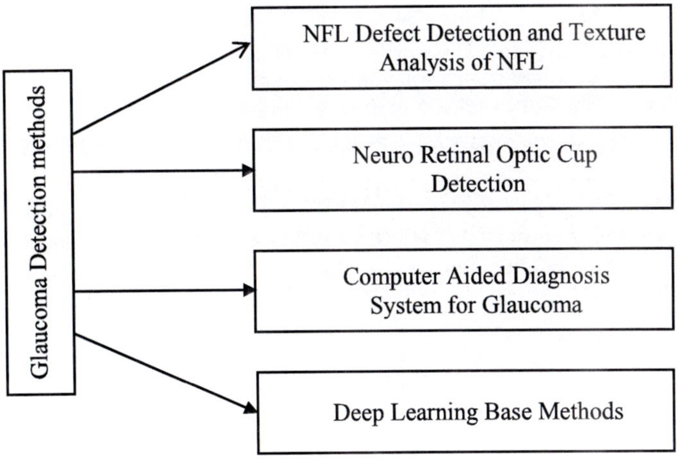

Fig. 1 Detection of glaucoma techniques

stratum pathological disorder [14]. Mainly, NFL is used as this technique is most popular to diagnosis of glaucoma [15]. NFL technique in this domain is discussed below.

For the detection of RNFL defect automatically from fundus images, NFLD is the most useable parameter for diagnoses of glaucoma for ophthalmologist. For the automatically detection of nerve fibre layer defect, the CAD system is used because it takes less time to find number of disease like diabetes, hypertension and glaucoma. CAD system helps to find these types of disease more accurately [16]. The flow chart of proposed algorithm is shown in Fig. 2.

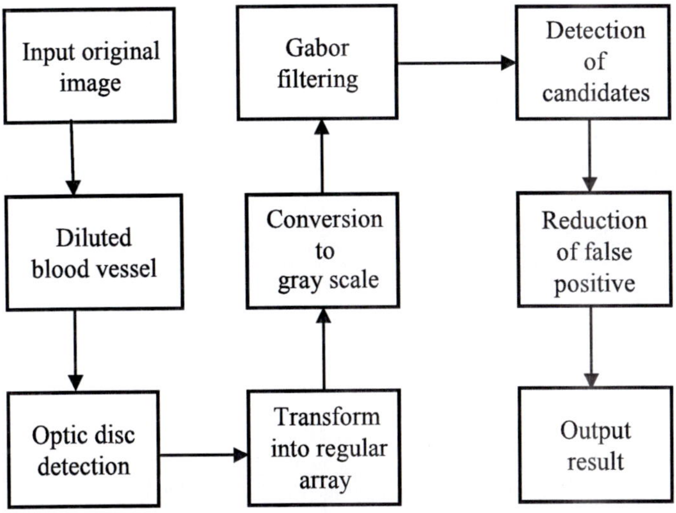

Fig. 2 Flow chart of proposed methodology [17]

K. A. Vermeera et al. show that automatic classification process addressed the global parameters but important local information is ignored. To search the local strong edges with a preference for straight edges, author has classified these local edges in term of strength and came out with results as combined into wedge defects with sensitivity and specificity by 73% and 90% on set of 65 images [19, 20].

Texture analysis is done by using local binary pattern. By detection of RNFL loss, author detected important region around the ONH. This technique is used for spatial interaction among nearby pixel in the textural images. Author differentiated between healthy and glaucomatous by description of RNFL [21].

2.2 *Neuro Retinal Optic Cup Detection in Glaucoma Diagnosis*

This class of technique is called a multimodality fusion for optic cup detection, which can provide the more accurate way for the extraction of neuro retinal optic cup (Table 1).

Glaucoma is the primary cause of blindness. Furthermore, the early detection of glaucoma with the calculation of CDR is more accurate, but automatically, calculation of CDR is very difficult job. Recently, KA Vermeer et al. proposed a technique which can easily find out area of interest (ROI) from fundus image, by using the least fitting algorithm and result showed that the detected boundary of optic cup very accurately and outperform in comparison to other exiting algorithms [20].

Later, Novotný et al. [21] suggested a novel method to find optic cup detection, assess the approach for more precise appraisal of neuro retinal optic cup diagnosis founded on labelling, convex hull and ellipse flitting methods as depicted in Fig. 3. Automatic calculation of cup boundary is not easy task because of interweavement

Table 1 Comparison of nerve fibre layer techniques for detection of Glaucoma

Author Name	Sensitivity	Specificity	Accuracy	Description
Vermeer et al. [19, 20]	80	93	–	Discloser of wedge is assisting the clinician for diagnosis
FM Vosa et al	73	90	–	Classification of local edges based on wedge defect
Hayashi et al. [18]	–	71	71	Using morphological filtering, blood vessel is erased, and Gabor filtering is used for enhancing the vertical dark bands
Novotny et al. [21]	–	–	–	This method has 3% classification error with distinguished normal and glaucoma eye

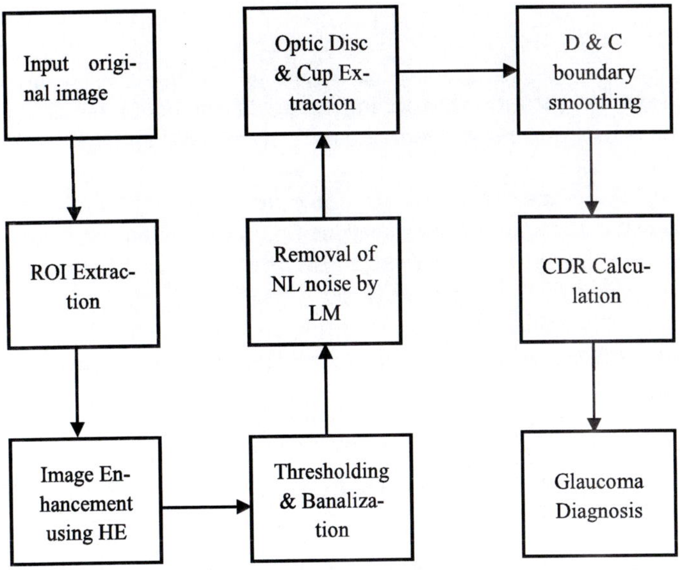

Fig. 3 Steps of CDR measurement using CAD system [21]

of blood vessel around the cup. These neuro retinal detection techniques improved the accuracy of the boundary calculation. Furthermore, Zhang et al. [22] present the methods of detection and segmentation of cup, and mixed approach is present on multimodalities including level set segmentation convex hull and ellipse fitting boundary smoothing.

Recently, Cheng et al. [23] proposed a method using the RetCam for automatically differentiate closed and open angle of glaucoma. In this method, author presents the two approaches for RetCam is arc amount and angle width-based. This block diagram explains the working of RetCam system (Fig. 4).

Liul et al. [24] proposed to describe the ARGALI system that calculated CDR automatically from non-stereographic retinal fundus images. By this system, author first extracted the region of interest via pixel intensity analysis, and then, level set algorithm is used for segmentation of the optic disc. The complexity of this method is segmentation of cup because of interweavement with blood vessels. Therefore,

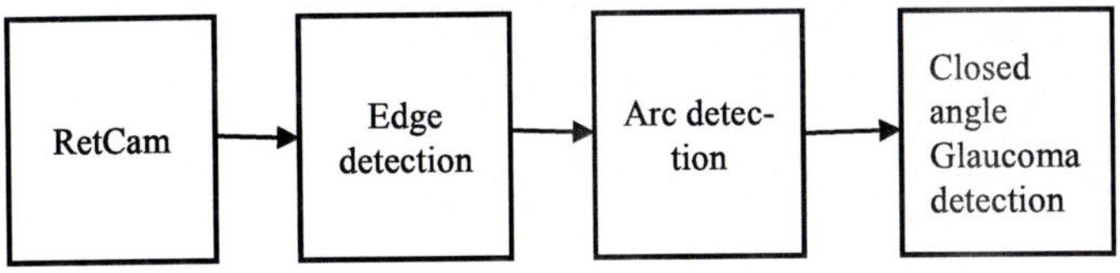

Fig. 4 Block diagram of RetCam system [23]

multimodal techniques are used for optic cup segmentation. The result demonstrated that the proposed system is a cost effective, efficient in screening also.

Gopal Datt et al. [25] presented a method in which area of neuro retinal rim continuously changes with cross-sectional study of rim loss for the estimation of the glaucoma. Authors demonstrated that most rim is decreased in the super temporal sector. The severity of glaucoma expressed as incensement in the outer nasal area than in the nasal inferior area. Jost et al. [26] proposed algorithm in which extracted multidimensional feature space around the optic disc. In this algorithm based on the anatomical evidence, blood vessels bend irregularly at the cup boundary. The severity of glaucoma is expressed in terms of bending irregularity around optic cup. This algorithm is evaluated on dataset of the 138 images with 33 normal and 105 glaucomatous. The comparison of NROC techniques results is shown in Table 2.

Table 2 Comparison of NROC techniques for detection of glaucoma

Author Name	Sensitivity	Specificity	Accuracy	Description
Zhuo Zhang et al.	97.2	97.2	97.2	The good performance achieved by the multimodalities including level set segmentation convex hull and ellipse flitting boundary smoothing method
Cheng et al.	97.8	92.6	–	This system intelligently clearly differentiated the open- and close-angle glaucoma
A. Murthi et al.	–	–	97.5	Calculation of CDR is the way the diagnosis of glaucoma by some feature extraction
J. Liul et al.	–	–	–	Non-stereographic retinal fundus photographs used for automatically calculate the CDR
Jost B et al.	–	–	–	Detection glaucoma by the progression of visual field defects and the morphology of lamina cribrosa
Gopaldatt Josi et al.	–	–	–	Parameterization technique used for OD and monocular retinal images used for cup detection

2.3 *Computer Added Diagnoses System for Diagnosis of Glaucoma*

This section discusses computer-based system for glaucoma diagnosis. CAD system is used for diagnosis of glaucoma as the ratio of optic disc to optic cup. A novel, active contour model is presented was proposed by GD Joshi et al. [27] to get robust OD segmentation. This has been achieved by enhancing the C-V model by including image information at the support domain around each contour point. Causes of glaucoma are high amount IOP inside the optic nerve.

In proposed system, glaucoma is detected by automatic calculation of CDR which is calculated by extraction of optic disc and cup. In this paper, used algorithm is convex null-based neuro retinal optic cup ellipse optimization for accuracy upgrade of the boundary [27]. The laid down procedure the author adopted has been shown in Fig. 5.

Zhang et al. proposed a novel algorithm to diagnose glaucoma very quickly using computer-based system. This system is normally focused on optic nerve defection, simulation of visual field and the special rule which increases the sensitivity and specificity of the system. This method is cost effective and effectively works on fast detection of glaucoma screening system [28]. Xiaoyang et al. proposed to detect various features which are used to calculate the CDR. In the system, input is images data set, and output is feature extracted. The main feature extracted are CD ratio, length of medially OD center, optic nerve head and ISNT ratio and finally glaucoma is detected [29]. Nayak et al. [30] proposed statistical model-based method use for extraction of cup and disc. This method is combination of novel optimal channel selection and knowledge based circular Hough transform for extraction of Optic Disc and achieved 0.92 and 0.81 as dice coefficient for optic disc and cup detection.

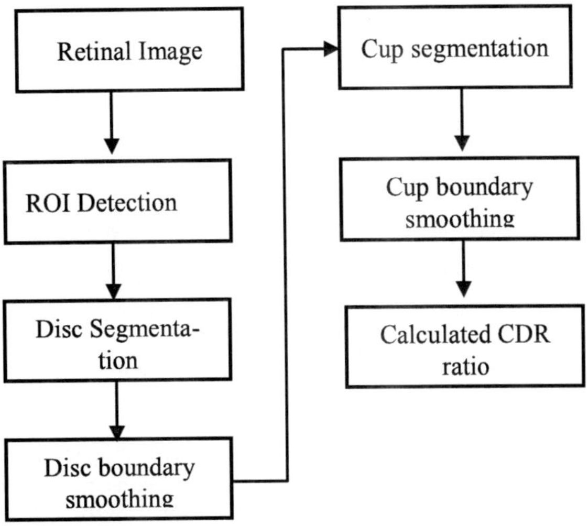

Fig. 5 Flow chart of methodology [27]

Table 3 Comparison of CAD system techniques for detection of glaucoma

Method	Sensitivity	Specificity	Accuracy	Description
Xiaoyang et al. [29]	96.2	96.6	–	Optic nerve defects and visual field simulation using CAD system
Nayak et al. [30]	100	80	–	Find features of cup-to-disc ratio. Calculate length between optic disc centre and optic nerve head diameter of the OD
Hatanaka [31]	80	85	–	This method used for the optic disc glaucoma examination
Noronhha [32]	100	92	84.72	High order spectra system can be used for mass screening
Salam [33]	100	87	–	This system designed for rural area specialists

Cheung et al. [31] proposed an algorithm to find CDR using Canny edge detector to detect edge of optic disc after removing vessel of image.

Then, cup is determined by using thresholding techniques on vertical profiles followed by the cup-to-disc ratio. Yuji Hatanaka et al. proposed system for diagnosis glaucoma using higher order spectra (HOS) cumulates on fundus images. Firstly, the third-order HOS cumulate features have been extracted to linear discriminate analysis to downgrade number of features, and then, the features are fed to SVM for classification images to diagnosis of glaucoma [32]. This system is used for the mass screening. Salam et al. use structural and non-structural features to enhance the accuracy of system. Proposed algorithm was tested using dataset of 100 patients fundus image. Machine learning is used to classify the image features like texture and intensity, vertical height of cup and disc [33]. Table 3 gives the implementation results of CAD system.

2.4 *Detection of Glaucoma by Deep Learning*

Algorithm based on classification methods reported in the literature has different issues due to which the accuracy is not high. Most of algorithms are not independent of datasets. Researchers hence shifted the focus on deep learning-based algorithm in recent years. With the deep neural networks, very large number of local or global features can be extracted, and then, by using deep classifier, glaucoma can be screened with higher accuracy [11]. Some of algorithm-based deep networks are discussed below:

Noronhaa et al. [34] proposed support vector machine (SVM) as well as advanced deep learning algorithm using hybrid feature set to classify the glaucoma. In this algorithm, author extracted colour and texture in fundus image along with other features of CDR for detecting glaucoma disease. This algorithm was validated on 100 patients, and obtained sensitivity and specificity are 100% and 87%, respectively.

In this paper [11], author proposed convolution neural network to extract the features from raw pixel intensities. Deep belief network (DBN) and softmax are used in this algorithm. DBN is allowed to select the different deep features, and softmax is used as a linear classifier to distinguish glaucoma and non-glaucoma image. Evaluated result of glaucoma deep system achieved sensitivity, specificity and accuracy as 84.50%, 98.01% and 99%, respectively. The methodology of glaucoma deep described in Fig. 6. Kevin et al. [34] proposed method by deep learning classifier like deep feedforward neural network with machine learning classifier like random forest and neural network. These classifiers are used to extract the feature like standard deviation values, mean deviation and other 52 features for the diagnosis of glaucoma. Chen et al. [35] developed the convolutional neural network architecture by using the four convolutional layers and two fully connected layers for diagnosis of glaucoma. Author used hierarchical representation of feature for detection of glaucoma.

Asaoka et al. [36] developed an algorithm using support vector machine and detect colour and texture feature for the image and some other severity-level to calculate the cup-to-disc ratio. 100 images of a dataset are evaluated by author and get the 100% and 87% sensitivity and specificity. Salam et al. [33] proposed robust algorithm addressing images effected by abnormalities like hemorrhages in which CDR calculation is very complex. So, the author proposed the end to end supervised model for informative feature of optic disc extracted in the retinal image. Alghamdi Hanans et al. developed a method by combining hybrid deep learning method and single wide-filed OCT protocol diagnosis of glaucoma. Convolutional neural network was used to extract rich features, and random forest classifier was used to train model

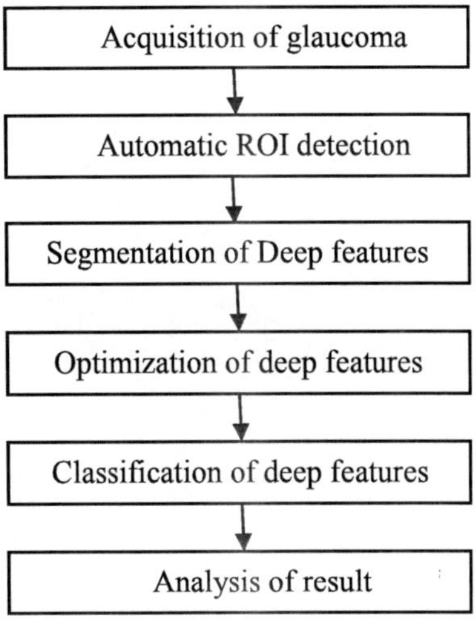

Fig. 6 Block diagram of glaucoma deep system

Table 4 Comparison of deep learning techniques for detection of Glaucoma

Method	Sensitivity	Specificity	Accuracy	Area under curve	Description
Chen et al. [35]	–	–	–	83.1%	CNN architecture is developed for detection of glaucoma with area under of curve 0.831
Asaoka et al. [36]	–	–	–	92.6%	Perperimetric glaucoma of visual field segmented the glaucoma and normally eye
Salam et al. [33]	100%	87%	–	–	For high sensitivity, author makes suspect class of glaucoma
Alghamdihaanhs et al. [37]	96.42%	86%	86.52%	–	This method shows fast and power full OD localization and sensitivity in the evaluated data set

on these features [37]. Some of implementation of deep learning algorithms is listed in Table 4.

3 Conclusion

Glaucoma is the wildly advanced research topic of biomedical field in image processing. Screening test for glaucoma presently available is very cost effective. In view of rapid increases of glaucoma patient over the global, it is very much necessary to have a screening method or screening algorithm easily accessible to all label of society. In this survey paper, different detection techniques of glaucoma have been analysed and compared. The main purpose of this paper is to highlight the entire glaucoma detection algorithm in a single reach. Deep learning is most advanced techniques of the subarea of artificial intelligence. In glaucoma deep system, convolution neural network is an unsupervised architecture. CAD system for glaucoma screening is realistic option to lower the workload of the ophthalmologists and reduce the costs of health care. Thus, this work has comparative studies of detection techniques for glaucoma in term of accuracy. After the study of all these existing methods, it is found that the existing algorithm is not insensitive to databases. As a future work, an effort may be made to design a screening algorithm based on deep neural network are to be exploited to have high accuracy rate and higher efficient in terms of speed and cost effect and database insensitive.

References

1. Greco, A., Rizzo, M. I., De Virgilio, A., Gallo, A., Fusconi, M., de Vincentiis, M. (2016). Emerging concepts in Glaucoma and review of the literature. *The American Journal of Medicine* (2016), https://doi.org/10.1016/j.amjmed.2016.03.038.
2. Dervisevic, E., Pavljasevic, S., Dervisevic, A., & Kasumovic, S. S. (2016). Challenges In early Glaucoma detection. *Med Arch., 70,* 203–207.
3. Spanò, A., Cedrone, C., Floris, R., Simonetti, G., & Nucci, C. (2009). Optic nerve and optic radiation neurodegeneration in patients with Glaucoma. In Vivo analysis with 3-T diffusion-tensor MR imaging.*Radiology, 252*(2), 496–501.
4. Casson, R. J., Chidlow, G., Wood, J. P., Crowston, J. G., & Goldberg, I. (2012). Definition of glaucoma: clinical and experimental concepts. *Clin Exp Ophthalmol, 40*(4), 341–349. PMID22356435
5. Cvenkel, B., & ŠketKontestabile, A. (2011). Correlation between nerve fibre layer thickness measured with spectral domain OCT and visual field in patients with different stages of glaucoma. *Graefe's Archive for Clinical and Experimental Ophthalmology, 249*(4), 575–584.
6. Mwanza, J.-C., Oakley, J. D., Budenz, D. L., Chang, R. T., O'Rese, J. K., & Feuer, W. J. (2011). Macular ganglion cell–inner plexiform layer: automated detection and thickness reproducibility with spectral domain–optical coherence tomography in glaucoma. *Investigative Ophthalmology & Visual Science, 52*(11), 8323–8329.
7. Lam, D. Y., Kaufman, P. L., Gabelt B'Ann, T., To, E. C., & Matsubara, J. A. (2003). Neurochemical correlates of cortical plasticity after unilateral elevated intraocular pressure in a primate model of glaucoma. *Investigative Ophthalmology & Visual Science 44*(6), 2573–2581.
8. Ystemucel, Y. H., Gupta, N., Zhang, Q., Mizisin, A. P., Kalichman, M. W., & Weinreb, R. N. (2006). Memantine protects neurons from shrinkage in the lateral geniculate nucleus in experimental glaucoma. *Archives of Ophthalmology 124*(2), 217–225.
9. Nouri-Mahdavi, K., Zarei, R., & Caprioli, J. (2011). Influence of visual field testing frequency on detection of glaucoma progression with trend analyses. *Archives of Ophthalmology, 129*(12), 1521–1527.
10. Chandrika, S., & Nirmala, K. (2013). Analysis of Cdr detection for Glaucoma diagnosis. *International Journal of Engineering Research and Applications (IJERA).* ISSN: 2248-National Conference on Advanced Communication & Computing Techniques (NCACCT-19 March 2013, 23–27.
11. Gangwani, R. A., Chan, J., Lee, J., Kwong, A., & Lai, J. S. M. (2013). Detection of Glaucoma in a cohort of Chinese subjects with systemic hypertension. *Journal of Ophthalmology,* 5 (Hindawi Publishing Corporation)
12. Abbas, Q. (2017). Glaucoma-deep: Detection of Glaucoma eye disease on retinal fundus images using deep learning. *(IJACSA) International Journal of Advanced Computer Science and Applications, 8*(6).
13. Takayama, K., Ooto, S., Hangai, M., Ueda-Arakawa, N., Yoshida, S., Akagi, T., Ikeda, H. O., et al. (2013). Highresolution imaging of retinal nerve fiber bundles in Glaucoma using adaptive optics scanning laser ophthalmoscopy. *American Journal of Ophthalmology, 155*(5), 870–881.
14. Sharma, P., Sample, P. A., Zangwill, L. M., & Schuman, J. S. (2008). Diagnostic tools for glaucoma detection and management. *Survey of Ophthalmology, 53*(6), S17–S32.

15. Golzan, S. M., Avolio, A., & Graham, S. L. (2011). Minimising retinal vessel artefacts in optical coherence tomography images. *Computer Methods and Programs in Biomedicine, 104*(2), 206–211.
16. Jan, J., Odstrcilik, J., Gazarek, J., & Kolár, R. (2012). Retinal image analysis aimed at blood vessel tree segmentation and early detection of neurallayer deterioration. *Computerized Medical Imaging and Graphics, 36*(6), 431–441.
17. Hayashi, Y., Nakagawa, T., Hatanaka, Y., Aoyama, A., Kakogawa, M., Hara, T., Fujita, H., & Yamamoto, T. (2007). Detection of retinal nerve fiber layer defects in retinal fundus imagesusing Gabor filtering. In*Medical Imaging, International Society for Optics and Photonics* (pp. 65142Z–65142Z).
18. Vermeer, K. A., Vosa, F. M., Lemijb, H. G., & Vossepoela, A. M. *Detecting Glaucomatous wedge shaped defects in polarimetric images.* 1361–8415/03. https://doi.org/10.1016/S1361-8415(03)00041-0.
19. Vermeer, K. A., Reus, N. J., Vos, F. M., Vossepoel, A. M., & Lemij, H. G. Automated detection of wedgeshaped defects in polarimetric images of the retinal nerve fibre layer. In *2nd International Conference on Nature Publishing Group All rights reserved 0950-Informatics, BMEI'09* (pp. 1–4). IEEE, 2009222X/06.
20. Novotný, A., Odstrčilík, J., Kolář, R., & Jan, J. (2010). Texture analysis of nerve fibre layer in retinal images via local binary patterns and Gaussian Markov random fields. In*20th Biennial International EURASIP Conference BIOSIGNAL 2010 Analysis of Biomedical Signals and Images* (pp. 308–315).
21. Zhang, Z., Liu, J., Wong, W. K., Tan, N. M., Lim, J. H., Lu, S., Li, H., Liang, Z., & Wong, T. Y. (2009). Neuro-retinal optic cup detection in glaucoma diagnosis. In *2nd International Conference on Biomedical Engineering and Informatics, Tianjin* (pp. 1–4).
22. Murthi, A., & Madheswaran, M. (2012). Enhancement of optic cup to disc ratio detection in glaucoma diagnosis. In*International Conference on Computer Communication and Informatics (ICCCI)* (pp. 1–5), IEEE.
23. Zhang, Z., Liu, J., Wong, W. K., Tan, N. M., Lim, J. H., Lu, S., et al. (2009). *Neuro-retinal optic cup detection in Glaucoma diagnosis.* 978-1-4244-4134-1/09/$25.00 ©2009 IEEE.
24. Cheng, J., Liu, J., Lee, B. H., Wong, D. W. K., Yin, F., Aung, T., et al. (2010). Closed angle glaucoma detection in RetCam images. In*Annual International Conference of the IEEE, Engineering in Medicine and Biology Society (EMBC)* (pp. 4096–4099). IEEE.
25. Liu, J., Wong, D. W. K., Lim, J. H., Li, H., Tan, N. M., Zhang, Z., et al. (2009). An automatic cup-to-disc ratio measurement system for glaucoma analysis using level-set image processing. In C. T. Lim, & J. C. H. Goh (Eds.).,*ICBME 2008, Proceedings* (Vol. 23, pp. 559–562).
26. Jonas, J. B., Fernández, M. C., & Stürmer, J. Pattern of Glaucomatous neuroretinal rim loss.*Ophthalmology, 100(1)*, 63–68. doi:10.1016/s0161.
27. Joshi, G. D., Sivaswamy, J., & Krishnadas, S. R. (2011). Optic disk and cup segmentation from monocular color retinal images for Glaucoma assessment. *IEEE Transactions on Medical Imaging, 30*(6).
28. Zhang, Z., Liu, J., Cherian, N. S., Sun, Y., Lim, J. H., Wong, W. K., Tan, N. M., Lu, S., Li, H., & Wong, T. Y. (2009). Convex hull based neuro-retinal optic cup ellipse optimization in glaucoma diagnosis. In*Engineering in medicine and biology society, 2009. EMBC 2009 Annual International Conference of the IEEE* (pp. 1441–1444). IEEE.
29. Song, X., Song, K., & Chen, Y. (2006). A computer-based diagnosis system for early glaucoma screening. In*27th Annual International Conference of the Engineering in Medicine and Biology Society* (p. 66086611). IEEE.
30. Nayak, J., Acharya, R., Subbanna Bhat, P., Shetty, N., & Lim, T.-C. (2009). Automated diagnosis of Glaucoma using digital fundus images. *Journal of Medical Systems 33*(5), 337–346.
31. Cheung, C., Baskaran, M., Aung, T., & Wong, T. Y. (2012). *Automated segmentation of optic disc and optic cup in fundus images for Glaucoma diagnosis.* 978-1-4673-2051-1/12/$31.00 2012 IEEE.

32. Hatanaka, Y.,, Noudob, A., Muramatsuc, C., Sawadad, A., Harac, T., Yamamotod, T., et al. (2010). Vertical cup-to-disc ratio measurement for diagnosis of glaucoma on fundus images. In *Medical Imaging 2010: Computer-Aided Diagnosis, edited by Nico Karssemeijer, Ronald MSummers, Proc. of SPIE*, Vol. 7624, 76243C © 2010 SPIE CCC code: 1605-7422/10/$18 doi: https://doi.org/10.1117/12.843775.
33. Salam, A. A., Khalil, T., Akram, M. U., Jameel, A., & Basit, I. (2016). Automated detection of glaucoma using structural and non structural features. *Springer Plus, 5,* 1–22.
34. Noronhaa, K. P.,Rajendra Acharya, U., Prabhakar Nayaka, K., Martis, R. Y., Sulatha, V. B. (2013). Automated classification of glaucoma stages using higher order cumulant features. 1746–8094/$ – see front matter ©2013 Elsevier Ltd. All rights reserved.
35. Chen, X., Xu, Y., Wong, D. W. K., Yin Wong, T., Liu, J. (2015). Glaucoma detection based on deep convolutional neural network. *IEEE EngMeSoc* (pp. 715–718), Annual Conf., August 2015.
36. Asaoka, R., Murata, H., Iwase, A., & Araie, M. (2016). Detecting preperimetric Glaucoma with standard automated perimetry using a deep learning classifier. *Ophthalmology, 123,* 1974–1980.
37. Alghamdi, H. S., Tang, H. L., Waheeb, S. A., & Peto, T. (2016). Automatic optic disc abnormality detection in fundus images: A deep learning approach. In X. Chen, M. K. Garvin, J. Liu, E. Trucco, Y. Xu (Eds.)., *Proceedings of the Ophthalmic Medical Image Analysis Third International Workshop, OMIA 2016, Held in Conjunction with MICCAI 2016*, Athens, Greece, October 21, 2016. 17–24.

Security Risks and Challenges in IoT-Based Applications

Ashwin Perti, Alka Singh, Amit Sinha, and Prabhat Kr. Srivastava

Abstract Internet of things (IoT) is a network having very large scale of heterogeneous devices, sensors and equipment connected together with the communication capability. IoT is capable of establishing the interface between things (digital or physical surrounding objects) and human. Cloud servers are getting used to manage, control and service the IoT devices. Internet is the backbone of IoT which establishes the communication system between IoT devices beyond geographical limit. The diverse and heterogeneous structure of IoT phenomenon introduces variety of new security-related risks and challenges. Many threats like botnets, home intrusion, remote control of the IoT devices and man-in-the-middle attacks are emerging and need a stronger security implementation to stop IoT devices being compromised. This paper surveys the different kind of IoT network technologies, security-related challenges and solutions to these challenges to form more secure IoT environment for trustful adoption of services through industrial or personal use.

Keywords Internet of things · IoT · Intrusion · RFID · Wi-Fi · Heterogeneity · Cloud · Malware · Virus · Botnet

A. Perti · A. Sinha
Department of Information Terchnology, ABES Engineering College, Ghaziabad, India
e-mail: ashwinperti@abes.ac.in

A. Sinha
e-mail: amit.sinha@abes.ac.in

A. Singh (✉)
Department of CSE, NIET Engineering College, Greater Noida, India
e-mail: alka@niet.co.in

P. Kr. Srivastava
Department of CSE, Dr. KNMIET, Modinagar, Ghaziabad, India
e-mail: sri_prabhat@rediffmail.com

S. Tiwari et al. (eds.), *Proceedings of International Conference on Big Data, Machine Learning and their Applications*, Lecture Notes in Networks and Systems 150,
https://doi.org/10.1007/978-981-15-8377-3_9

1 Introduction

The Internet of things (IoT) is a complex network architecture consists of variety of devices, sensors and equipment following different communication protocols so-called heterogeneous network. IoT is getting more attention of researchers and industries from last two decades. The main objective of IoT is free flow of information by connecting various types of digital or physical objects having different communication protocols. Devices connects in IoT platform through an Internet connection to deliver specific type of services using real-time communication. IoT includes communication technologies like radio frequency identification (RFID), cloud computing, wireless sensor network (WSN), near-field communication (NFC), machine-to-machine (M2M) communication, and worldwide interoperability for microwave access (WiMAX), low-power wireless personal area network (LoWPAN), etc. IoT concept was initially given by Kevin Ashton in 1982 to establish an interface between human beings and virtual environment to make their life easier [1]. The growth rate of connected devices in IoT is highly tractive today. According to an article published on Forbes, the global market of IoT will grow this predicted from $157 billion in 2016 to $457 billion by 2020, attaining a compound annual growth rate (CAGR) of 28.5% [2]. The number of connected devices in IoT will grow up to 50 billion in 2020 and will surge up to 125 billion by 2030 [3].

IoT is capable to serve in any application area but due to dependency on Internet, we could not neglect the security- and privacy-related issues. Security vulnerabilities cyber-attacks are more advance and improved than before. IoT devices are widely used in electronic health monitoring systems, smart cars, home appliances, military and other type of personal or industrial objects. These devices could be vulnerable to external threats like malware, viruses, hackers, physical damages and theft. A hacker can attempt to launch phishing, SQL injection, cross site scripting, DDoS attacks to hack, performance downgrade or damaging devices used in IoT. Some of the popular attacks on IoT have been discussed below.

Stuxnet is a malicious computer worm which affected the industrial programmable logic controllers (PLCs) in Iran's nuclear-fuel enrichment project. Although Stuxnet was not a type of IoT attack, but it was a sign that smart devices can be compromised [6]. The Mirai botnet attack was launched to infect older routers, DVD players or IP cameras especially in 2016 [4]. Mirai used these compromised IoT devices to launce the http flood attack (DDoS attack) to the Dyn server. This IoT botnet is made by Mirai malware and causes Twitter, New York Times, Netflix, GitHub, and CNN like networks to get affected by this DDoS attack. The Reaper (IoTroop) was another botnet which stunned everyone in 2017 and more dangerous than Mirai botnet [5]. Reaperbotnet came in spotlight in September 2017 and infected over 1 million wireless networks. Reaper is an evolution of Mirai and uses more sophisticated hacking tools and software than Mirai. Another very harmful malware called BrickerBot came in existence and is capable of killing any unsecured IoT device. The most awful thing about BrickerBot is that consumers of IoT devices could never know that their device is affected by this bot. BrickerBot finds an unsecured IoT device on the network and performs series of Linux commands to corrupt the device storage or disturbs the connectivity to affect the device performance [10]. We have discussed security threats in detail in Sect. 3 (Fig. 1).

The world of IoT is still full of security loopholes and needs a strong and trustful security mechanism that can eliminate these loopholes. A clear and complete structure and design of IoT is yet to be defined, and this could be a reason that above threats are still capable to harm devices and applications in IoT. This paper surveys about these security threats and their proposed cure in detail.

The rest paper is organized in following six sections. Section 2 describes the communication strategies in IoT. In Sect. 3, we will discuss existing security threats and vulnerabilities that can harm IoT. Section 4 describes major technologies and mechanisms to secure IoT. In Sect. 5, we have given the summary related to the topic that can help to identify all the risks and challenges before adaptation of IoT. In Sect. 6, we have extracted the conclusions about IoT security solutions and future scope for better security.

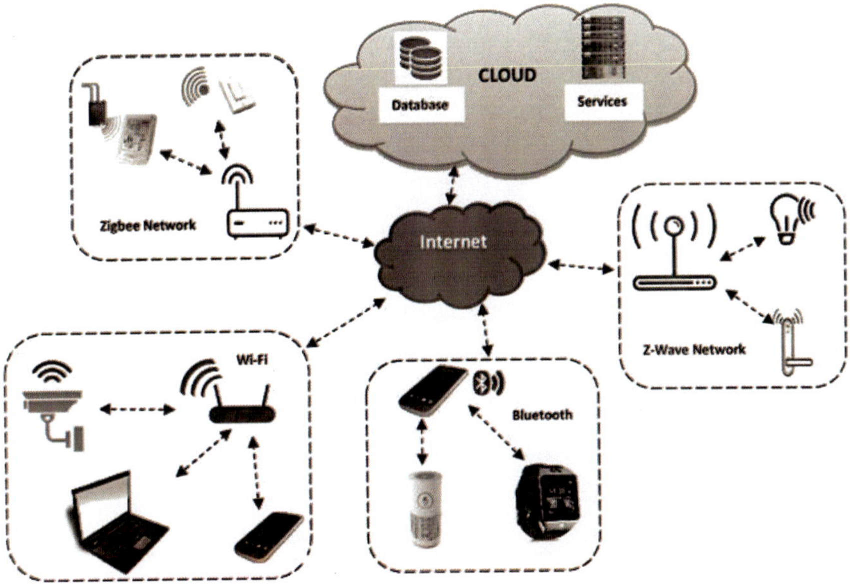

Fig. 1 IoT architecture

2 Communication Technologies Used in IoT

(1) Different communication technologies have been adopted in IoT. This section describes these communication strategies with their operative background and architecture. Table 1 summarizes the characteristics of different IoT network architectures and a comparison chart on different criteria like frequency (Fig. 2a), data rate (Fig. 2b) and range (Fig. 2c) have been picturized in Fig. 2.

(2) **IEEE 802.15.4 (ZigBee)** is a low data rate, low power consumption and low-cost wireless networking protocol, which is used to defining the operation of WSNs and currently uses IEEE 802.15.4 MAC and PHY layers. IEEE and ZigBee combined their technological research regarding communication in devices inbuilt with Bluetooth technology and having low power and low data rate. These devices require long battery life and does not require the high-speed data rate. The range of these devices may vary to 10–75 m. The data rates are 250 kbps at 2.4 GHz, 40 kbps at 915 MHz, 20 kbps at 868 MHz [12]. ZigBee provides interoperable data networking which operates on upper level of protocol stack as network layer to application layer. This will eliminate the consumer's dependency on product manufacturer and ensures the working between different manufacturer devices [11].

Table 1 Different communication strategies in IoT

Technology	Frequency (GHz)	Stream data rate (kbps)	Approximate range (m)	Vulnerabilities
ZigBee	2.4	250	10	Device tempering, key secrecy required
RFID	2.45	640	100	MITM attack, sniffing, denial-of-service attack, cloning & spoofing
NFC	0.13	424	0.10	Low range, low security
WiMAX	2.5	55,296	50,000	Jamming attack, scrambling attack, water torture attack
BLE (Bluetooth)	2.4	2048	30	Bluejacking
Wi-Fi	5	55,296	36	Vulnerable to passive attacks, jamming and scrambling
6LoWPAN	2.4	250	30	Low security in multihop

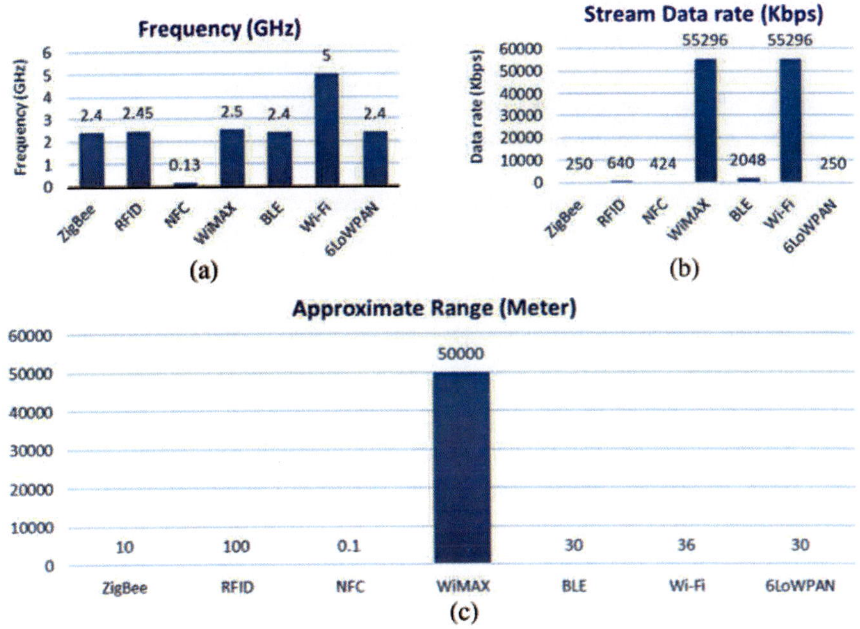

Fig. 2 IoT communication strategy comparison

(3) **The radio frequency identification (RFID)** technology allows a computing device to read identity of RFID tags from a distance and is a replacement of barcode technology. RFID devices can be categorized in two classes. First is active class, in which devices use power either from an integrated power source battery or connected to a powered infrastructure. The second class is passive RFID which contains antenna, a semi-conductor chip attached to antenna and some form of encapsulation [8].

(4) **Near-field communication (NFC)** uses magnetic field induction to establish communication among short-range high-frequency wireless devices. NFC devices uses a peer-to-peer network to perform data exchange. NFC is an upgrade to the RFID technology and has been developed by Philips and Sony jointly [13]. NFC operates in three different modes. The first mode is read/write mode in which interaction is made with an NFC-enabled device and reads data in from device or writes data out. In peer-to-peer mode, a two-way communication is established between NFC-enabled devices. In card emulation mode, system acts as a contactless smart card [22].

(5) **Machine-to-machine (M2M)** system establishes direct communication between two IP-based IoT machines or sensors over wired or cellular networks to send the data to gateways or cloud servers in IoT network. Human interaction is not required for communication and data transferring between devices [14].

(6) The worldwide interoperability for **microwave access (WiMAX)** allows the high-speed data transfer (30–40 mbps) and belongs to IEEE 802.16 wireless

family. WiMAX is much faster than Wi-Fi, and its range for connectivity and data transfer is up to 40 km. Thousands of users or devices can be connected simultaneously through this network with security-level implementation which lacks in Wi-Fi network [15].

(7) **Bluetooth low energy (BLE)** uses IEEE 802.15.4 for communication between ultra-low power IoT devices. BLE may use one of the topology formations like tree, mesh, cluster or star for the connectivity. BLE implements frequency hopping over 37 channels for bidirectional and three channels of unidirectional [16].

(8) IEEE 802.11 is a set of some technical specifications related to communication between Wi-Fi devices. These specifications are related to physical layer and media access control (MAC) and connect devices like printers, scanners, smartphones and laptops without wire. These network connections are easy target for passive attacks, and active attacks can also be performed by exploiting the hardware security loopholes and protocol vulnerabilities [17].

(9) **Low-rate and low-power wireless personal area networks (LoWPAN)** send the data in the form of packets and uses IPV6 over wireless network. Internet engineering task force (IETF) defines the 6LoWPAN which later defines the compression and encapsulation mechanisms that enables the IPV6 over low-power wireless LAN (WLAN). 6LoWPAN is being used in application areas of industrial monitoring, smart grid, general automation, home automation. 6LoWPAN utilizes the IEEE 802.15.4 to provide low layers for low-power wireless network and uses 128-AES link layer security defined by IEEE 802.15.4. IPV6 is applied to PHY and MAC layer in 6LoWPAN communication of the existing 802.15 standard [18].

3 Security Vulnerabilities and Threats in IoT

(1) IoT Botnets (Thingbot)

IoT botnets are compromised independent Internet-connected IoT devices like wearables, medical instruments, industrial systems infected with malwares. These compromised devices and sensors are Internet enabled and able to transfer data automatically. Devices infection increases from one infected device to another without the knowledge of device owner. Attackers can use these compromised devices as botnet to launch DDoS attacks. Aidra, Bashlite, Linux/IRCTelnet, Hajime, Linux. Wifatch, BrickerBot and Mirai are some popular IoT botnets. Mirai is one of the biggest destructive botnets [7]. The first Mirai botnet attack (DDoS attack) was traced on September 20, 2016, against the Web site of journalist Brian Krebs at the 620 Gbps. Over 24,000 systems infected in this massive attack [10]. Table 2 summarizes the popular botnets with their attack techniques.

Table 2 IoT botnet discovery timeline and attack techniques

IoT botnets	Year	Attack technique/Impact
Aidra	2012	Telnet-based attacks on IoT devices
Bashlite	2015	Infects Linux system to launch DDoS attacks
Linux/IRCTelnet	2016	Sends UDP and TCP floods in both Ipv4 and Ipv6 protocols
Hajime	2016	Targets devices via Telnet and gains access by brute-forcing default credentials
Wifatch	2014	Removes other malware and disables telnet access
BrickerBot	2017	Uses exploit code to gain access and rewrites the device's flash storage with random data
Mirai	2016	DDoS attacks, GRE floods and water torture attacks

(2) Spam
Spam is a messaging system which sends unrequested bulk messages to a target device. Spam filters are the option to identify and stop these unwanted messages. Spammers can use 2D bar codes to flood the physical side of the IoT and mislead users to reach unsolicited and unrelated content over the Internet [9]. Digital signature system can be used to overcome this problem. Mass flooding, Web site referrals, and redirection hiding technique are the spamming techniques used by spammers.

(3) Advanced Persistent Threats (APTs)
Advanced persistent threat (APT) is a type of attack in which a unauthorized user gets foothold through malware, physical malware infection or external exploitation to execute future continuous attacks for long time period to achieve his malicious objective without being detected. There are many activities performed in this attack like network hacking, detection avoidance, determining the target area, collecting important information to gain access. This attack is basically targeted, goal-oriented, persistent and unnoticed in nature [19] (Fig. 3).

(4) *Ransomware*
Ransomware is a type of malware that encrypts all data of your computer and sells the decryption keys to decrypt. The damage made by ransomwares is irreversible, and decryption key is required for getting data back. The ransomware is more serious threat for IoT because its action cannot be reset with our own and will have to pay for that.

(5) *Data Theft*
Data and identity theft could be a more serious security-related problem in the IoT. Suppose that all information got by your smart watches, fitness tracker, GPS location data, and social media is combined together and may be sufficient to reveal your identity. Thus, identity and data theft are one of the biggest threats to the IoT.

(6) *Home Intrusions*
Smart home corresponds to a heterogeneous network structure having variety of devices, applications and technologies connected together. The globally

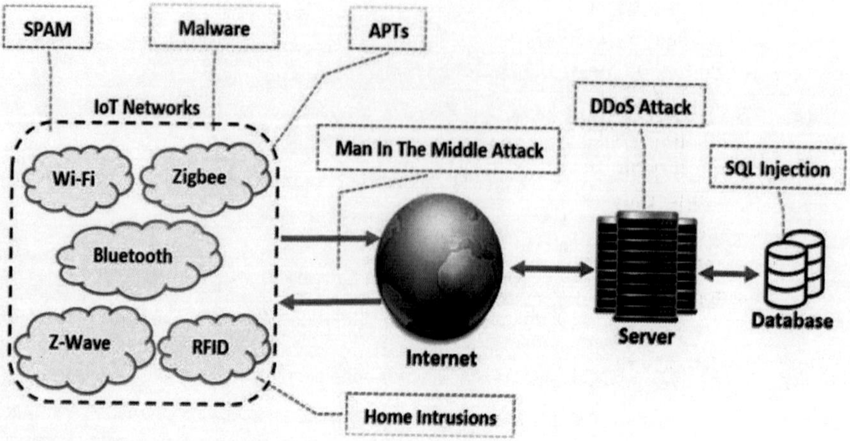

Fig. 3 Security threats in IoT

available smart home ecosystem data may prone to security vulnerability. This electronic data needs to be protected from external intrusions which may cause several security issues like denial-of-service attacks. Home intrusion could be launched through several attacks like DDoS attack, device hijacking, and phlashing (PDoS). Intrusion detection systems (IDS) are highly required for the safety of electronic data of smart home.

(7) *Remote Vehicle Control*
Current security measures of connected vehicles in IoT are in poor state today. Connected remote vehicles may face several security-related issues like vehicle sensor attack, wireless carjacking, GPS jamming and spoofing, backdoor attacks, frontdoor attack, hacking of remote vehicle control application.

(8) Man-in-The-Middle Attack
Intercepting a communication channel with malicious intention between two systems without acknowledgement of sender and receiver is called man-in-the-middle attack. The man-in-the-middle attack can be launched through several techniques like address resolution protocol (ARP), DNS spoofing, session hijacking, and sniffing. Once a communication channel is compromised, an attacker can hear all communication as well as can transmit false messages too. In the scenario of IoT, this attack can be more effective and saboteur. An interceptor can track your daily activities through compromised IoT devices like health monitoring system, smart cars, mobile devices, cameras, GPS navigation system and many others. Smart cars can be misguided, false health monitoring system data can be transferred to show emergency like situation. A strong encryption mechanism like RSA, AES, and Blowfish can be used in IoT to get protected from this threat [20].

(9) RFID Skimming
RFID skimming is the process of stealing the data or information through a chip reading device from RFID chips. Most of new debit cards, credit cards and

identity cards contain RFID chips inside them. These RFID chips use radio waves to read and capture the information from several feet away, and this facility can be used to hack the RFID chips for malicious intentions. Hacked information can be used to create duplicate cards or chips and use them for illegal financial benefits.

(10) Insecure Cloud Interface
The unencrypted data traveling to cloud interface from IoT devices can be intercepted by attackers. The cloud computing introduces potential security-related risks to IoT devices connected with cloud. Although cloud has many strong security implementations but when IoT devices with insecure credentials, unencrypted data transmission, and weak authentication mechanism connects with the cloud computing, this type of insecure connection possesses many security vulnerabilities for IoT-cloud collaboration.

(11) *Insecure Mobile Interface*
IoT devices with mobile interface having weak or no security implementations are one of the biggest threats for the IoT. Information can be hacked from the wearables, remote vehicle control system, remotely controlled home appliances, and other computing devices and sensors connected with insecure mobile interface. An attacker can trace anyone's health-related information, identity, banking details easily through intercepting insecure mobile interface.

(12) *Insecure Software/Firmware*
Insecure software and firmware are easy target for botnets or malwares. IoT devices firmware falls under two categories: embedded and OS-based firmware. Non-encrypted communication to the firmware of IoT devices is vulnerable to external threats like botnets. Access to these devices' firmware must be password protected and regular updates must be performed for better security. The easiest targeted devices are with default passwords. Default passwords must be changed as soon as possible to save device from botnets and malwares.

(13) *Insecure Mobile Interface*
IoT devices with mobile interface having weak or no security implementations are one of the biggest threats for the IoT. Information can be hacked from the wearables, remote vehicle control system, remotely controlled home appliances, and other computing devices and sensors connected with insecure mobile interface. An attacker can trace anyone's health-related information, identity, banking details easily through intercepting insecure mobile interface.

(14) *Insecure Software/Firmware*
Insecure software and firmware are easy target for botnets or malwares. IoT devices firmware falls under two categories: embedded and OS-based firmware. Non-encrypted communication to the firmware of IoT devices is vulnerable to external threats like botnets. Access to these devices' firmware must be password protected, and regular updates must be performed for better

security. The easiest targeted devices are with default passwords. Default passwords must be changed as soon as possible to save device from botnets and malwares.

4 Security Implementations for IoT

IoT security threats are the major cybersecurity challenges in current IT ecosystem. In previous Sects. 2 and 3, we have discussed about major IoT communication technologies and threats. Table 3 summarizes the different IoT threats with their attach techniques and identity. There should be a stronger security mechanism to stop IoT devices being compromised. It's quite complex to implement stronger security to IoT devices due to their low computational capability and low memory. Still there are some anti-threat implementations for harmful IoT threats. IoT network is getting larger as it includes more devices day by day. IoT devices must include stronger security techniques before communicating with other devices.

Table 3 IoT threats and their security mechanism

Threat type	Threat identity	Year	Security technique
Botnet	DDoS attack	2001	Authentication, encrypted device identity
Spam	Sends bulk messages	1994	Change passwords frequently, web application firewall (WAF), DDoS mitigation system
APTs	Monitors network activity and steals data with no damage	2006	Beware of trojans, suspicious emails, and malicious port scanning, install patches to prevent previously known vulnerabilities
Ransom ware	Encrypts all data on victim's computer	1989	Data backup, use cryptolocker software, disable RDP, be careful when an email has files with '.exe' extension
Man-in-the-middle attack	Intercepting and interrupting an interconnection between two separate network devices	2003	Analyze the response time in the web traffic, authentication, use SSL/TLS certificates for Web sites, PKI technology, WEP/WPA encryption
RFID skimming	Stealing information from RFID cards	2006	RFID blocking using RFID shield. Disable the RFID chip in your credit card

Cryptography with secure encryption and decryption keys can be used to determine device identity and could make a hurdle between user data and threats. SSL certificates can play a vital role to facilitate the device identification and authentication process. Authentication process must be enforced before any software or firmware update to save IoT devices from being compromised as a botnet (Thingbot). There should be a periodic examination of the IoT network by anti-malware utility to detect any malicious activity. Network devices like routers, printers, security cameras and other IoT devices having default passwords must be changed to new one so that Mirai like botnets could not harm them.

Spam filters can be used to stop flood of spam. Spammers can flood the physical side of IoT devices to increase traffic for specific page. As the problem of spamming explained in Sect. 3, a possible solution to the spam problem is to digitally sign the 2D barcode and embedding the digital signature in QR code.

Advance persistent threats include persistent behavior of attackers as they have patient until getting their target complete. Solutions to mitigate APTs may include secure the entry point of the network, be careful to the outgoing traffic, install new security patches, and aware of any unusual activity being occurred in the network traffic. Utilities like firewall, intrusion prevention system (IPS), antivirus, botnet or command detection system and sandboxing should be used to ensure better security through these threats [19].

The strongest possible way to save IoT from man-in-the-middle attack is strong encryption system between IoT devices and communicating server of these devices. IoT devices communicate also with each other without involvement of server. So, encryption schemes should also be applied between IoT devices because MITM attack is also possible between two communicating IoT devices. Another possible way to mitigate the MITM attack by using encrypted virtual private network (VPN). This method ensures that everything comes in and goes out is encrypted and secured [20] (Table 4).

Table 4 IoT network security mechanisms [21]

Network type	Security mechanism
ZigBee	Link layer encryption using 128-bit AES, EAP, TLS
BLE	Secure pairing
WiMAX	Sends UDP and TCP floods in both Ipv4 and Ipv6 protocols
Wi-Fi	WEP, AES, TKIP, WPA, WPA2, and 802.1x
6LoWPAN	Access control list, 802.15.4 link layer encryption
NFC	Cryptographic methods and hardware-based security (TDES, AES, RSA, ECC)
RFID	Cryptography such as Advance Encryption Standard (AES)

5 Summary

This paper shows that the future of the IoT is still full of security loopholes. We find through our survey that identity of IoT devices must be secured and input output traffic must be examined in real-time basis for any malicious activity. Device firmware must be protected from being compromised through malwares. IoT devices need stronger security mechanisms for new emerging threats. ZigBee, LoWPAN, RFID, Bluetooth and other type of IoT networks are suffering from various types of threats like malwares, MITM attack, RFID skimming, APTs and SPAM. IoT security-related challenges are getting attention of researchers and inspire them to discover stronger security technique to mitigate these threats. Although most of IoT devices have low hardware configuration, that implementation of stronger security mechanism is not possible on them. This weakness makes them vulnerable to security threats and needs further research to overcome this problem.

6 Conclusion

The Internet of things reveals vulnerabilities exists in it and requirement of research work to secure the communication between IoT devices. Although IoT network has faced several massive threats in recent years, still it needs to be researched to mitigate these threats. There are several threats present which can cause damages to the IoT devices security and world of computing has no any full proof plan or security technology to trace or eliminate these threats completely. New security techniques must be continued to research until we can completely secure the IoT.

References

1. Shen, G., & Liu, B. (2011). The visions, technologies, applications and security issues of Internet of Things. In *2011 International Conference on E-Business and E-Government (ICEE)*. IEEE.
2. Howell, J. (2017, October 24). *Number of connected IoT devices will surge to 125 billion by 2030, IHS Markit says.* Retrieved from IHS Markit: https://technology.ihs.com/596542/number-of-connected-iot-devices-will-surge-to-125-billion-by-2030-ihs-markit-says.
3. Nordrum, A. (2016, August 18). *Popular Internet of Things forecast of 50 billion devices by 2020 is outdated.* Retrieved from IEEE Spectrum: https://spectrum.ieee.org/tech-talk/telecom/internet/popular-internet-of-things-forecast-of-50-billion-devices-by-2020-is-outdated.
4. Fruhlinger, J. (2018, March 9). *The Mirai botnet explained: How teen scammers and CCTV cameras almost brought down the internet*. Retrieved from CSO. https://www.csoonline.com/article/3258748/security/the-mirai-botnet-explained-how-teen-scammers-and-cctv-cameras-almost-brought-down-the-internet.html.
5. Goodin, D. (2017, October 28). *Assessing the threat, the Reaper botnet poses to the Internet—What we know now*. Retrieved from ArsTechnica. https://arstechnica.com/information-techno

logy/2017/10/assessing-the-threat-the-reaper-botnet-poses-to-the-internet-what-we-know-now/.

6. Kushner, D. (2013, February 26). *The real story of Stuxnet*. Retrieved from IEEE Spectrum. https://spectrum.ieee.org/telecom/security/the-real-story-of-stuxnet.
7. Constantinos Kolias, G. K. (2017). DDoS in the IoT: Mirai and other Botnets. In *CYBERTRUST* (pp. 40–44) (published by the IEEE computer society).
8. Kaur, M., Sandhu, M., Mohan, N., & Sandhu, P. S. (2011). RFID technology principles, advantages, limitations & its applications. *International Journal of Computer and Electrical Engineering, 3*(1), 151.
9. Razzak, F. (2012). Spamming the Internet of Things: A possibility and its probable solution. *Procedia Computer Science, 10*, 658–665.
10. Kolias, C., Kambourakis, G., Stavrou, A., & Voas, J. (2017). DDoS in the IoT: Mirai and other botnets. *Computer, 50*(7), 80–84.
11. Ergen, S. C. (2004). ZigBee/IEEE 802.15. 4 Summary. UC Berkeley, September, 10, 17.
12. Ray, P. P. (2018). A survey on Internet of Things architectures. *Journal of King Saud University-Computer and Information Sciences, 30*(3), 291–319.
13. Hong, Y. G., Choi, Y. H., Youn, J. S., Kim, D. K., & Choi, J. H. (2015). *Transmission of IPv6 packets over near field communication*. draft-ietf-6lo-nfc-00.
14. Iqbal, M. A., Olaleye, O. G., & Bayoumi, M. A. (2017). A review on Internet of Things (IoT): Security and privacy requirements and the solution approaches. *Global Journal of Computer Science and Technology*.
15. Papapanagiotou, I., Toumpakaris, D., Lee, J., & Devetsikiotis, M. (2009). A survey on next generation mobile WiMAX networks: Objectives, features and technical challenges. *IEEE Communications Surveys & Tutorials, 11*(4).
16. Narendra, P., Duquennoy, S., & Voigt, T. (2015, October). BLE and IEEE 802.15. 4 in the IoT: Evaluation and interoperability considerations. In *International Internet of Things Summit* (pp. 427–438). Cham: Springer.
17. Mendez, D. M., Papapanagiotou, I., & Yang, B. (2017). *Internet of things: Survey on security and privacy*. arXiv preprint arXiv:1707.01879.
18. Ech-Chaitami, T., Mrabet, R., &Berbia, H. (2011). Interoperability of LoWPANs Based on the IEEE802. 15.4 Standard through IPV6. *International Journal of Computer Science Issues (IJCSI), 8*(2), 315.
19. Hudson, B. (2014). *Advanced persistent threats: Detection Protection and Prevention*. US February: Sophos Ltd.
20. Cekerevac, Z., Dvorak, Z., Prigoda, L., & Cekerevac, P. (2017). Internet of Things and the man-in-the-middle attacks–security and economic risks. *Journal of MEST, 5*(2), 15–25.
21. Madakam, S., & Date, H. (2016). Security mechanisms for connectivity of smart devices in the internet of things. In *Connectivity frameworks for smart devices* (pp. 23–41). Cham: Springer.
22. Rahul, A., Gokul Krishnan, G., Unni Krishnan, H., & Rao, S. (2015). Near Field Communication (NFC) technology: A survey. *International Journal on Cybernetics & Informatics (IJCI), 4*(2), 133–144.

An Assessment of Type-2 Diabetes Risk Prediction Using Machine Learning Techniques

Hardeep Kour, Munish Sabharwal, Shakhzod Suvanov, and Darpan Anand

Abstract Diabetes is a metabolic chronic disease that due to the lack of insulin makes a serious problem with the transfer of blood all over the body. Nowadays, this epidemic has been expanding steadily everywhere throughout the world. Furthermore, negligence can lead to visual deficiency, amputations, cardiovascular breakdown, heart failure, and stroke. Diabetes generally occurs in two types (diabetes Type-1, diabetes Type-2) through the body, diabetes Type-2 shows the acute condition in the patient's body. The present study reviews the existing research studies and their methods for predicting diabetes using machine learning techniques, these methods use previously stored patient information to predict the next steps or the next events. During this research that using machine learning methods in addition to high speed and accuracy in the prediction of diabetes, it can improve the efficiency and importance of this process.

Keywords Diabetes risk prediction · Diabetes Type-2 · Ensemble methods · Machine learning

1 Introduction

Diabetes disease has become one of the main sources of death in the world. Diabetes mellitus is an incessant disease and caused due to high-sugar level the International Diabetes Foundation (IDF) [1], until now, India had a greater number of diabetic patients than any other nation of insulin or the body cells not reacting appropriately to insulin. Diabetes of Type-1 [1] and Type-2 are the normal types of this disease,

H. Kour (✉) · D. Anand
Department of Computer Science and Engineering, Chandigarh University, Mohali, Punjab, India
e-mail: erhardeepkour7@gmail.com

M. Sabharwal
School of Computing Science and Engineering, Galgotias University, Greater Noida, Uttar Pradesh, India

S. Suvanov
Faculty of Digital Technologies, Samarkand State University, Samarkand, Uzbekistan

S. Tiwari et al. (eds.), *Proceedings of International Conference on Big Data, Machine Learning and their Applications*, Lecture Notes in Networks and Systems 150,
https://doi.org/10.1007/978-981-15-8377-3_10

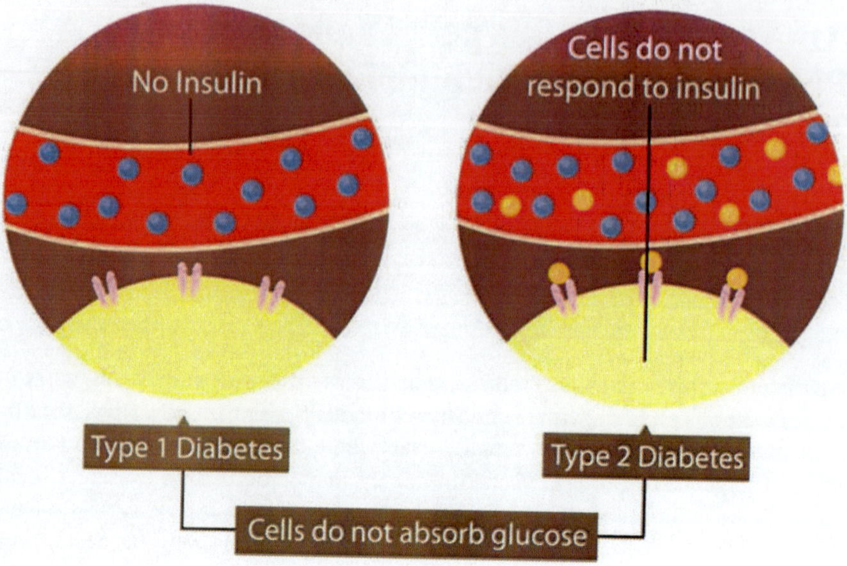

Fig. 1 Type-1 and Type-2 diabetes mellitus

yet some other sorts are also there, like gestational diabetes, which happens during pregnancy, just as different types. Type-2 diabetes [3], as indicated by Canadian Diabetes Association (CDA) for 10 years [4], somewhere in the range of 2010 and 2020, expected to increase from 2.5 million to 3.7 million [5]. As shown in Fig. 1, the diabetes Type-1 and 2, in diabetes Type-1 that is the primary state, the body does not try to make insulin, but in diabetes Type-2, which is in the critical stage that the body cells do not respond to the insulin that is available in the blood [6].

In the circulatory system. It is a significant general medical issue, one of four priority non-communicable diseases (NCDs) directed for activity by the world [7]. As per the World Health Organization (WHO) report [8], about 422 million individuals worldwide have diabetes, especially in low and middle-income nations and approximately 98 million population in India [9] might have Type-2 diabetes by 2030. One of the other reports of IDF center of education demonstrates that 82 million people in the SEA region and this will ascend to 151 million by 2045 [10]. A report by association (American Diabetes Association) states that India would see the best increment in people diagnosed to have diabetes by 2030 [11]. According to reviewing different medical sciences papers, we have discovered that there could be numerous reasons which lead to Type-2 diabetes [12]. Some of the important reasons are related to an individual's lifestyle, absence of the activity, sustenance propensities, weight, high cholesterol (hyperlipidemia) [13], smoking, hypertension (Hyperglycemia), and so forth, which on a very basic level augmentation the danger of treating diabetes [14]. As per the Indian Heart Association India is anticipated to be home to 109 million people with this disease by 2035 [15]. A report published in the 'Lancet Diabetes and Endocrinology' journal [16] found that the measure of insulin expected to viably treat Type-2 diabetes will rise by more than 20 percent worldwide in the coming 12 years [17].

Due to the importance of the topic, the use of modern technology such as artificial intelligence that can predict the different status of patients in less time and with high accuracy can make help the doctors [18], as well as the patients with itself, the present research study with consideration of the importance of diabetes [19], try to have a review on the resent work-based machine learning methods and show the status and efficiency [20].

2 Machine Learning Techniques

Due to the significant regulation of machine learning models and its valuable role in the healthcare area, which leads to the identification and treatment in a short time and with high accuracy. It should also be noted that machine learning models for doctors and specialists help to follow the patient's past information and change the patient's symptoms well through the machine learning model, which will worsen if the patient's condition is not considered. In this section, I will briefly discuss the models that are more important in this learning than the others [21].

2.1 Support Vector Machine

The SVM model is implemented to prevent over fit of training and testing samples and also with the choice of Gaussian kernel, the technique can make more similarity in whole classes [22]. One of the most important drawbacks of SVM is that it does not know the combination of features most influence a prediction, but an important method that works with medical images and plays the best role in classifying. As per the capability of SVM in the classification of medical data, various research works used to predict and detect the disease [23].

2.2 Naïve Bayes

In this manner, another technique for data classification is Naïve Bayes, which is an effective technique for constructing classifiers and an efficient technique to forecast the possibility of various modules based on different attributes [24]. The genetic techniques rank attributes that contribute more toward the arrangement. Least positioned attributes are expelled and the algorithms of classification are manufactured dependent on the evaluated attributes [25].

Table 1 Summary information about the performance of each model

Algorithm	Type	Class	Restriction bias	Performance bias
Support vector machine (SVM)	Supervised	Decision boundary	Works where there is definite decision between two classification	Prefers binary classification problems
Naïve Bayes	Supervised	Probabilistic	Work on problems where inputs independents from each other	Prefer problems where the probability will always greater than zero for each class
Decision tree	Supervised	Tree	Become less useful on problems with low covariance	Prefer problems with categorical data

2.3 Decision Tree (DT)

Based on DT technique, it tries to model decision-making in the form of rules [26]. It can be useful also for predicting a specific value when measuring a patient's result. It can apply also even when the difference value between the two classes is small. Basically, the decision tree with having great problem solving can be used with different data, and also it possible to have advantages of different algorithms together to expand the excellence of the result [27].

Besides this, to evaluate the performance of each method here as shown in Table 1 the details of each method are described as below:

3 Objective of Study

1. To review the previous research work done on Type-2 diabetes prediction using machine learning technique.
2. To analyze and find out the best way for prediction of the disease.

4 Related Work

The motive of this assessment is to show the comparison of previous studies using different machine learning techniques and models used to predict Type-2 diabetes risk and to discover the best options for achieving the highest output accuracy for the same. The various studies use different machine learning techniques like decision trees, SVM, KNN, Naïve Bayes, genetic algorithms, ANN, etc. [28–30].

The comparative chart of all the previous studies related to machine learning techniques and models used to predict Type-2 diabetes depicting the author and publication details, dataset, and classifier used in the model and conclusions showing prediction accuracy as shown in Table 2.

The review of several studies shows that different machine learning algorithms were applied on various types of medical datasets and different machine learning techniques have diverse power when applied on the various datasets, and individual algorithms provided less accuracy than ensemble one, ensemble methods prompt more efficient outcomes. The datasets are shown in Table 3.

5 Conclusion

Diabetes is a metabolic disease that can be forestalled through the way of life alteration, diet control, and control of overweight. This paper is focusing towards reviewing previous research work to predict Type-2 Diabetes, and suggested methods that can be used for better prediction after observing previous research work. Based on this it is better to use a hybrid model using an ensemble approach for classification than an individual classifier. The second approach is the author found that different research studies apply their prediction model only on one dataset to predict accuracy, the author tries to with consideration the importance of dataset on research try to apply on several datasets to show a cumulative accuracy when applied on different datasets. And eventually, the present with consideration of the machine learning method on prediction shows that can improve the effectiveness as well as can predict in the further state of the patient.

Table 2 Comparative studies of related work on machine learning techniques

Ref	Dataset	Method	Conclusion
[1]	University of California(UCI)/Irvine Repository Dataset	Naïve Bays, N-CART	The algorithm named DMP_MI has achieved an accuracy of 87.10%
[2]	San Antonio Heart Study (SAHS) Dataset	SVM, tenfold cross-validation	This model gave an accuracy of 84.1% having recall rate as 81.1% averagely over 100 iterations
[3]	Local dataset	Single-layer perceptron, RF, PCA	The present research study, with the help of several ML-methods, try to predict the performance of ML models using PCA and RF methods
[4]	Canadian Diabetes Association (CDA), Diagnostic of Medical Center Chittagong (MCC), Bangladesh	SVM, K-nearest, NB, decision tree (C4.5)	The paper shows that C4.5, DT achieved a better accuracy of 73.5%
[5]	UCI, ML repository: 1-cancer dataset 2-PIMA diabetes	Hybrid method, K-fold cross-validation	RF plays out the best in any case of cancer data, however, it was the slowest one too. KNN has minimal precision but the quickest one. This is likewise valid for the diabetes dataset
[6]	Pima Indian dataset	Backpropagation, J48 algorithm, NB, SVM	The backpropagation algorithm gives 83.11% predictive accuracy
[7]	PIMA Indian dataset	DNN, PCA, LDA, K-component	The best accuracy of 77.86%. The future work of this can be to and for hidden layers into DNN
[8]	PIMA Indian dataset	RF, DT, NN, SVM, NB	The model trained by various algorithms provides an accuracy of 94.5%
[9]	Pima Indians of Arizona, diabetes dataset, National Institute of Diabetes	Logistic regression, SVM, RF	RF was the most ideal algorithm giving an accuracy of around 84%
[10]	University of Wisconsin hospital, Madison	DT, ANN, logistic regression, SVM, NB	Decision trees and artificial neural networks produce better results in comparison with other models

(continued)

Table 2 (continued)

Ref	Dataset	Method	Conclusion
[11]	UCI/Irvine, ML repository	Deep learning (DL), Distributed- RF, GLM, GBM, NB	The meta-learner algorithm was deep learning algorithm which gave the best results
[12]	Canadian Primary Care Sentinel Surveillance Network (CPCSSN)	No sampling, random under/over, K-meanoids under-sampling, DT, NB	J48 (C4.5), NB, DT model have used the relevant risk factor to generate balanced training sets for improving accuracy
[13]	UCI ML repository	KNN, NB, SVM, DT, LR, RF	It can be seen that SVM and KNN give the highest accuracy (77%) for predicting diabetes
[14]	PIMA Indian Diabetes dataset	Sequential regression multiple imputation	The RF algorithm performed very well giving a precision of 79.7%
[15]	Pima Indian Diabetes dataset	NB, SVM, RF, simple cart	The performance of SVM is better than NB, RF, and simple cart by running on the WEKA tool
[16]	Local Dataset	Big Data, ANN, time series, association, clustering	The available study, with the help of ANN and time sequence-based, estimates the diabetes that with ANN got 19% to 22% shows improvement
[17]	PIMA Indian dataset, UCI	SVM, KNN, ANN, GNB, Min-Max scaling	The overall accuracy of 76.25% with the Gaussian NB Algorithm, ANN provides the maximum accuracy of the scaling method named Min-Max
[18]	Pima Indian dataset	LR, DT, RM, SVM, adaptive boosting	The prediction accuracy is 87.1% in heart by LR 85.71% in diabetes by SVM and 98.57% by (AdaBoost classifier) for breast cancer detection
[19]	PIMA Indian dataset	Neural network-based method, K-means-based methods, multilayer classifier	The process that includes neural network offers high efficiency (85%) with less computation
[20]	Pima Indians diabetes	K-Mean, DT	It provides better accuracy than the previous classification models with data preprocessing and reduction steps
[21]	Database of NID and digestive and kidney diseases	DT, KNN	KNN and C4.5 provide 76.96% and 90.43% accuracy, respectively

(continued)

Table 2 (continued)

Ref	Dataset	Method	Conclusion
[22]	Prima Indian/UCI	Multilayer perceptron	The built model achieved an accuracy of 88.7%

Table 3 Related medical dataset of diabetes

S. No	Dataset
1	PIMA Indian dataset
2	Korean-Health and Genome Epidemiology Study (KHGES) Database
3	Canadian Primary Care Sentinel Surveillance Network (CPCSSN)
4	San Antonio Heart Study (SAHS) dataset
5	University of California (UCI)/Irvine Repository Dataset

References

1. Ramírez, J. C., & Herrera, D. (2019). *Prediction of diabetic patient readmission using machine learning* (pp. 78–88).
2. Faruque, M. F., Sarker, I. et al. (2019). *Performance analysis of machine learning techniques to predict diabetes mellitus* (pp. 1–4).
3. Al Helal, M., Chowdhury, A. I., Islam, A., Ahmed, E., Mahmud, M. S., & Hossain, S. (2019). *An optimization approach to improve classification performance in cancer and diabetes prediction* (pp. 1–5).
4. Sarwar, M. A., Kamal, N., Hamid, W., & Shah, M. A. (2018). *Prediction of diabetes using machine learning algorithms in healthcare* (pp. 1–6).
5. Woldemichael, F. G., & Menaria, S. (2018). *Prediction of diabetes using data mining techniques* (pp. 414–418).
6. Perveen, S., Shahbaz, M., Keshavjee, K., & Guergachi, A. (2018). Metabolic syndrome and development of diabetes mellitus: Predictive modeling based on machine learning techniques. *IEEE Access, 7,* 1365–1375.
7. Rubaiat, S. Y., Rahman, M. M., & Hasan, M. K. (2018). *Important feature selection & accuracy comparisons of different machine learning models for early diabetes detection* (pp. 1–6).
8. Dey, S. K., Hossain, A., & Rahman, M. M. (2018). *Implementation of a web application to predict diabetes disease: An approach using machine learning algorithm* (pp. 1–5).
9. Mir, A., & Dhage, S. (2018). *Diabetes disease prediction using machine learning on big data of healthcare* (pp. 1–6).
10. Rani, S., & Kautish, S. (2018). *Association clustering and time series based data mining in continuous data for diabetes prediction* (pp. 1209–1214).
11. Kohli, P. S., & Arora, S. (2018). *Application of machine learning in disease prediction* (pp. 1–4).
12. Alahmar, A., Mohammed, E., & Benlamri, R. (). *Application of data mining techniques to predict the length of stay of hospitalized patients with diabetes* (pp. 38–43).
13. Barhate, R., & Kulkarni, P. (2018). *Analysis of classifiers for prediction of type ii diabetes mellitus* (pp. 1–6).
14. Dutta, D., Paul, D., & Ghosh, P. (2018). *Analysing feature importances for diabetes prediction using machine learning* (pp. 924–928).
15. Indoria, P., & Rathore. (2018). A survey: Detection and prediction of diabetes using machine learning techniques. *Int. J. Eng. Res. Technol, 7,* 2278–0181.
16. Hema & Palanivel. (2018). A study on prediction of diabetic disorder using classification based approaches *International Journal of Computer Science and Mobile Computing, 7,* 53–60.
17. Deepika & Kalaiselvi. (2018). *A empirical study on disease diagnosis using data mining techniques* (pp. 615–620).
18. Wei, S., Zhao, X., & Miao, C. (2018). *A comprehensive exploration to the machine learning techniques for diabetes identification* (pp. 291–295).
19. Mareeswari, S., Mahalakshmi, & Preethi. (2017). Prediction of diabetes using data mining techniques. *Research Journal of Pharmacy and Technology, 10,* 1098–1104.

20. Suresh, K., & Pranavi. (2017).*Performance analysis of machine learning algorithms on diabetes dataset using big data analytics* (pp. 508–513).
21. Khalil, R., & Jumaily, A. (2017). *Machine learning based prediction of depression among type 2 diabetic patients* (pp. 1–5).
22. Kaur, H., & Batra, S. (2017). *HPCC: An ensembled framework for the prediction of the onset of diabetes* (pp. 216–222).
23. Khan, N. S., Muaz, M. H., Kabir, A., & Islam, M. N. (2017). *Diabetes predicting mhealth application using machine learning* (pp. 237–240).
24. Jahangir, M., Afzal, H., Ahmed, M., Khurshid, K., & Nawaz, R. (2017). *An expert system for diabetes prediction using auto tuned multi-layer perceptron* (pp. 722–728).
25. Hashi, E. K., Zaman, M. S. U., & Hasan, M. R. (2017). *An expert clinical decision support system to predict disease using classification techniques* (pp. 396–400).
26. Chen, W., Chen, S., Zhang, H., & Wu, T. (2017). *A hybrid prediction model for type 2 diabetes using k-means and decision tree* (pp. 386–390).
27. Negi, A., & Jaiswal, V. (2016). *A first attempt to develop a diabetes prediction method based on different global datasets* (pp. 237–241).
28. Lee, B. J., & Kim, J. Y. (2015). Identification of type 2 diabetes risk factors using phenotypes consisting of anthropometry and triglycerides based on machine learning. *IEEE Journal of Biomedical and Health Informatics, 20,* 39–46.
29. Gandhi, K., & Prajapati, N. (2014). Diabetes prediction using feature selection and classification. *International Journal of Advance Engineering and Research Development, 1.*
30. Wild, S., Roglic, G., Green, A., Sicree, R., & King, H. (2004). Global prevalence of diabetes: Estimates for the year 2000 and projections for 2030. *Diabetes Care, 27,* 1047–1053.

Comparative Study of Different Reduced Precision Techniques in Deep Neural Network

Rishi Raj Gupta and Virender Ranga

Abstract There has been rising interest in reduced precision in the training of deep neural networks (DNNs) that is used from the single-precision (FP32) to different precision format (FP16, FP8, bfloat16) due to the rapid increase in model sizes, which require less representational space when stored in lower precision. However, training a DNN in reduced precision format (FP16, FP8, and bfloat16) is challenging, because the data format may be inadequate for representing the gradients during backpropagation. In this research paper, we compare the various novel approaches to train a DNN using the different reduced precision formats and explore the challenges that we get during the training of DNN in reduced precision. Besides, we also examine various layers in the neural network which make a significant reduction in the backpropagation and also observe that when should get sufficient precision.

Keywords Low precision · FP-16 training · FP-8 training

1 Introduction

Over the last few decades, deep learning has achieved near human-level efficiency for various tasks including computer vision [4], machine translation [5], speech recognition [9, 11], image processing [4]. These facts have led to the development of even deeper neural networks to achieve higher accuracy by training them over larger and more diversified datasets. As a result, the computational demands for training of deep models have been evolving exponentially, and in the last few years, it beats the Moore's law by a broad margin [11]. Training and inferencing of the model in the

Equal contribution

R. R. Gupta · V. Ranga (✉)
Computer Engineering Department, NIT, Kurukshetra, Haryana, India
e-mail: virender.ranga@nitkkr.ac.in

R. R. Gupta
e-mail: rishirajgupta2094@gmail.com

S. Tiwari et al. (eds.), *Proceedings of International Conference on Big Data, Machine Learning and their Applications*, Lecture Notes in Networks and Systems 150,
https://doi.org/10.1007/978-981-15-8377-3_11

reduced precision become prominent an area of research to manage these increases of computational demands.

Many studies on inferencing using DNNs have shown that the bits for inferencing can sharply scale down to few bits(2–4) while maintaining the good accuracy; however, training of DNNs is more difficult due to the necessity to preserve the consistency of the gradients during the backpropagation, and thus, training is primarily continued to be done in single precision (FP32).
Current research shows that at least 16-bit precision is required to train a DNN without affecting the model accuracy. Consequently, various hardware targeted training of DNNs support half-precision (16-bit), which gives excellent performance over the single-precision (32-bit). There have been numerous attempts to train DNNs at further reduced precisions, which are lesser than 16-bit. Recently, it has been shown that the training of a DNN can also be done with 8-bit [1, 2, 5, 25, 27]. In the case of improving the efficiency of computing, researches have done various related works like training the DNN with a 16-bit fixed point [21] with insignificant loss in accuracy followed by the studies on half-precision floating point [22] and dynamic fixed point [5, 23]. Several efforts have been done to decrease the precision of DoReFa-Net [24] which is trained with 1-bit and 2-bits to represent the weights while incurring a small loss in accuracy. On the image classification [26] showed the mixed precision training (MPT) of ConvNet using INT16 tensors for GEMM operations the remaining part of the paper is as followed: Section 2 discusses the training of DNN and various precision formats. Section 3 shows the training neural network. Section 4 discusses the training in reduced formats. Result analysis is shown in Sect. 5. Section 6 concludes the paper.

2 Various Precision Formats

2.1 Fixed Point

The fractional number is the representation of the fixed-point format. Fractional number in the computer memory stored as the signed integer in two's complement format. We use the split notation, putting the radix point, which separates the integer and the fractional part. In fixed point, the format has a fixed number of bits for the signed part and fractional part. Though there is a deficiency of fixed point that is shown in Fig. 1.

2.2 Floating Point

Floating point format is a method to describe pretty big or tiny numbers accurately using the logical representation in binary. To accomplish this, floating-point format

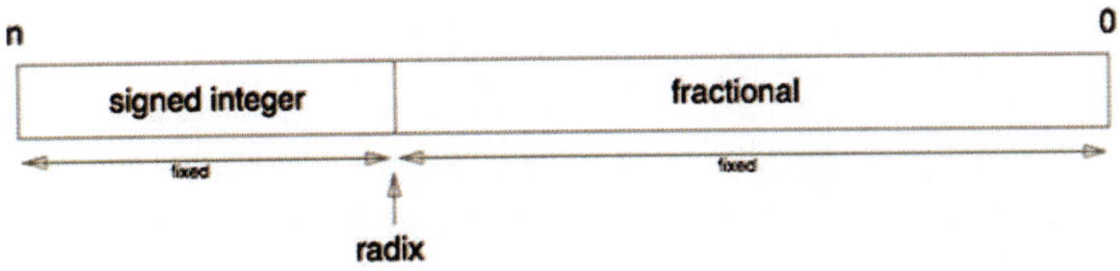

Fig. 1 Fixed-point representation

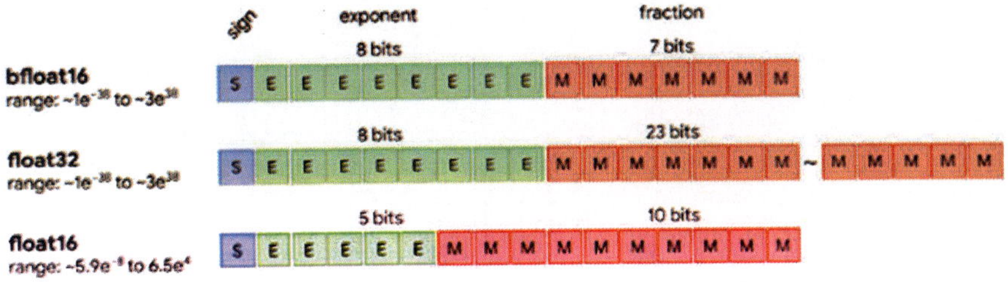

Fig. 2 Different floating formats

gives differing degrees of precision based on the order of the numbers that you are using. There are various floating-point representations like Float32, Float16, Bfloat16, Float8 [2] which have a dynamic range of numbers. Figure 2 shows the representation of Float32, Float16, and Bfloat16.

Training the DNNs requires a dynamic range of numbers during forward-propagation and backpropagation to store weights, activation, gradients. However, due to the very small magnitudes of gradients (compared to weights and activations), there is a high-precision requirement in backward propagation during the gradient calculation. Therefore, we compare all the different precision formats: Float32, Float16, Bfloat16, Float8 based on the range of the number that above precisions format can represent. Table in Fig. 3 shows the comparison among single-precision (FP32), half-precision (FP16), 8-bit floating-point and Bfloat16 where S, E, and M represent the number of bits used to store the sign, exponent, and mantissa of a number.

3 Training Neural Network

Neural network (NN), also termed as an artificial neural network (ANN), is a software, which is encouraged by a biological neuron. The biological neuron is responsible for solving a tiny part of the problem, and collectively these neurons led to solving the significant issue. The brain consists of several neurons connected. A biological neuron consists of four parts Dendrites, Soma, Axon, and Synapses (Fig. 4).

Data type	Bit Format (S, E,M)	Max Normal	Min Normal	Min Sub-Normal
IEEE-754 float	[1,8,23]	3.40e38	1.17e-38	1.40e−45
IEEE-754 Half-float	[1,5,10]	6.55e4	6.10e-5	5.96e−8
Bfloat16[20]	[1,8,7]	3.38e38	1.17e−38	N/A
FP-8[2]	[1,5,2]	57344	6.10e-5	1.52e−5

Fig. 3 Different floating formats

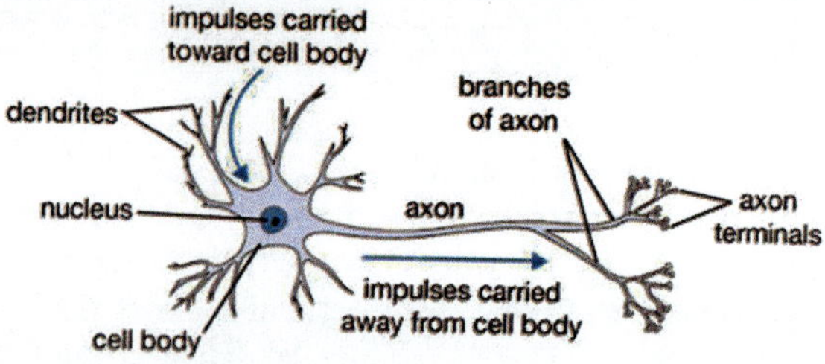

Fig. 4 Single biological neuron

3.1 Architecture of Neural Network

3.1.1 Artificial Neuron(AN)

"Artificial neurons" were designed with a similar structure to mimic the brain. An artificial neuron takes the input values from the other neuron that is connected using dendrite and after that the neuron merges those activations. The connection between the two biological/artificial neurons can be weak and strong. In Fig. 5, Wi denotes the intensity of the neuron, and the product of xiwi gives the amount of information received from the neuron. Combining all the information xiwi with a bias b forms $xiwi + b$, and after that this equation is passed to the activation function (nonlinear) $f(xiwi + b)$, and the value from the activation is passed to different neurons.

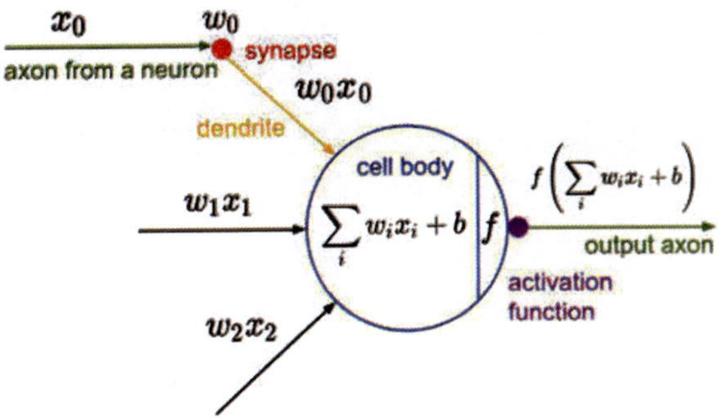

Fig. 5 Function of single artificial neuron

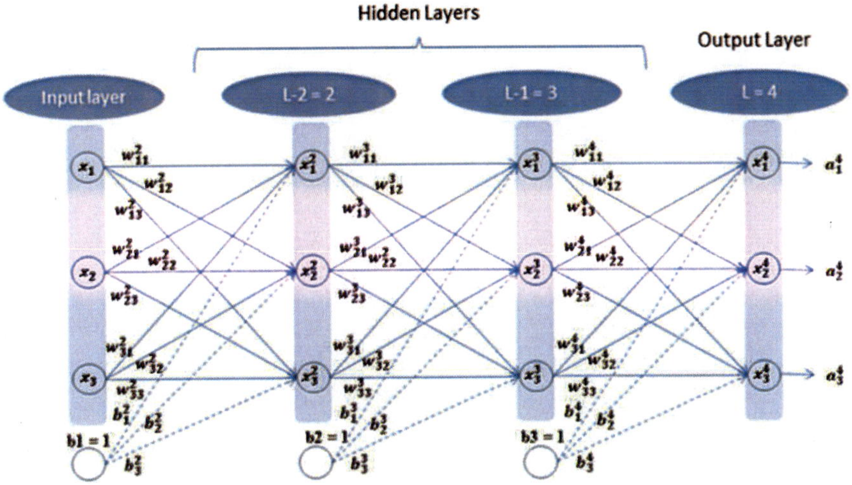

Fig. 6 Multilayer neural network

3.1.2 Architecture of Neural Network

In the neural network, there are thousands of neural networks connected together, and a connection link has weights that show the strength of the link. In Fig. 6, we have shown architecture feed-forward neural network (input-hidden-output).

3.2 Training

Training a neural network is adjusting the trainable variables in the model so that we can reduce the loss function and get the right prediction. There are two methods to train the neural network.

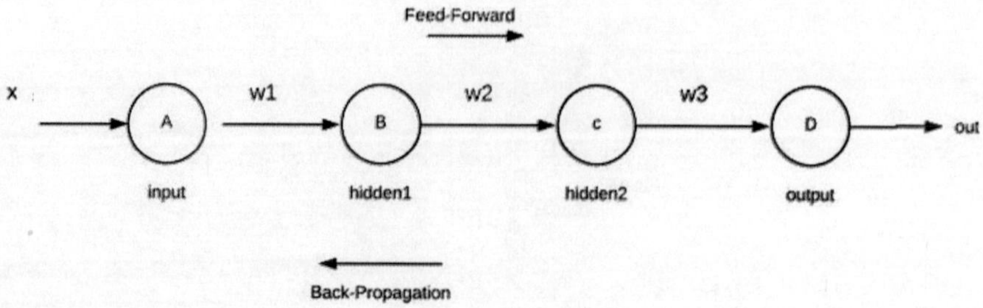

Fig. 7 NN with forward and backward pass

3.2.1 Forward-pass

The "training" stage begins by randomly initializing all weights, i.e., weights linked with all AN. Input is passed into the neural network and after passing the data to the activation, the middle layer (hidden) is calculated, and after calculating the activation of the last layer (output), we get the predicted output and this procedure is called as forward pass.

Forward propagation equation for NN in Fig. 7:

$$L_1 = \sigma_3(\sigma_2(\sigma_1(w_1 x)w2)w3) \tag{1}$$

where σ = Activation function, w_i = weights, x = input

A loss is then calculated based upon the actual and predicted output, and initially, the loss value is high because we have not adjusted our trainable variables. A high loss says that the final output is far from the actual output. Here, our aim is to finally get an optimum value of weights, such that NN can predict the output with some reasonable accuracy. To improve the trainable variables in the neural network, we can use a backpropagation algorithm.

$$Loss = L(\text{actual, predicted}) \tag{2}$$

3.2.2 Backward Pass

After getting the loss from the loss function, this loss is then backpropagated to a network to adjust the trainable variables. To improve the trainable variables, we use optimizers which decide based on the loss, which trainable variable to change and how much it should be changed. Gradient descent algorithm is used to minimize the loss function.

3.2.3 Weight Updation

$$
\begin{aligned}
W_1^+ &= w_1 - \eta * \frac{\partial E}{\partial W_1} \\
W_2^+ &= w_2 - \eta * \frac{\partial E}{\partial W_2} \\
W_3^+ &= w_2 - \eta * \frac{\partial E}{\partial W_3}
\end{aligned}
\qquad (3)
$$

So finally, we do forward and backward pass for many iterations so that a reasonable accuracy can be achieved.

4 Training in Reduced Precision

4.1 16-Bit Training

4.1.1 Mixed Precision Training

In mixed precision training (MPT) [1] , the authors proposed a methodology for training a deep neural network using 16-bit floating-point numbers while preserving the efficiency and without altering the hyperparameters. By using the IEEE 16-bit, floating-point format authors have nearly halved the memory requirement, and using the latest GPU accelerated the arithmetic operation. In DNNs, we generally need to store the weights, activations, gradients in the MPT [1]. These three trainable variables are stored in floating-point16 (FP-16). To overcome the problem created due to the reduction in precision from single (32-bit) to half (16-bit), MPT uses three techniques to solve those problems created due to the reduced precision:

1. Maintain a copy of weights in single FP-32, and then the copy of weights is downgraded to FP-16 for the forward and backward propagations.
2. MPT proposes the loss-scaling technique to prevent gradient values from a significant reduction.
3. The third technique is to use FP-16 arithmetic that accumulates into FP-32 before storing into memory converted back to FP-16.

Referring the Fig. 8 from MPT [1] for saving the master copy of weights in (FP-32).

4.1.2 Loss Scaling

In the loss scaling technique, one optimal way is to shift the gradient value into half-precision (FP-16) range to scale loss estimation evaluated in the forward propagation

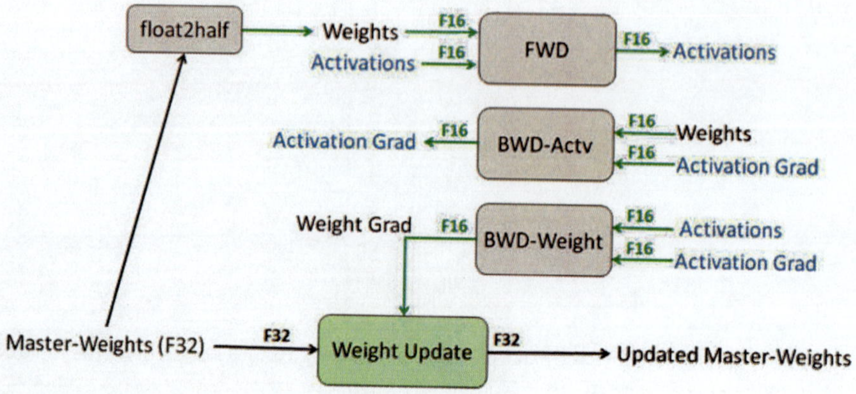

Fig. 8 Steps for doing the MPT with 16-BIT

before the backward propagation. During the backpropagation, the chain rule is used, which assures that all the gradients are scaled to an equal amount. By using this, during the backpropagation gradients, values can be preserved from becoming zero. Weight gradients should be unscaled before the weight update to sustain the gradients in FP-32. One most straightforward way is to pick a fixed scaling factor when the gradient values are known, and we can directly select a factor for scaling.

4.1.3 Arithmetic Precision

To sustain the accuracy of the network in MPT, the authors find that some network requires vector-dot product addition in FP-16, and partial products in FP-32, which is changed to FP-16 while writing to memory. We need to do the FP-32 accumulation because some model in FP-16 did not match the accuracy compare to FP-32 model. While training some layer make the large reduction across the vector so this should be carried out in FP-32, because FP-16 does not have the dynamic range of numbers compared to FP-32. Generally, batch normalization and SoftMax do the large reduction, so we need to do the arithmetic operation in FP-32.

4.2 8-Bit Training

4.2.1 Training of DNN with FP-8

Today, most of the hardware platforms are moving from single to half-precision (FP-16) due to high performance and smaller storage. Reducing below 16-bit is a challenging task due to maintaining the computation of the gradient during backpropagation. In training of DNN with FP-8, authors have done successful 8-bit training

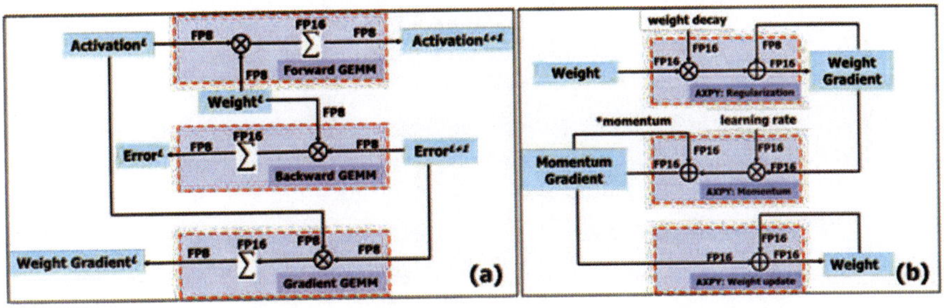

Fig. 9 GEMM operations architecture [3]

while completely maintaining the accuracy compare to models in FP-16 and FP-32 [1]. In this paper, the authors reduce the data and computation to 8-bit and also reduce the accumulation from FP-32 to FP-16 through the idea of the chunk-based calculation. The three primary challenges while training in 8-bit precision as firstly, reducing the weights, error, activation, gradients in 8-bit shows the accuracy degradation. Secondly, reducing the accumulation precision form 32-bit to 16-bit significantly affects the convergence of DNN training. Reducing to 16-bit precision weight update affects the accuracy while 32-bit requires a copy of weight to be saved in the memory that is costly operation.
In the paper, authors present following techniques to solve the above problems.

1. Proposed a novel 8-bit floating-point format which is defined as 1 bit-sign, 5-exp, 2-mantissa. In this approach, all the GEMM operations of the DNN are performed in 8-bit without any accuracy degradation.
2. Developed a computation technique that can be applied hierarchically to allow all matrix operation to be calculated in 8-bit and 16-bit.
3. Stochastic rounding is used while updating the weight in 16-bit precision.

In Fig. 9, the author shows the precision setting for three general matrix multiplications during the forward pass and backward pass. In the new 8-bit reduced precision, which has a precision setting (1,5,2), the author has chosen the precision setting carefully to get a dynamic range of numbers for the weights, gradients, activation, error, and Fig. 9 shows the 8-bit computation on weights and 16-bit accumulation. In the paper, the authors successfully demonstrated FP-8 training with 2–4 speedups in computation. To reduce the problem of swamping, authors proposed a chunk-based computation technique with stochastic rounding for accumulation which reduces the bits down to 16-bits.

4.3 Mixed-Precision Training in 8-Bit

In the mixed-precision training with 8-bit [3], authors proposed training of the DNN applying 8-bit precision for weights, gradients, errors, activation, and they also trying to reduce the master copy weights from 32-bit to 16-bit. Paper also illustrates the accuracy over multiple datasets like WMT, ImageNet, and various machine learning models like ResNet-18/34/50, GNMT, Transformer. To improve the error propagation, the authors introduced an improved loss scaling approach to exchange the subnormal range of 8-bit. To reduce the gradient noise, authors used stochastic rounding technique. Applying all these techniques, they get a slightly higher accuracy then the baseline.

In the paper, authors extended the 8-bit floating-point training [2] with the below-given contribution:

1. To develop the FP8 compute primitives and eliminating the necessity of hardware stochastic rounding, a scalable method is proposed, with an 8-bit floating point [3].
2. Illustrate SOTA accuracy using 8-bit for weights gradients, errors, activation on various datasets like ImageNet, WMT and on numerous models like ResNet (18–50) [14], GNMT [16], Transformer [17].
3. By diminishing the precision of a master copy of single-precision weights to half-precision weights, and the memory requirement is reduced by half.
4. Introduced an improved loss scaling approach to neutralize the reduced 8-bit floating-point range, which improves the error propagation and produces good accuracy.
5. After a comprehensive study about the quantization noise and its impact on model generalization, a method is proposed for gradient noise in the preceding epochs for more reliable generalization using stochastic rounding technique.

Figure 10 illustrates the precision setup of different estimate procedures used in the mixed-precision organization. The general matrix multiplication [GEMM] operator presented in the above Fig. 10 denotes the key compute kernel utilized by DNN in forward and backward, and the gradient step. Q is the quantization node that defines the don conversion of the 32-bit floating point to 8-bit floating point before proceeding to the subsequent layer.

In Fig. 10, all the weight, error, gradient are updated to 8-bit format for the forward and backward pass during the training.

5 Result Analysis

We have compared the outcomes of various reduced precision training techniques based on the accuracy, and Bilingual Evaluation Understudy (BLEU) score after doing the precision setting for weights, errors, gradient, activation on the different reduced precision technique as we have discussed in the paper like mixed precision training in FP-16 [1], Training with FP-8 [2], mixed precision training in FP-8 [3]. In the given Table 1, we have shown the various reduced precision training techniques and we have defined a baseline training (FP-32) on ILSVRC [18] ImageNet validation set and model used is AlexNet, and we have compared with different techniques like

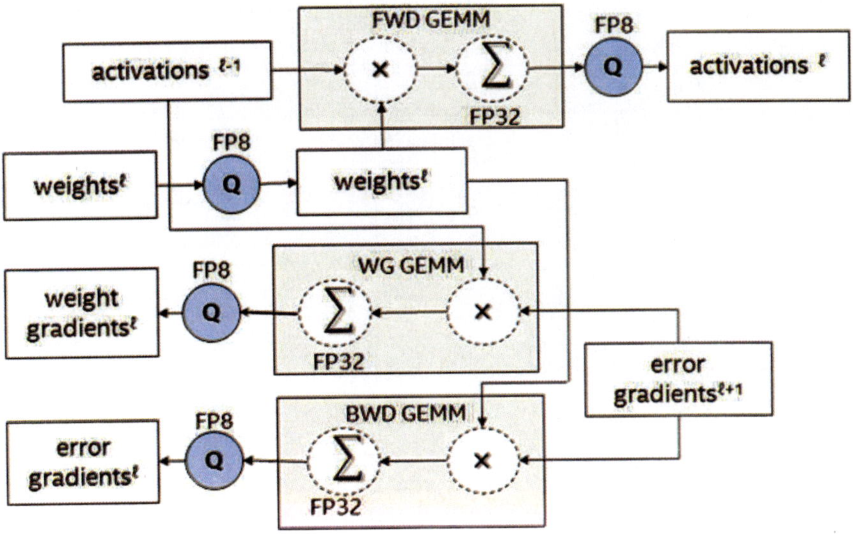

Fig. 10 GEMM operations architecture [4]

Table 1 BLEU [8] Score on WMT2016

Training scheme	Dataset	Model	Bit precision X W dw dx acc Mw	FP-32	Reduced precision
Mixed precision FP-8	WMT 2016 (E → G)	GNMT	8 8 8 8 32 16	24.3	24.6
Mixed precision FP-8	WMT 2016 (E-→G)	Transformer	8 8 8 8 32 16	23.6	23

mixed precision FP-16 [1], MPT FP-8 [3], training with 8-bit precision. We have also examined it on different model like ResNet-18/34/50 [14] with respect to the baseline accuracy, and we have found that we can reduce the precision for training below 16-bit without losing the accuracy in model; for lowering this, we have previously looked into techniques in [1–3] how we can we achieve the similar accuracy to baseline (FP-32). We have also seen that in paper [3] that the authors have successfully done the training for the machine translation task like Google machine translation task (GNMT) [16] and transformer language translation model (T-LT) [17] on WMT-16 [19] Eng. to German dataset.

So, the papers [1–3] show results on various image classification task and on the machine, translation tasks for the image classification, it shows the Top 1 accuracy and for the machine translation, it shows the (BELU) score in Table 2.

Table 2 Top-1 accuracy on ImageNet dataset for different convolution neural networks

Training scheme	Dataset	Model	Bit precision X W dw dx acc Mw	FP-32 (%)	Reduced precision (%)
FP-32 training	Image-net-1k	AlexNet [13]	32 32 32 32 32 –	58.1	-
Mixed precision FP-16	Image-net-1k	AlexNet	16 16 16 16 16 32	57.1	56.93
Training FP-8	Image-net-1k	AlexNet	8 8 8 8 16 16	58.0	57.5
Mixed precision FP-8	Image-net-1k	ResNet-50	8 8 8 8 32 16	75.67	75.70
Mixed precision FP-16	Image-net-1k	ResNet-50	16 16 16 16 16 32	75.92	76.04
Mixed precision FP-16	Image-net-1k	Inception v3	16 16 16 16 16 32	73.85	74.13
Mixed precision FP-8	Image-net-1k	ResNet-34	8 8 8 8 32 16	72.96	75.95

6 Conclusion

In the paper, we have demonstrated the various precision formats like fixed point and floating point and showed the dynamic range of numbers, which is needed to represent weights and gradients during the calculations in backpropagation and also described how the training happens in a neural network in the forward pass and in the backward pass. Based on the understanding of training and formats, we have shown a comparative study on various reduced precision training techniques. All the comparisons are based on the ILSVRC ImageNet dataset processed on various neural network models like ResNet 18/34/50 and AlexNet and apart from that, MPT with 8 bits [3] also showed result for machine translation on WMT-16. During the comparisons, we figure out that reducing the precision requires a lot of arrangements for weights, error gradients, so our research work introduces various techniques to overcome the problems like swamping and vanishing of gradients due to the large reduction on the layer like batch normalization [9]and SoftMax which lead to NAN and model will not converge. Today, GPU supports 16-bit arithmetic operation, and Nvidia also provides mixed precision support library and documentation [10] so training in mixed precision is possible compared to other lower precision training.

References

1. Micikevicius, P., Narang, S., Alben, J., Diamos, G., Elsen, E., Garcia, D., et al. (2017). *Mixed precision training*. arXiv preprint arXiv:1710.03740.
2. Wang, N., Choi, J., Brand, D., Chen, C.-Y., & Gopalakrishnan, K. (2018). Training deep neural networks with 8-bit floating point numbers. In *Advances in neural information processing systems* (pp. 7675–7684).
3. Mellempudi, N., Srinivasan, S., Das, D., & Kaul, B. (2019). *Mixed precision training with 8-bit floating point*. arXiv preprint arXiv:1905.12334.
4. He, K., Zhang, X., Ren, S., & Sun, J. (2016). Deep residual learning for image recognition. In *Proceedings of the IEEE conference on computer vision and pattern recognition* (pp. 770–778).
5. Wu, Y., Schuster, M., Chen, Z., Le, Q. V., Norouzi, M., Macherey, W., et al. (2016). *Google's neural machine translation system: Bridging the gap between human and machine translation*. arXiv preprint arXiv:1609.08144.
6. Graves, A., Mohamed, A.-R., & Hinton, G. (2013). Speech recognition with deep recurrent neural networks. In *2013 IEEE international conference on acoustics, speech and signal processing* (pp. 6645–6649). IEEE.
7. Hannun, A., Case, C., Casper, J., Catanzaro, B., Diamos, G., Elsen, E., et al. (2014). *Deep speech: Scaling up end-to-end speech recognition*. arXiv preprint arXiv:1412.5567.
8. Matt Post. (2018). *A call for clarity in reporting bleu scores*. arXiv preprint arXiv:1804.08771.
9. Ioffe, S., & Szegedy, C. (2015). *Batch normalization: Accelerating deep network training by reducing internal covariate shift*. arXiv e-prints.
10. Training With Mixed Precision. https://docs.nvidia.com/deeplearning/sdk/mixed-precision-training/index.html.
11. Amodei, D., & Hernandez, D. *AI and compute*. https://openai.com/blog/ai-and-compute.
12. Neural Network Terminology. https://www.cs.toronto.edu/~lczhang/360/lec/w02/terms.html.
13. Krizhevsky, Sutskever, A., Hinton, I., & Geoffrey. (2012). ImageNet classification with deep convolutional neural networks. *Neural Information Processing Systems,25*. https://doi.org/10.1145/3065386.
14. He, K., Zhang, X., Ren, S., & Sun, J. (2016). Deep residual learning for image recognition. In *Proceedings of the IEEE conference on computer vision and pattern recognition* (pp. 770–778).
15. Szegedy, C., Vanhoucke, V., Ioffe, S., Shlens, J., & Wojna, Z. (2016). Rethinking the inception architecture for computer vision. In *Proceedings of the IEEE conference on computer vision and pattern recognition* (pp. 2818–2826).
16. Wu, Y., Schuster, M., Chen, Z., Le, Q.V., Norouzi, M., Macherey, W., et al. (2016). *Google's neural machine translation system: Bridging the gap between human and machine translation*. arXiv preprint arXiv:1609.08144.
17. Vaswani, A., Shazeer, N., Parmar, N., Uszkoreit, J., Jones, L., Gomez, A.N., Kaiser, Ł., et al. (2017). Attention is all you need. In *Advances in neural information processing systems* (pp. 5998–6008).
18. Russakovsky, O., Deng, J., Su, H., Krause, J., Satheesh, S., Ma, S., et al. (2015). Imagenet large scale visual recognition challenge. *International Journal of Computer Vision, 115*(3), 211–252.
19. Sennrich, R., Haddow, B., & Birch, A. (2016). *Edinburgh neural machine translation systems for WMT 16*. arXiv preprint arXiv:1606.02891.
20. Kalamkar, D., Mudigere, D., Mellempudi, N., Das, D., Banerjee, K., Avancha, S., et al. (2019). *A study of bfloat16 for deep learning training*. arXiv preprint arXiv:1905.12322.
21. Gupta, S., Agrawal, A., Gopalakrishnan, K., & Narayanan, P. (2015). Deep learning with limited numerical precision. In *International conference on machine learning* (pp. 1737–1746).
22. Micikevicius, P., Narang, S., Alben, J., Diamos, G., Elsen, E., Garcia, D., et al. (2017). *Mixed precision training*. arXiv preprint arXiv:1710.03740.
23. UrsKöster, Webb, T., Wang, X., Nassar, M., Bansal, A. K., Constable, W., et al. (2017). Flexpoint: An adaptive numerical format for efficient training of deep neural networks. In *Advances in neural information processing systems* (pp. 1742–1752).

24. Das, D., Mellempudi, N., Mudigere, D., Kalamkar, D., Avancha, S., Banerjee, K., et al. (2018). *Mixed precision training of convolutional neural networks using integer operations*. arXiv preprint arXiv:1802.00930.
25. Banner, R., Hubara, I., Hoffer, E., & Soudry, D. (2018). Scalable methods for 8-bit training of neural networks. In *Advances in neural information processing systems* (pp. 5145–5153).
26. Das, D., Mellempudi, N., Mudigere, D., Kalamkar, D., Avancha, S., Banerjee, K., et al. (2018). *Mixed precision training of convolutional neural networks using integer operations*. arXiv preprint arXiv:1802.00930.
27. Cambier, L., Bhiwandiwalla, A., Gong, T., Nekuii, M., Elibol, O. H., & Tang, H. (2020). *Shifted and squeezed 8-bit floating point format for low-precision training of deep neural networks*. arXiv preprint arXiv:2001.05674.

Design of a Predictive Measure to Enhance Neural Network Architecture for Plant Disease Detection

Swati Vashisht, Praveen Kumar, and Munesh Chandra Trivedi

Abstract Agriculture is the foundation of Indian economic framework and financial exercises. It adds to national incomes, international trades, and overall industrial development. Indian population significantly depends upon agriculture for their livelihood. The key responsible reason in the decay of crop production is infected crops, which leads to noteworthy financial losses consistently. It is in this way absolutely critical to pursue the detection of leaf diseases. As per reports, the human populace is anticipated to reach at 9 billion by 2050, and food consumption is expected to increase by 60 percent. Improving and expanding the crop yields is hence a significant area of interest. Irresistible abiotic and biotic infections have essentially influenced potential yields and decreased it by a normal of 40%, a large number of population engaged in farming in this developing world suffering losses of up to 100%. Farmers around the globe are dealing with the diagnosis of plant leaf diseases and their legitimate treatments. The basic approach for classification process flows from data pre-processing to creating classifiers. Data pre-processing plays an important role in achieving higher classification accuracy while reducing performance time. In this paper, we analyzed the predictive measures of various artificial neural networks used in detection and classification of plant diseases, to better understand the existing architectures, and also proposed Gaussian filter as one of the predictive measure to increase the efficiency of trained model at optimal computational costs, while giving some insights about the future research work as well.

Keywords Neural networks architectures · Predictions · Disease detection · Deep learning · Gaussian filter · Machine vision

S. Vashisht (✉) · P. Kumar
Computer Science & Engineering, Amity University, Uttar Pradesh, India
e-mail: svashisht@gn.amity.edu

P. Kumar
e-mail: pkumar3@amity.edu

M. C. Trivedi
Computer Science & Engineering, NIT Agartala, Agartala, India
e-mail: munesh.trivedi@gmail.com

S. Tiwari et al. (eds.), *Proceedings of International Conference on Big Data, Machine Learning and their Applications*, Lecture Notes in Networks and Systems 150,
https://doi.org/10.1007/978-981-15-8377-3_12

1 Introduction

Recently, A good number of researches have proposed for the improvement of current procedures for early identification of plant disease another, automated image recognition models based upon deep learning [1]. Transfer learning is a significantly used method in machine learning which provides solutions to the real-world problems by gaining knowledge from its past experiences while applying and implementing in the same or related domain [2]. In this survey, an illustrative study has been done and introduced, to make it simpler for the peruser to comprehend AI structures executed upon crop health area with the goal that unusual agriculture ecosystems are better comprehended by observing, estimating, and analyzing it. At present, all the research conducted in the field of crop disease identification, the standard methodology is to use a deep convolutional network combined with a cross-entropy function. Some of the approaches have been executed to recognize healthy and unhealthy crop, to distinguish weed from crop or creating multiple class classification to detect and classify more than one type of infected leaves. An existing architecture used the apple plant leaves image dataset which comprises 2462 diseased leaves of six different categories for data modeling and method evaluation, and this model achieved better results than multiple classification system based on cross-entropy loss with 92.29 percent accuracy while achieving 93.51 percent, 93.31 percentm and 93.71 percent accuracy on the test set [3].

2 Convolutional Neural Network

We have witnessed that artificial intelligence-based machines are replacing human interventions in real world. Scientists and researchers alike work on a wide variety of facets of the field to make great things possible, and computer vision is one of them. A convolutional neural network takes an input image, accordingly assigned weights and biases to various aspects in the image, and be able to distinguish between them. ConvNet is proved to give efficient models because pre-processing is not required here but in other classification algorithms. Although filters are hand-engineered in primitive methods, ConvNets have the ability to learn these filters/characteristics.

3 Existing Architectures and Predictive Measures

Several models have been proposed for plant disease identification and treatment within the machine learning domain [4] which have been measured in terms of their area of use, techniques used for pre-processing which includes data augmentation, segmentation, digital image processing [5] task, etc. In addition, it has been claimed that DenseNets require a sensible computation time and less number of parameters to perform best in the classification models, and it achieves 99.75 percent test accuracy for 30th epoch and gives better results than other architectures [6]. Some of the included measures are as follows:

3.1 Areas of Use

15 area has been identified where machine learning algorithms have been be applied, and these algorithms can be implemented to detect and recognize beetle species from plant in a food inspection [7]. Begomovirus infected tomato [8], pepper, and cucurbits which were investigated in Thailand for continuous monitoring of viral diversity in these economic crops [9]. Tea leaves have been identified as healthy or unhealthy using conditional deep convolutioanl generative adversial network (C-DCGAN) [10]. Onion [11] and tomato are consumed all over the world and in every kitchen, whereas India ranked second in tomato production [12]. Tomato crop have been monitored in two ways: first approach has been proposed to identify the external defects on tomatoes [13], and the second one is to detect and recognize the diseases in tomato crop [14]; only one article proposed a multi-classifier approach to implement on a small real-time dataset to classify capsicum disease [15]; one of the approaches used five fine-grained cassava leaf disease categories obtained from conducting a survey in Uganda; the dataset contains 10,000 labeled images [16]; two approaches have been proposed to identify healthy and infected rice plant [17, 18].

Disease identification model for coffee plant has been designed as coffee is one of the commodities that has been consumed worldwide, but its quality and productivity are affected by diseases such as rust and cercospora, and an article has been written to elaborate how to apply statistical methods to identify major coffee plant leaf diseases, with the goal of introducing an expert system to provide a solution to the coffee producers in the early identification of diseases [19].

One of the approaches has been given for tulip breaking virus (TBV) as Tulip is again a commodity in Netherlands whose production significantly affects the country's economy, and TBV infected plants can spread the disease through aphids, which must be eliminated from the field in early growth days [20]. Deep learning algorithms do not only apply for the recognition of disease but also to identify other characteristics such as fruit cultivars in apricots; hence, one approach has been proposed to determine apricot cultivars considering their shape features [21]. On the other hand, discrimination of weed and crop can also be done by implementing local binary pattern (LBP) and support vector machine (SVM) method [22].

3.2 Data Sources

While implementing these deep learning approaches, a large dataset must be used just to make the proposed algorithm more accurate. Observing the sources of data, it can be said that most of the time data has been used either from the laboratories or from online available data repositories, and if real-time data is used to train a classifier, it was very less in number, number of images used to train, and test the classifier ranges from minimum 60 images to maximum 82,000 images approximately. Some of the available datasets are discussed here.

3.2.1 Available Laboratory Datasets:

The beetle elytra images (BEI) dataset was collected from the Arkansas Laboratory, which was further used for the research work, and this dataset contains 69 images for 15 beetle species [7].

Fujian Institute of Subtropical Botany, Xiamen, China, provided leaf image dataset which includes 466 maize images and 500 rice images, captured under different lighting strengths and background conditions [17]. Images were also obtained from Digipathos, and a database of eight different typical diseased plants of soybean was provided by the Brazilian Agricultural Research Agency consisting of 354 images [23].

A total of 486 RGB soybean color images are obtained from PDDB, a publicly accessible benchmark database. Each image consists of only one disease with a complex history, and also these images are of different sizes. There are three groups primarily referred to as safe, contaminated, and unknown. Not all images are of the same size [24].

3.2.2 Online Data Repositories:

The Kaggle database [25] accommodates approximately 174,000 images of both types of healthy and infected leaves of tomatoes, oranges, and maize; the subset used includes four separate categories with 3076 image dataset and 12,332 training set images [25].

Dataset was also obtained from plantvillage.org and divided into equal quantity of images in each category to ensure that the model did not store any specific class over another [4]. Data used for training and testing has also been collected from (https://challenger.ai/) which is a library of plant leaf diseases [26]. 43,843 photos were used to train the classifier, and the data was made available online. The dataset is highly imbalanced against a stable class [13]. The PlantVillage dataset has 82,161 images having 55 distinct groups of stable and infected plants [27].

3.2.3 Manual Image Acquisition:

Begomovirus-infected leaves were collected from the production areas in Thailand, and symptoms of these viruses include leaf mosaic, leaf curling, and leaf yellowing [9]. The image of diseased leaf with clear symptoms is taken by a tablet or smartphone [28].

Tianjingshan National Forest Park of China provided tea leaf image dataset. Image acquisition systems include a handheld video camera and an aerial vehicle. Another part of the picture of the disease of the tea leaf comes from the agricultural graphics database of the Anhui Provincial Agricultural Committee on Pest and Disease [10].

Onion images were obtained from National Institute of Crop Science, Muan, South Jeolla Province. These images were captured using a field-monitoring device which was mounted at a distance of 10 m [11]. 62 capsicum images are trained and checked manually [15], fruit market apricot images [21]. Some of the images are directly taken from farmland [18], some of the photos are taken by mobile phone in actual field conditions [29], dataset has been arranged from a tulip field planted at a higher production density that results in images with overlapping plants [20], and a dataset of 600 images or 200 images of various different crop varieties are used for experiment which are labeled with ten different types of crop diseases which are healthy and rusty [30].

3.3 Data Pre-processing

Data pre-processing plays a vital role in increasing the accuracy of classification algorithms, and some of the techniques used for pre-processing in state-of-art architectures are described below.

The training set imbalances can be treated using random over sampling (ROS) and random under sampling (RUS). In the ROS process, data images are replicated to make the available dataset efficient enough to give accurate results, while in RUS, some of the samples are arbitrarily discarded from the majority classes until the quantity of samples corresponds to that of the minority classes [25]. Also to remove the complex background of images, RGB images can be converted to hue–saturation–value (HSV) images [18]. Image segmentation is also a technique to be applied to diseased part of the plant image and discards the history, and then, the HSV transformation is applied to the segmented image [28, 31]. Convolutional neural network gives better results if segmented image is fed to the network for disease detection/classification even when limited dataset is available for experiments [24].

For data pre-processing, relief algorithm and least square estimator are used to input the database, and this pre-processed data is fed to teaching-learning-based optimization (TLBO)-trained radial basis function network (RBFNs) to produce a brief and meaningful definition for each class to identify subsequent instances without a known class label [32]. If smart process automation is used in combination with machine learning methods, training models are used to reduce the performance time and to predict the cultivar; different variables were used as inputs to these training models [21].

3.4 Data Augmentation

Image data augmentation is a method that can be used to artificially enhance the size of the training dataset by producing new versions of images in the dataset. Noise injection, image flipping, color augmentation, gamma correction, principal component analysis, scaling, and rotation are basic augmentation methods used in existing classification models [33, 24].

3.5 Technical Analysis

Some of the existing approaches based upon neural network architectures have been discussed here:

Approach 1: Transfer learning approach has been applied on a pre-trained network model, and pre-trained weights of five convolution blocks in the VGG16 architecture were frozen, allowing only the final dense block to obtain gradient updates during the training phase. This final block consists of four fully connected hidden layers. One dropout unit was connected after the first hidden layer with a drop rate of 0.8. Rectified linear units (ReLU) was considered to be the activation function for each fully connected layer; the final block provided the last output layer with Softmax activation for the classification of the 15 species of beetle. Network weights in these layers have been equipped over 1000 epochs where each epoch has a batch size of 128 for each training and modeling cycle, just to align the vectorization capability [7].

Approach 2: In order to make these models easily accessible to the end user, researcher has chosen convolutional neural network as it is appropriate for the representation of real-world data. This model is developed using keras API with GPU enabled version of tensor flow as a backend, training dataset contains total 643 images per class, a total of 6 gigabytes of graphics memory was used, and the proposed model was trained on a Raspberry pi system; this technique has proved to be abortive because it has no processing power to train high resolution images of a large database against its limited memory [4].

Approach 3: Segmentation has been implemented as one of the data pre-processing tasks, and then, GrabCut algorithm provided monochromatic versions of the segmented image. LBP is an approach that labels image pixels and provides a binary result, and finally, SVM classifies the dataset images into being healthy or diseased [28].

Approach 4: Pytorch 0.4 and Python 3.6 are used to simulate deep neural networks for the automatic detection of plant diseases. VGG16 is used as the basic architecture, and by shuffling the network layers, total four models were built just to compare the effects of changing connections of intermediate layers to achieve the highest accuracy [11].

Approach 5: Deep residual neural network was trained to detect external defects of the tomatoes, and it can be used upon any other food too [13].

Approach 6: The VGGNet pre-trained ImageNet and inception modules are chosen in this method, so that we could skip randomly initializing the weights in the classification model, and these initial weights can be assigned using pre-trained network on massive labeled dataset, ImageNet [17].

Approach 7: Classification model basically uses ResNet50 architecture and shuffle units as the auxiliary architectures [34].

Approach 8: To overcome the issues of traditional system, CNN is used for classification purpose, and this study focused on strategies to obtain a precision score of more than 93% with class weight, SMOTE, and focal loss with deep CNN from scratch, but achieving high precision was very difficult because of limited dataset and a high-class disparity with a heavy bias toward cassava mosaic disease (CMD) and cassava brown streak virus disease (CBB) classes [16].

Approach 9: VGG16 and AlexNet are two deep learning architectures which are used to classify the images of tomato leaves. This model is basically focused on the significance of bias learning rate, minibatch size, and weight [14].

Approach 10: A model is designed to identify the limited number of disease using sufficient amount of a dataset. In this study, local binary patterns and statistical attributes were considered, results were compared against the efficiency of a CNN, it has been observed that LBP proved to achieve surprisingly higher accuracy, the best kappa coefficient achieved was 0.970, and sensitivity was 0.980 [19].

Approach 11: Pre-trained models VGG16, MobileNet, and Inception V3 are used as basic architectures, the final classification model consists three convolutional layers, three max pooling layers, and two fully connected layers for disease detection and classification, and proposed model claims 91.2 percent accuracy for one healthy and nine diseased class [12].

Approach 12: A faster R-CNN network is applied on part of the data, and the outcome showed that the results are almost same compared to the previous methods using only RGB data [20].

Approach 13: To gain better results than other traditional techniques, different convolutional filters and different pool sizes are used, max pooling with a filter size of 32*32*3 demonstrates 92% accurate results, and average pool size with a convloutional filter size of 64*64*3 gives accuracy of 93.7% and is proved to be better than other traditional machine learning techniques [30].

4 Integration-Based Models

It has been observed that integration-based models claim to be more accurate in terms of detection and recognition of diseased plants, some existing models have been discussed thoroughly, and one of the efficient approach is: combining VGG16 and VGG19 (pre-trained networks) together to make the final prediction; this model demonstrates the significance of hyperparameters which are optimized using the orthogonal learning particle swarm optimization (OLPSO) algorithm [25].

Integration-based model basically focuses on two measures, i.e., number of hyperparameters used for classification and training time, and this integrated model achieved 91.7% accuracy to determine the infected areas. In order to reduce the above-mentioned measures, the conventional CNN has been enhanced by integrating the structure of the inception module to boost the accuracy of dataset, the squeeze-and-excitation (SE) module, and the global pooling layer to reduce the number of model parameters [26].

Another efficient approach developed an integration-based model in which feature extraction is done using SVM considering color and texture attributes as significant features for the classification process and conditional-DCGAN (C-DCGAN) for data augmentation; these segmented disease spot images are used to train the VGG16 to recognize the diseased leaves with an accuracy of 90%. Purpose of using SVM is to spot disease image segmentation on low-level learning while preserving the quality of edge information, enhanced C-DCGAN can implement image augmentation without changing the distribution of real disease spot images, and VGG16 equipped with increased spot images can accurately identify the diseases of the tea leaf [10].

While designing a classifier for capsicum plant leaf diseases, K-means clustering technique is used to extract the infected area, and then, classification is done by SVM [15]. Feature extraction can also be done by LBP Algorithm to extract textual features and help vector machine method for multiple plant classification [22]. One of the proposed approaches explains pre-processing as well as post-processing to precise the stability, where pre-processing includes a process of background removal of RGB images, and then, these images have been converted into HSV images; these binary images are classified based upon saturation and hue parts of images. Then, segmentation is done by a clustering method, and classification is done by optimized deep neural network along with Jaya optimization algorithm; a feedback loop is then generated to precise the stability [18]. LBP algorithm can also be used in combination with the bag of visual words (BoVW) and SVM [23].

Another efficient approach is focused on images with complex background, and the proposed model is a combination of two passes, the first pass removes the leaf component from the entire image by removing the complex context, while the second one implements a deep learning transformation neural network (CNN) using those segmented images. Identification accuracy of 98.14 percent is being achieved which prove it to be better than many conventional deep learning models [24].

5 Overall Performance

In total 38 research papers, which are discussed in this article, the most recent and advanced work aims to create an easy handheld application for the identification of tomato diseases using deep learning algorithms where convolutional neural network was able to remove selected features from the images automatically and gives 99.01 percent validation accuracy. Land owners can effectively use this model to automatic recognition of diseased leaves, while readings from distant lands can differ due to the image dataset being used in a supervised environment [4]. So that a farmer does not need to take the leaf samples to some laboratory and wait for the results to find out the severity of diseases, he can only use his own device to detect and recognize the disease and treat his crop as soon as possible to increase the production.

One of the approaches designs a model for automatic recognition of 15 species of beetle products that are frequently identified in food inspection. This approach trained a convolutional neural network for 6900 microscopic image dataset and achieved a calculative accuracy of 83.8 percent in cross-validation [7].

When feature extraction is done by LBP and classifier is implemented further for the detection of healthy/infected leaves, the findings have shown that the model has been successful in most cases. The model has focused to implement upon 44 of the 46 plant disease combinations, and it achieved overall 95% accuracy. It is observed that achieving 100 percent identification capability for more than 50 percent cases is critical through conflict resolution. Through addition of new images to the existing database, it expands modal's ability of recognition and also enriches its database. The proposed application is capable of recognizing a list of health conditions in plant species [28]. We have identified some of the neural networks which have been used as basic architecture in classification models.

6 ResNet

The ResNet classifier model proves to achieve of 94.6% average precision on the test set while the optimal classifier achieved a recall of 86.6% while maintaining a precision of 91.7% [13]. ResNet50 architecture is also used as basic architecture, and an excellent performance (overall accuracies of 0.99, 0.91, and 0.98 for plant species recognition, disease severity estimation, and plant disease classification) is experienced in the process [34].

7 VGGNet

When used to design a classification model, VGGNet achieved a validation accuracy of 91. 83%, and on overall accuracy of 92% in the complex background [17].

8 Support Vector Machine

SVM has given 100% accurate results when applied on the small dataset of 62 images collected manually on capsicum plant [15]. SVM in coperation with SPA demonstrates 90.7% accuracy in test set [21]. The SVM algorithm in addition with LBP achieves classification accuracy of 91.85% [22]. Support vector machine demonstrates approximately 17% improvement in results and reaches a success rate of 75.8% [23]. With the help of image processing techniques, 22 visual features have been extracted from each of the plant leaves, and then, a number of algorithms have been applied for the classification, and concluded that, SVM model has given the most efficient and accurate results [35].

9 Optimized Deep Neural Network

A classification model achieves an accuracy of 90.57 percent for normal leaf images, significantly higher 98.9 percent for the blast affected, 95.78 percent for bacterial bright, 94 percent for brown spot, and 92 percent for sheath rot, and this model used Jaya algorithm with optimized deep neural network for multi-classification process [18].

10 RandomForest

RandomForest-based model has claimed 98.4% classification accuracy implemented upon a segmented dataset, segmentation is used to evaluate the diseased part, and it has been done by a novel fuzzy set extended from neutrosophic logic-based segmentation technique [31].

11 MobileNet

Reduced MobileNet demonstrated 98.34% accurate results, as it uses significantly lesser number of parameters than VGG and MobileNet [27].

Compared to common machine learning approaches, transfer learning methods achieve better performance and a higher validation accuracy of 96.46 percent [33]. Hence, it can be seen that an overall performance ranges from 75% success rate to 100% accuracy, while considering the various distinct neural network architectures, pre-processing methods, evaluation tools, and performance time.

12 Proposed Measure

Predictive measure is a property (already existing in the dataset being observed or manipulated after applying different processing techniques) which can be considered to predict the class value for each data instance. We have proposed Gaussian filter as a predictive measure for classification.

It is also a part of data pre-processing, which is an important task to be carried out to increase efficiency of a trained model. My trained model has given 78% and 97% accuracy before and after data pre-processing, respectively. Data pre-processing also significantly increases the reliability of the analysis in order to facilitate the extraction of useful information from the data. In any machine learning process, data pre-processing is the step that transforms or encodes the data to such a state that the machine can easily parse it. In other words, the data features can now be easily interpreted by an algorithm. According to our studies, data augmentation and data segmentation have been carried out on the dataset for improved efficiency, and if we apply image filtering on the image dataset, it would give better results. Image filtering is used to reduce noise from the image dataset, and Gaussian filter is best suited here for the process. Gaussian blur is a filter used to blur the given image where Gaussian function is used to measure the transformation to be applied to each pixel of the image. The Gaussian function formula for 1D is given below

$$g(x) = \frac{1}{\sqrt{2\pi}\sigma_1} \exp\left(\frac{-x^2}{2\sigma^2}\right)$$

where x represents distance from the origin(horizontal axis), y represents distance from the origin (vertical axis), and Σ is the standard deviation of the Gaussian distribution. Two-dimensional Gaussian function is given as

$$G(x, y) = \frac{1}{2\pi\sigma^2} e^{-\frac{x^2+y^2}{2\sigma^2}}$$

The Gaussian blur effect is usually created by convolving an image with the finite impulse response (FIR) kernel of Gaussian values. Gaussian blur process can be divided into two phases; image blurring is done by a one-dimensional kernel in any of the vertical or horizontal direction, and second phase blurs the remaining direction of that image.

13 Comparison with Other Approaches

An analysis has been done on different deep learning and machine learning approaches in which an image processing technique is applied upon 637 distinct images of 32 leaf species and 22 visual features that are extracted from each leaf. Training data and test data contain 510 and 127 images, respectively. Simulation proved that SVM technique achieves higher success rate over K-nearest neighbor (KNN) technique with 94.2% and 88.6% accuracy, respectively. In the event that artificial intelligence processing times are found, KNN is known to be the fastest technique. Artificial neural network (ANN) yields similar results to SVM, with a processing speed approximately seven times longer than SVM. Hence, ANN does not seem feasible for classification. It is also inferred from the study that visual feature selection is considered to be the most efficient technique to improve the efficiency and accuracy of classification models [35].

14 Conclusion

Number of articles have been studied, and findings can clearly state that integration-based models are more efficient; these models enhance success rate of diseased crop symptoms identification while reducing the training time; SVM gives best results when used in integration models; data pre-processing proves to be an important task while designing a classification model; moreover, choosing the appropriate hyperparameters can give best results.

We have also proposed image filtering as one of the predictive measures while data augmentation and data segmentation were already considered, and various techniques were also applied on the dataset being observed. Among three of the image filtering techniques, Gaussian filter has proved to be best in terms of computational costs. It has also been observed that computer vision applications are an excellent solution to the unhealthy crop recognition; it is also been observed that yet farmers does not have any handheld device for the detection; in order to make this task easier and convenient, some device must be designed that would be available at feasible prices or there can be some application that can detect the disease.

References

1. Leee, S. H., Goëaua, H., Bonnet, P., Joly, A. (2020). New perspectives on plant disease characterization based on deep learning. *Computers and Electronics in Agriculture, 170*, 105220.
2. Kayaa, A., Keceli, A. S., Catal, C., Yalic, H. Y., Temucin, H., & Tekinerdogan, B. (2019). Analysis of transfer learning for deep neural network based plant classification models. *Computers and Electronics in Agriculture, 158*, 20–29.
3. Kamilaris, A., & Prenafeta-Boldú, F. X. (2018). Deep learning in agriculture: A survey. *Computers and Electronics in Agriculture, 147*, 70–90.
4. Liang, S., & Zhang, W. (2020). Accurate ImagetRecognition of plant diseases based on multiple classifiers integration. In Y. Jia, J. Du, & W. Zhang (Eds.)., *Proceedings of 2019 Chinese Intelligent Systems Conference*. CISC 2019. Lecture Notes in Electrical Engineering (vol. 594). Singapore: Springer.

5. Khairnar, K., Goje, N. (2020). Image processing based approach for diseases detection and diagnosis on cotton plant leaf. In P. Pawar, B. Ronge, R. Balasubramaniam, A. Vibhute, & S. Apte (Eds.)., *Techno-Societal 2018*. Cham: Springer.
6. Tooa, E. C., Li, Y., Sam, N., & Liu, Y. (2020). A comparative study of fine-tuning deep learning models for plant disease identification. *Computers and Electronics in Agriculture.*
7. Tyagi A.C. Towards a second green revolution. *Irrigation and Drainage, 65*(4), 388–389.
8. Barbedo, J. C. A. (2019).Plant disease identification from individual lesions and spots using deep learning. *Biosystems Engineering, 180*, 96–107.
9. Platt, J. C., Cristianini, N., & Shawe-Taylor, J. (1999). *"Large margin DAGs for Multiclass classification" fast training of Support Vector Machines using sequential minimal optimization.* Cambridge, MA: MIT Press.
10. Hu, G., Wu, H., Zhang, Y., Wan, M. (2019). A low shot learning method for tea leaf's disease identification. *Computers and Electronics in Agriculture, 163*, 104852.
11. Kim, W.-S., Lee, D.-H., Kim, Y.-J. (2020). Machine vision-based automatic disease symptom detection of onion downy mildew. *Computers and Electronics in Agriculture.*
12. Agarwal, M., Abhishek Singh, Arjaria, S., Sinha, A., & Gupta, S. (2020). ToLeD: Tomato leaf disease detection using convolutional neural network. *Procedia Computer Science, 167*, 293–301.
13. Da Costa, A. Z., Figueroa, H. E. H., Fracarolli, J. A. (2020). Computer vision based detection of external defects on tomatoes using deep learning. *Biosystems Engineering, 190*, 131–144.
14. Rangarajan, A. K., Purushothaman, R., Ramesh, A. (2018). Tomato crop disease classification using pre-trained deep learning algorithm. *Procedia Computer Science, 133*, 1040–1047.
15. Anjana, M. S., & Singh, P. K. (2019). Hybrid system for detection and classification of plant disease using Qualitative texture feature analysis. In *International Conference on Computational Intelligence and DataScience (ICCIDS 2019).*
16. Sambasivam, G., Opiyo, G. D. A predictive machine learning application in agriculture: Cassava disease detection and classification with imbalanced dataset using convolutional neural network. *Egyptian Informatics Journal.*
17. Chena, J., Chena, J., Zhanga, D., Sunb, Y., Nanehkarana, Y. A. (2020). Using deep transfer learning for image-based plant disease identification. *Computers and Electronics in Agriculture, 173*, 105393.
18. Ramesh, S., Vydeki, D. Recognition and classification of paddy leaf diseases using optimized deep neural network with Jaya algorithm. *Information Processing in Agriculture.*
19. Sorte, L. X. B., Ferraz, C. T., Fambrini, F., Goulart, R. D. R., & Saito, J. H. (2019). Coffee Leaf disease recognition based on deep learning and texture attributes. *Procedia Computer Science, 159*, 135–144.
20. Polder, G., van de Westeringh, N., Kool, J., Khan, H. A., Kootstra, G., Nieuwenhuizen, A. (2019). Automatic detection of Tulip Breaking Virus (TBV) using a deep neural network. *IFAC PapersOnLine, 52–30*, 12–17.
21. Yang, X., Zhang, R., Zhai, Z., Pang, Y., & Jin, Z. (2019). Machine learning for cultivar classification of apricots (*Prunus armeniaca* L.) based on shape features. *Scientia Horticulturae, 256*, 108524.
22. Nguyen, V., Le, T., Apopei, B., Alameh, K. (2019). Effective plant discrimination based on the combination of local binary pattern operators and multiclass support vector machine methods. *Information Processing in Agriculture, 6*, 116–131.
23. Araujo, J. M. M., Peixoto, Z. M. A. A new proposal for automatic identification of multiple soybean diseases. *Computers and Electronics in Agriculture.*
24. Karlekar, A., Seal, A. (2020). SoyNet: Soybean leaf diseases classification. *Computers and Electronics in Agriculture, 172*, 105342.
25. Darwish, A., Ezzat, D., Hassanien, A. E. An optimized model based on convolutional neural networks and orthogonal learning particle swarm optimization algorithm for plant disease diagnosis. In *2020 Swarm and Evolutionary Computation.*
26. Han, J., Zhang, D., Chen, P., Zhang, J., & Wang, B. (2019). Classification of plant leaf diseases based on improved convolutional neural network. *Sensors, 19*, 4161.

27. Kamal, K. C., Yin, Z., Wu, M., & Wu, Z. (2019). Depthwise separable convolution architectures for plant disease classification. *Computers and Electronics in Agriculture, 165*, 104948.
28. Pantazi, X. E., Moshou, D., Tamouridou, A. A. (2019). Automated leaf disease detection in different crop species through image features analysis and One Class Classifier. *Computers and Electronics in Agriculture,156*, 96–104.
29. Picona, A., Seitzc, M., Alvarez-Gila, A., Mohnke, P., Ortiz-Barredo, A., Echazarra, J. Crop conditional convolutional neural networks for massive multi-crop plant disease classification over cell phone acquired images taken on real field conditions. *Computers and Electronic in Agriculture.*
30. Khamparia, A., Aman Singh, Luhach, A. K., Pandey, B., Pandey, D. K. Classification and identification of primitive Kharif crops using supervised deep convolutional networks. *Sustainable Computing: Informatics and Systems.*
31. Dhingra, G., Vinay Kumar, & Joshi, H. D. (2019). A novel computer vision based neutrosophic approach for leaf disease identification and classification. *Measurement 135*, 782–794.
32. Dash, C. S. K., Behera, A. K., Dehuri, S., & Cho, S.-B. Building a novel classifier based on teaching learning based optimization and radial basis function neural networks for non-imputed database with irrelevant features. *Applied Computing and Informatics.*
33. Geetharamani, G., Arun Pandian, J. (2019). Identification of plant leaf diseases using a nine-layer deep convolutional neural network. *Computer and Electrical Engineering, 76*, 323–338.
34. Liang, Q., Xiang, S., Hu, Y., Coppola, G., Zhange, D., & Suna, W. (2019). PD2SE-Net: Computer-assisted plant disease diagnosis and severity estimation network. *Computers and Electronics in Agriculture, 157*, 518–529.
35. Yigit, E., Sabanci, K., Toktas, A., & Kayabasi, A. (2019). A study on visual features of leaves in plant identification using artificial intelligence techniques. *Computers and Electronics in Agriculture, 156,* 369–377.

Design and Analysis of Smart Automatic Street Light System

Prakhar Maheshwari, Monika Agrawal, and Vishal Goyal

Abstract Smart automatic street light system helps in controlling the street lights intelligently and automatically to minimize power consumption and reduces man work by fluctuating the street lights time to time and in certain conditions toggling as well as turning OFF of some lights concerning the motion observed on the roads. The interfacing of the real-time clock (RTC) module is to control the street light on exact timing, so this phenomenon reduces excess power consumption by up to 50%. On the other hand, the proposed system uses an ARM controller which has a high clock speed. All controlling processes work faster as compared to all other typical controllers like arduino, Avr, 8051, etc.

Keywords Smart street lights · RTC module · Power consumption · Time stamp · PIR sensor

1 Introduction

In past, halogen light was used in street light whose element heating time is high to glow and it takes much power and these light harm the environment such as emission of carbon dioxide (CO_2) [1, 2]. After this, halogen lights were replaced with LED lights which glows up in a millisecond as well as consumes less power [3]. For automatization, LDR interfaces with street light which automatically start and stop the lights after sunset and before sunrise [4–6]. But this is not sufficient for reducing the power consumption. Then, PIR sensor was introduced which senses the motion of living being near the PIR sensor; if the sensor does not detect any motion, then,

P. Maheshwari (✉) · M. Agrawal · V. Goyal
Department of Electronics Communication and Engineering, GLA University, Mathura 281406, India
e-mail: prakhar1505@gmail.com

M. Agrawal
e-mail: agrawalmonika1999@gmail.com

V. Goyal
e-mail: vishal.glaitm@gmail.com; vishal.goyal@gla.ac.in

S. Tiwari et al. (eds.), *Proceedings of International Conference on Big Data, Machine Learning and their Applications*, Lecture Notes in Networks and Systems 150,
https://doi.org/10.1007/978-981-15-8377-3_13

controller dimming/toggling the light [6–8]. This reduces the power consumption upon a certain mark, but this is not enough. So, this paper shows the advancement in the existing system.

For the sample, 8 light shows the automatic working of the street lighting system, further it will extend as per requirement. Nowadays, in this fast pace world, no one even concerned about the wastage of electricity and forgot to switch the lights OFF which is not in use which leads the lights to be continuously ON and each of the lights in the colonies, expressway, and roads continuously ON for the vehicles during whole night and continuously stay ON till the very next morning till sunset. Due to which lot of power is wasting on lighting the roads for vehicles. So, this involves an experimental idea to reduce power consumption more than 50%, also automated the street light.

As soon as sunlight falls of LDR sensor cadmium sulfide then sensor automatically switches OFF the lights which depends on its light-intensity, otherwise switches ON the lights. Now, the main modification is after 10 p.m. to 4 a.m. When there is any motion observed, then lights start toggling and if not detect, then only 20% of the total LED is ON. Otherwise, all lights depend on the previous state.

2 Proposed System Functionality

This proposed system requires some basic components, i.e., LDR Sensor, PIR sensor, RTC module, and TIVA controller. Arduino and Avr controller work on 16 MHz and 32 MHz, respectively, that is why Tiva is used for fast processing with 80 MHz clock frequency. LDR is used to check whether there is daylight or not. If it is day, then controller turns OFF all light; otherwise, turn ON the light-dependent on RTC and PIR. RTC module is used to get exact timing to start-/stop-specific task which is dependent on PIR. If PIR output is HIGH (It means motion is detected or presence of a human in the area), then LEDs start toggling with delay and if PIR output is LOW (it means no motion is detected), then only 20% of all LED's ON. The whole working is dependent on the values of LDR, RTC, and PIR. Thus, the control of LEDs will depend on Table 1.

Table 1 Different modes of operation

S. No.	LDR output	RTC timing	PIR output	Modes
1	<2000	Independent	Independent	Mode 1
2	≥2000	<10 p.m. and >4 a.m.	Independent	Mode 2
3	≥2000	≥10 p.m. and ≤4 a.m.	0	Mode 3
4	≥2000	≥10 p.m. and ≤4 a.m.	1	Mode 4

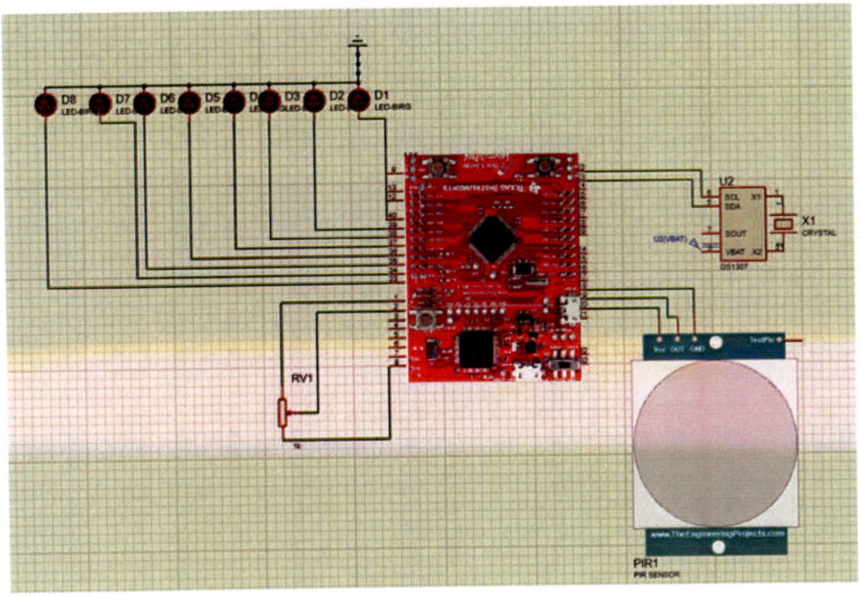

Fig. 1 All lights are OFF

2.1 Operating Modes

MODE 1: This is the condition when LDR gives an output less than 2000 (whatever be the value of LDR and PIR), means there will be daylight and automatically all the lights get OFF as shown in Fig. 1.

MODE 2: This situation comes when LDR gives an output more than 2000, and the switching ON/OFF of LED's depends on RTC and PIR. If the time (output of RTC) is less than 10 p.m. and greater than 4 a.m., then whatever the value of PIR will be, it switches ON all the lights till very next morning until there will be enough daylight on roads as shown in Fig. 2.

MODE 3: This condition comes in the picture when LDR gives an output more than 2000 and the switching ON/OFF of the LEDs depends on RTC and PIR. If the time (output of RTC) is more than 10 p.m. and less than 4 a.m., then it depends on the output of PIR. If PIR output is LOW, i.e., there is no vehicle on the road, so it switching ON two LEDs to provide enough lightning on the road as shown in Fig. 3.

MODE 4: This condition comes in the picture when LDR gives an output more than 2000 and the switching ON/OFF of the LEDs depends on RTC and PIR. If the time (output of RTC) is more than 10 p.m. and less than 4 a.m., then it depends on the output of PIR. If PIR output is HIGH, i.e., there is a vehicle on the road, then the toggling of LED's starts for that vehicle only, after passing the vehicle, it works in mode 3 as shown in Fig. 4 (Figs. 5 and 6).

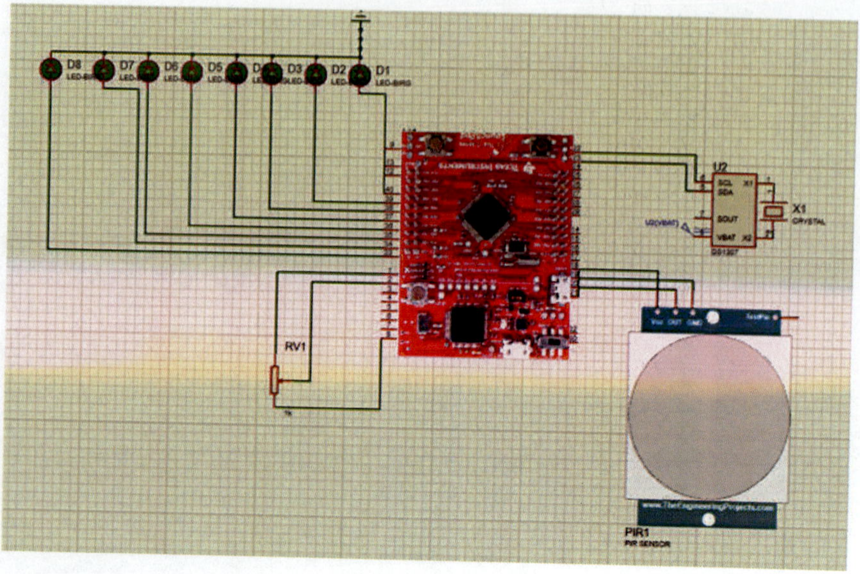

Fig. 2 All lights are ON

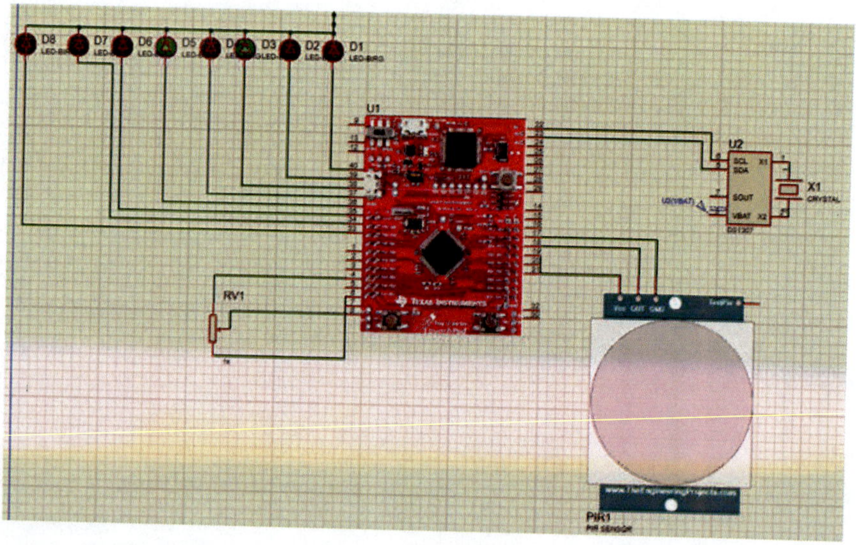

Fig. 3 Some lights ON

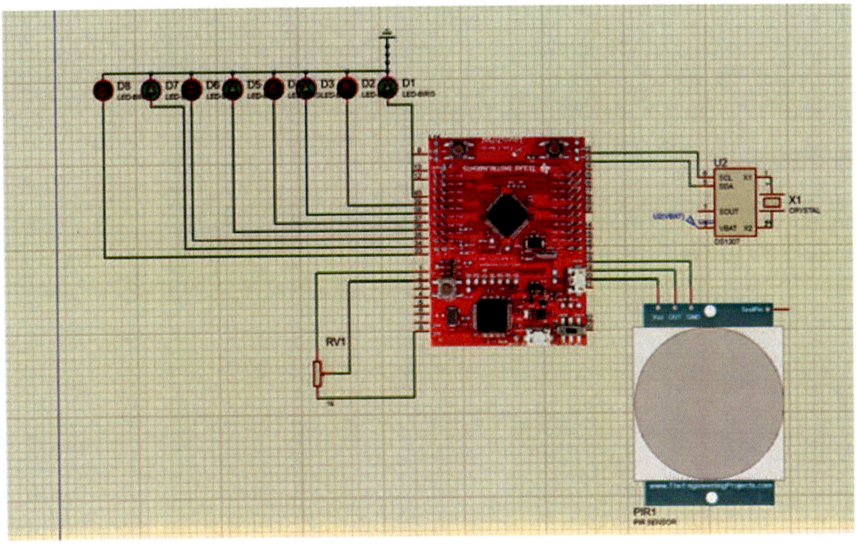

Fig. 4 When light toggles

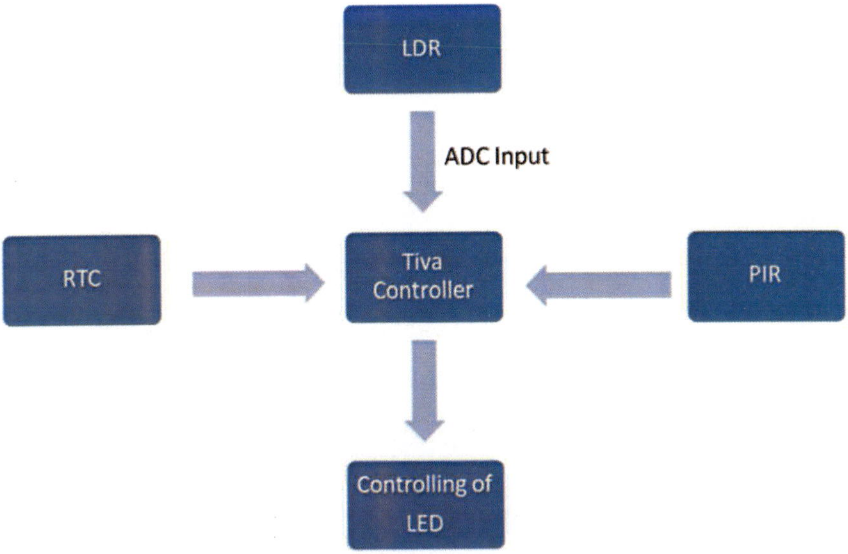

Fig. 5 Block diagram

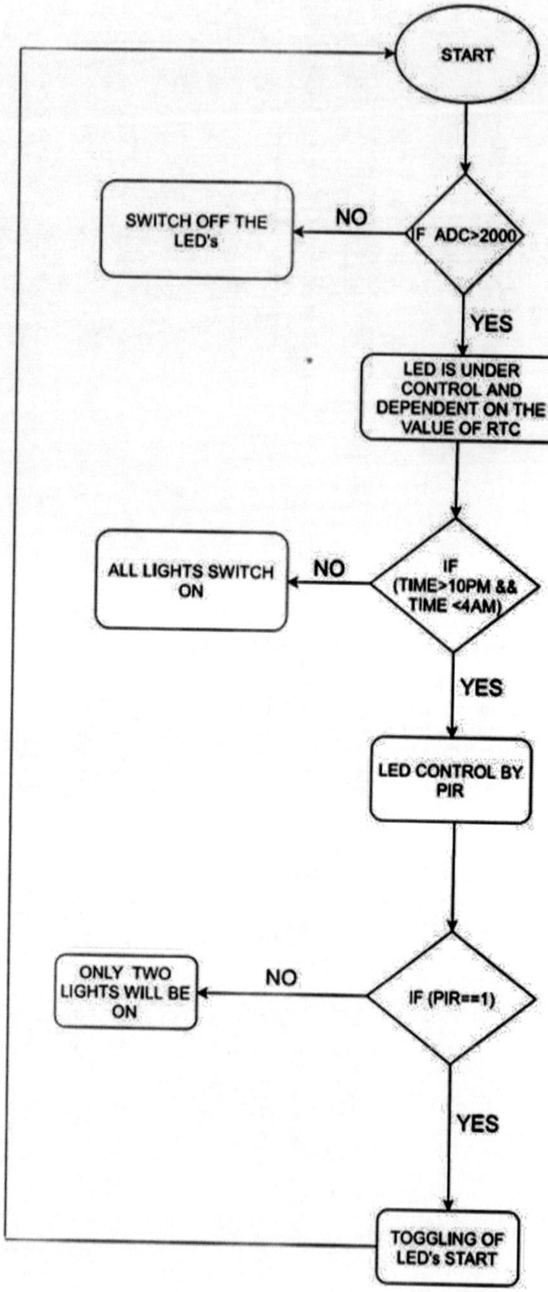

Fig. 6 Flowchart depicting system operations

3 System Implementation

(a) **RTC MODULE**: Real-time clock (RTC) module is the most widely used device in the current system that provides precise time and date information for various applications. It works on inter-integrated circuits (I2C) which is used to transmit the data. It provides hour, minute, and second along with day, date, and month.

(b) **LDR SENSOR**: A photoresistor (LDR) and cadmium sulphide track are used to measure the light intensity present in environment. Resistance changes according to light. When the light level decreases, the input analog voltage increases very sharply and vice versa. So, it works on light-sensitive devices.

(c) **PIR SENSOR**: Pyroelectric InfraRed sensor is an electronic sensor that senses motion and heat emitted by the human body in the range of 5–12 m and angle up to 170° depends upon sensitivity potentiometer.

4 Result and Conclusion

The main motive of the proposed system is to save energy and to make it eco-friendly which is implemented at a low cost in comparison to that system implemented earlier, i.e. ,using sodium vapor lamps. It successfully tackles the problem of manual switching (by making it automatically controlled) and the maintenance or primary cost that the world is facing. More than 50% of the light is utilized and the remaining will be saved. This system has a scope in various applications to provide lightning at highways, streets, hospitals, industries, universities, footpaths, agriculture field monitors, malls roads, etc. The experimental setup is as shown in Fig. 8 and the comparison of power consumption between the previous system [9] and the proposed system is shown in Fig. 7.

Fig. 7 Power consumption

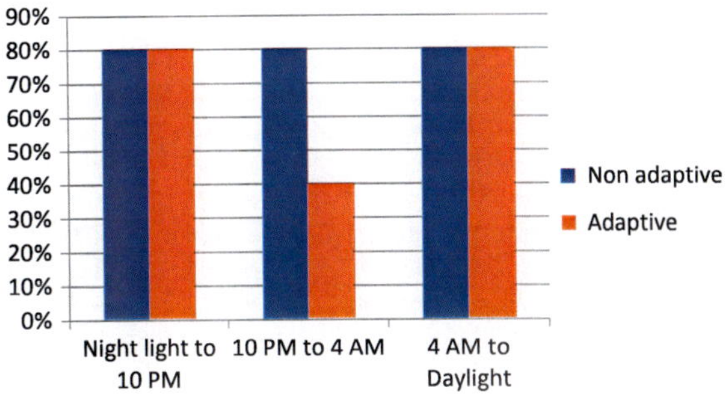

Fig. 8 Experimental setup of the proposed system

References

1. Yusoff, Y. M., Rosli, R., Karnaluddin, M. U., & Samad, M. (2013, September). Towards smart street lighting system in Malaysia. In *2013 IEEE Symposium on Wireless Technology & Applications (ISWTA)* (pp. 301–305). IEEE.
2. Caponetto, R., Dongola, G., Fortuna, L., Riscica, N., & Zufacchi, D. (2008). Power consumption reduction in a remote controlled street lighting system. In *International Symposium on Power Electronics, Electrical Drives, Automation and Motion (SPEEDAM 2008)* (pp. 428–433), Ischia, June.
3. Nuttall, D. R., Shuttleworth, R., & Routledge, G. (2008). Design of a LED street lighting system (pp. 436–440).
4. Sudhakar, K. S., Anil, A. A., Ashok, K. C., & Bhaskar, S. S. (2013). Automatic street light control system. *International Journal of Emerging Technology and Advanced Engineering, 3*(5), 188–189.
5. Divya, G., Lakshmi, G., Srinivas, S., Tejaswini, M., & Ashok, B. (2016). Design and implement of wireless sensor street light control and monitoring strategy along with GUI. *IJITR*, 78, 81.
6. Road lighting Part 4: Methods of measuring lighting performance. In BS EN 13201–4:2003 United Kingdom, 2003.
7. Ramli, N. L., Yamin, N. M., Ghani, S. A., Saad, N. M., & Sharif, S. M. (2015). Implementation of passive infrared sensor in street lighting automation system. *ARPN Journal of Engineering and Applied Sciences, 10,* 17120–17126.

8. Khade, D. R., Gajane, N. V., Gawade, S. N., & Metri, R. A. (2017). Intensity controller of LED street lights. In *2017 International Conference on Circuit, Power and Computing Technologies (ICCPCT)*, Kollam, pp. 1–4.
9. Rodi, P., Chandrakar, L., Sivanantham, S. S. G., & Sivasankaran, K. (2015). Energy conservation using automatic lighting system using FPGA. In *International Conference on Green Engineering and Technologies*.
10. Sheela, K. S., & Padmadevi, S. (2014). Survey on street lighting system based on vehicle movements. *International Journal of Innovative Research in Science, Engineering and Technology, 3*(2), 9220–9225.
11. Rath, D. K. (2016). Arduino based: Smart light control system. *International Journal of Engineering Research and General Science, 4*, March–April, .
12. Raghu, N., Yashaswini, N., Yashaswini, S., & Prathib Kumar, G. (2018). Automatic Street light control by detecting vehicle movement. In *2018 3rd IEEE International Conference*.
13. Shruthi, K., & Akhil, G. (2019). Arduino-based smart street lamp control system to save energy and improve lamp life. In *Innovations in electronics and communication engineering* (pp. 131–141). Singapore: Springer.
14. Abdullah, A., Yusoff, S. H., Zaini, S. A., Midi, N. S., & Mohamad, S. Y. (2019). Energy efficient Smart Street light for smart city using sensors and controller. *Bulletin of Electrical Engineering and Informatics, 8*(2), 558–568.

Combating DoS Attack on OpenStack Using Hypervisor Based Intrusion Detection System with the Help of Machine Learning

Smita Mondal and Arjun Choudhary

Abstract Evolution of Denial of Service(DoS) attacks are the critical issue for cloud over decades. The need of identifying DoS attack and preventing it at early stage becomes a challenge for cloud service providers and cloud hosting companies. OpenStack is a growing private cloud platform which ensures data security but like public cloud it is also vulnerable to various attacks like DoS attack. So there is still high need for an efficient Intrusion detection system (IDs) who can protect cloud hardware and resources from this attack. Currently OpenStack has only network based IDs which is implemented as a service and there is some ongoing research with OpenStack logs. In this paper a Hypervisor based Intrusion Detection System is implemented and tested for OpenStack cloud.

Keywords Cloud · OpenStack · DoS · IDPs · Machine learning

1 Introduction

With changing Cloud Computing mindset and an increased adoption among organizations, the Cloud computing market in India has grown beyond measure. Cloud is developed from the idea, Computer to be used by two or more people, simultaneously [1] Though development of cloud is started in early 1960's by developing "Intergalactic computer network" [2] but main cloud adaption by people started after 2002 when Amazon Web Service was introduced. In 2006, Amazon EC2 was introduced for providing small companies and individuals to rent virtual computers on which they can run their own computer applications. Unlike Amazon EC2 in 2008, Google cloud platform came up with the idea of offering Platform as a Service. Later Microsoft, with its Azure also came into picture with IaaS offerings, similar to EC2.

S. Mondal · A. Choudhary (✉)
Sardar Patel University of Police Security and Criminal Justice, Jodhpur, India
e-mail: a.choudhary@policeuniversity.ac.in

S. Mondal
e-mail: spu18cs10@policeuniversity.ac.in

S. Tiwari et al. (eds.), *Proceedings of International Conference on Big Data, Machine Learning and their Applications*, Lecture Notes in Networks and Systems 150,
https://doi.org/10.1007/978-981-15-8377-3_14

As of February 2020, canalys reports [3] AWS with 32.4% of the market, Azure at 17.6%, Google Cloud at 6%, Alibaba Cloud close behind at 5.4%, and other clouds with 38.5%. So, AWS is still a leader in public cloud market. But main disadvantage of public clouds is that services are owned and managed by third party which leads the big companies to think about data security. So, in 2010 OpenStack came up with its hybrid cloud concept.

Cloud computing models both public, private as well as hybrid are full of advantages but they are still susceptible to both inside and outside attacks. Therefore, cloud developers need to take security measures to protect their users' sensitive data from cyber-attacks. Some top most attacks are: Denial of Service, Malware attack, Side channel attacks, Man in the middle attack, SQL injection etc. which can lead to data theft, system infection by malware, service unavailability for intended users [4–6]. Though there are many ways of mitigation but in recent scenario deploying an Intrusion Detection System is most used and efficient technique to detect and prevent any type of attack. Intrusion Detection System provides network monitoring, log monitoring and notifies about the abnormal behavior. In current times, various machine learning techniques are being widely used to develop an Intrusion Detection System (IDS) for detecting and classifying cyberattacks. So, in this research paper a hypervisor based intrusion detection system using machine learning is deployed in XCP-ng which can detect anomalized network packets on OpenStack cloud scenario.

2 OpenStack DoS Vulnerability

Being a new comer OpenStack, the private cloud, is still facing many vulnerabilities [7, 8].

In Elia et al. [9] analyzes 5 years security vulnerabilities of OpenStack upto 2017 (Fig. 1).

OpenStack security guideline [10] also explained the recent attacks that can be possible in OpenStack [11, 12]. Most eminent and important attack is DoS attack, which is basically unavailability of services for intended users. It is relatively easy to create using some existing tool. DoS or DDoS both did physical resource damage. Hackers mostly launch this attack.

3 IDPs: The Solution

To prevent DoS attacks the only better approach is intrusion detection system [13]. Intrusion Detection system is an automated system that can monitor a system and issues alert in case of any anomaly is detected.

Types:

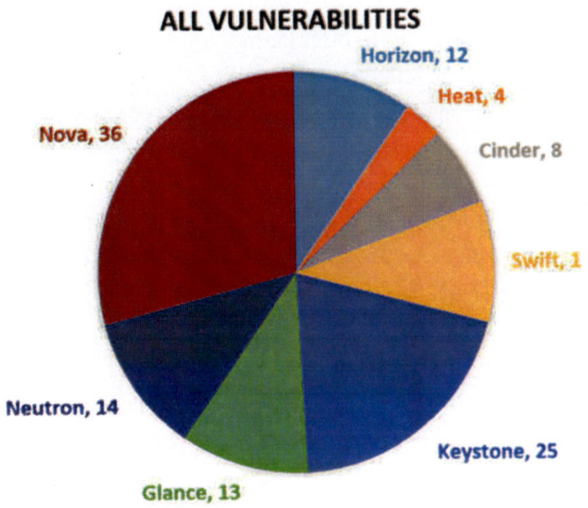

Fig. 1 OpenStack security issues till 2017

There are basically four types of IDS available. These are: hypervisor based, VM based, network based and host based IDs(Intrusion detection system).

Deploying Hypervisor based IDs allow the administrator to monitor the hypervisor and virtual machines built on that hypervisor. Host based IDS rely on the information available from various sources like host systems, that includes the contents of operating systems, system and application files. Network based IDS (NIDS) defends intrusions by monitoring network traffic generated through network end users and other network devices.

The common techniques or mechanism IDs uses are: Anomaly based detection, Signature based detection,behavioral analysis etc. Recently machine learning is used to reduce the number of false positive rate.

3.1 Recently Used IDPs on OpenStack

During literature review process, we found only log based and NIDS based intrusion detection system specifically for OpenStack.

Xu et al. [14] in the paper "Network intrusion detection system as a service on OpenStack cloud" proposed a lightweight approach of NIDS and implement it as a service for OpenStack. Limitation of this study is they only show CPU performance but they do not have information about what type of attack this NIDS is detecting. They also did not mention the OpenStack version which is an important parameter when some researchers want to compare it with other ones.

In second paper which is basically a survey paper of log based anomaly detection frameworks in cloud [15] by Meenakshi et al. This paper shows how researchers are using logs to detect network attacks. There are two more recent paper of Openstack log analysis. First is [16] which gives correlational analysis of OpenStack logs which can be helpful for model feature section. Second one is Log based cloud monitoring for OpenStack [17].

If we analyse large volume of logs, this can create problem in machine learning model as well for supervised learning we need labeled dataset. So far there is no processed labeled dataset of logs. So for this research use of logs was not undertaken as one of the research parameters.

Third paper is based on XEn. Dong et al. [18] which proposed a cache based countermeasures against DDoS attack for Xen. In this paper, structure of Xen is modified in such a way (they removed the linux bridge and integrate bridges function to cache queues), which eventually can only slowdown DDOS attack but cannot detect and prevent.

So we can see, currently no machine learning based IDPs is developed yet, which can detect DoS attack from captured packet and prevent it [19]. That is our major emphasis in this research paper.

4 Proposed Framework

In this below figure, to detect DOS attack on OpenStack a tool, which is acting like an IDPS, is implemented on control VM of XCP-ng server (Fig. 2).

5 Implementaton

5.1 *Used Setup Specifications*

Hardware Specification

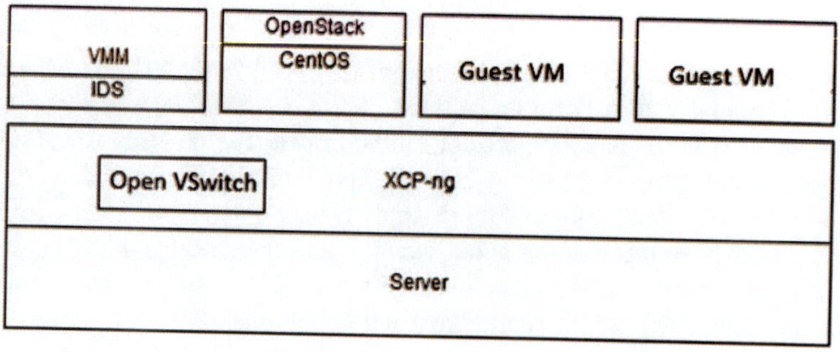

Fig. 2 Architecture of proposed framework

One server with 1TB Hard Drive, 64 GB of RAM One normal PC/laptop in order to access the server.

Software Specification

OS requirement (Cent Os 7) and XCP-ng bare metal hypervisor.

5.2 *Steps*

OpenStack cloud is an open source cloud service provider. It has much vulnerabilities [20] which we are trying to detect by deploying Intrusion detection prevention system. For this research, we intiallly installed OpenStack cloud infrastructure and then deployed the intrusion detection and prevention system on it.
The phases of implementation are:

1. Setup Type 1 hypervisor (XCP-ng).
2. Centos VM setup.
3. OpenStack (version-Train) setup.
4. Perform network capture from openvswitch.
5. Model building (Data preparation, Training and Testing).
6. Perform DOS attack on OpenStack and capture attacked traffic on hypervisor level.
7. Results taking phase.

The reason, we chose XCP-ng is that, it is a type 1 hypervisor or bare metal hypervisor which means it does not require an operating system as a base machine. The second reason is, it is an open source version of XEN hypervisor. Though some features are blocked like GPU integration etc. still for academic purpose we find it very useful and is easy to install. To install it, like any other OS installation process you have to create a bootable media with XCP-ng Iso. After installing, you have to give a static IP in order to access it from remote. In the same network, you have to browse the IP and download XCP-ng center on a windows machine. Now, by giving the static IP, username and password you can work upon your server. To enable openvswitch on XCP-ng you have to use command xe-switch-network-backend openvswitch in control VM.

Second step is Centos installation. For this you have to first make a local repository on the server using control panel of XCP-ng using xe sr-create command. And store the Iso images on that repository using wget command. Rest of the creation process of Centos is same like vmware or any other type 2 hypervisors.

For network capture, we use an automated script through which one can capture network packets using tcpdump command and save it in a directory with "pcap" extension.

In further step, we prepare our machine learning model. We use CICIDS 2018 dataset and more specifically the generated csv, files where DOS attacks are per-

Fig. 3 Result of python tool

formed. After cleaning the csvs, we simply apply Random forest algorithm on the dataset. To improve the model one can use other algorithms as well.

Now, we captured the live attacked packets using the script and transformed it into csvs using another script which use CICFlowmeter and test it in the model. For some limitations, we can not perform how much false positive it is giving. But yes the more accurate the model is the more you get the effective results. For prevention mechanism you just block the IPs against which the model gives output as DOS attack.

To automate the whole process a python based gui tool was made, which has two buttons. One is to capture packet and generate csv and the other one is to check whether packets are from DoS attack or benign packets. So, as button name suggests first button automates the pcap file making and csv generation process using tcpdump and CICFlowmeter and second process shows the output of the machine learning model.

6 Result Analysis

The desired result of the model is the python tool shown in below figure. The tool is tested on live captured packets i.e unlabeled unknown dataset of network packets (Fig. 3).

7 Conclusion and Future Scope

In this paper, we addressed the problem of most common OpenStack cloud attack i.e. DoS. Contribution of this research work is to provide a secure OpenStack platform by implementing hypervisor based intrusion detection system using machine learning which is capable of detecting DOS from features of network packets. And We have successfully implemented it on our laboratory. But, as we have used unlabeled data for result taking it is not possible for performance analysis.

No doubt this intrusion detection system can predict DoS attack and prevent it in real time better than network intrusion detection system. But there are many more vulnerabilities still exists in OpenStack which requires much more attention. Like web attacks, malware attacks, supply chain attacks. Like, we can develop another machine learning model which can take the memory dump of virtual machine from xen console and analyze if there is any malware in the machine. Real time monitoring using this model can be helpful to prevent malware attacks and system compromise. We can also develop a model to prevent SQL injection attack which can be capable of decrypting secure packets and monitor for SQL injection vulnerability. As application of IDPs is now everywhere improving efficiency and decreasing false positive rate can also be a possible research direction.

References

1. Mohamed, A. (2018). A history of cloud computing, Computer Weekly.com, 2018. [Online]. Available: https://www.computerweekly.com/feature/A-history-of-cloud-computing. Last Accessed 13 March 2020
2. Foote, K. D. (2020). Brief-history-cloud-computing, [Online]. Available: https://www.dataversity.net/brief-history-cloud-computing. Last accessed 13 March 2020.
3. Insight. Innovation. Impact.www.canalys.com Cloud market share Q4 2019 and full-year 2019.
4. Mishra, P., Pilli, E.S., Varadharajant, V., & Tupakula, U. (2016). NvCloudIDS: a security architecture to detect intrusions at network and virtualization layer in cloud environment. In *2016 International Conference on Advances in Computing, Communications and Informatics (ICACCI)*, (pp. 56–62) Jaipur.
5. Patil, R. N., Dudeja, H., & Modi, C. (2019). Designing an efficient security framework for detecting intrusions in virtual network of cloud computing. *Computers and Security*, *85*, 402–422.
6. Patil, R., Dudeja, H., & Modi, C. (2020). Designing in-VM-assisted lightweight agent-based malware detection framework for securing virtual machines in cloud computing. *International Journal of Information Security, 19*, 147–162. https://doi.org/10.1007/s10207-019-00447-w.
7. Suryateja, P. S. (2020). Experimental analysis of OpenStack effect on host resources utilization. In Khanna, A., Gupta, D., Bhattacharyya, S., Snasel, V., Platos, J., Hassanien, A. (Eds.), *International Conference on Innovative Computing and Communications. Advances in Intelligent Systems and Computing* (Vol. 1059). Singapore: Springer.
8. Anisetti, M., Ardagna, C. A., Gaudenzi, F., & Damiani, E. (2017). A security benchmark for openStack.
9. Elia, I. A., Antunes, N., Laranjeiro, N., & Vieira, M. (2017). An analysis of openstack vulnerabilities, pp. 129–134

10. Openstack: Openstack security guideline. https://docs.openstack.org/security-guide/introduction/security-boundaries-and-threats.html. Last accessed 03 March 2020
11. Bhardwaj, A., Sharma, A., Mangat, V., Kumar, K., & Vig, R. (2019). Experimental analysis of DDoS attacks on OpenStack cloud platform. In C. Krishna, M. Dutta, R. Kumar (Eds.), *Proceedings of 2nd International Conference on Communication, Computing and Networking. Lecture Notes in Networks and Systems* (Vol. 46). Singapore: Springer.
12. Khraisat, A., Gondal, I., Vamplew, P., et al. (2019). Survey of intrusion detection systems: Techniques, datasets and challenges. *Cybersecurity, 2*, 20 (2019). https://doi.org/10.1186/s42400-019-0038-7.
13. Bharati, M. P., & Tamane, S. C. (2020). Security vulnerabilities of openStack cloud and security assessment using different software tools. In Zhang, Y. D., Mandal, J., So-In, C., Thakur, N. (Eds.), *Smart trends in Computing and Communications. Smart Innovation, Systems and Technologies* (Vol. 165). Singapore: Springer.
14. Xu, C., Zhang, R., Xie, M., & Yang, L. (2019, October) Network intrusion detection System as a Service on OpenStack cloud. pp. 1–2. https://doi.org/10.1109/ICNP.2019.8888078.
15. Meenakshi, Ramachandra, A. C., & Bhattacharya, S. (2020). Literature survey on log-based anomaly detection framework in cloud. In A. K. Das, J. Nayak, B. Naik, S. Dutta, & D. Pelusi (Eds.), *Computational intelligence in pattern recognition* (pp. 143–153). Singapore: Springer Singapore.
16. Park, J., & Kim, E. (2017). Correlation analysis of openstack log using machine learning techniques.
17. Dong, X., & Du, S. (2020). A cache based countermeasure against ddos attacks in xen. In Y. Liu, L. Wang, L. Zhao, & Z. Yu (Eds.), *Advances in Natural Computation, Fuzzy Systems and Knowledge Discovery* (pp. 501–508). Cham: Springer International Publishing.
18. Agrawal, V., Kotia, D., Moshirian, K., & Kim, M. (2018). Log-based cloud monitoring system for OpenStack. In: *2018 IEEE Fourth International Conference on Big Data Computing Service and Applications (BigDataService)* pp. 276–281.
19. Tsai, S. C., Liu, I. H., Lu, C. T., Chang, C. H., & Li, J. S. (2017). Defending cloud Computing environment against the challenge of DDoS attacks based on Software defined network. In: J. S. Pan, P. W. Tsai, & H. C. Huang (Eds.), Advances in intelligent Information hiding and multimedia signal processing. Smart Innovation, Systems and Technologies (Vol. 63). Cham: Springer.
20. Elia, I. A., Antunes, N., Laranjeiro, N., & Vieira, M. (2017). An analysis of OpenStack vulnerabilities. In *13th European Dependable Computing Conference (EDCC)*, pp. 129–134.

Exploring Honeycomb Interconnection Networks

Akash Nigam, Pawan Jaiswal, Megha Malhotra, Neetu Faujdar, and Akash Punhani

Abstract Mesh topology is commonly used in system on chips. This paper proposes a variant of honeycomb mesh that is inspired by the mesh and the cube topology. The proposed topology 3D honeycomb mesh has high throughput and low latency in comparison to the traditional honeycomb topology. The proposed topology is compared with the existing 2D honeycomb mesh on the three parameters and four traffic patterns. From the results obtained after the execution of the topology it was observed that the latency of the network is reduced by 1.5 times and throughput was doubled by two times in comparison to the honeycomb mesh.

Keywords Honeycomb mesh · Traffic pattern · Latency · Throughput

1 Introduction

A mesh topology is of two types, complete mesh topology and partially connected mesh topology. In complete mesh topology, every node is connected every other node with the dedicated link. The total number of the links that we can have in the complete mesh topology is given by the relation as described in the equation (1):

A. Nigam (✉) · P. Jaiswal · M. Malhotra · N. Faujdar · A. Punhani
CSE Department, ASET, Amity Univesity, Noida, Uttar Pradesh, India
e-mail: nigakash05@gmail.com

P. Jaiswal
e-mail: pawanjaiswal131@gmail.com

M. Malhotra
e-mail: meghamalhotra1617@gmail.com

N. Faujdar
e-mail: nfaujdar@amity.edu

A. Punhani
e-mail: akashpunhani@gmail.com

S. Tiwari et al. (eds.), *Proceedings of International Conference on Big Data, Machine Learning and their Applications*, Lecture Notes in Networks and Systems 150,
https://doi.org/10.1007/978-981-15-8377-3_15

$$\frac{n(n-1)}{2} \tag{1}$$

In partially connected mesh topology, there are limited number of links to the adjacent nodes as this reduces the amount of wiring required in the connection and significantly reduces the cost incurred in designing complex routers by reducing the number of ports. It is preferred to have partial connected mesh for system on chip (SOC) due to limited area on the chip for the VLSI layout. From finding it was observed that tile-based representation is found to be more efficient in terms of area utilization chip. The importance of tile-based architecture can be understood from the fact that the latest Xenon processor developed by Intel is using the variant of Mesh Interconnection network. In past years the various researches have worked on different topologies of the network on chip like torus, which basically introduces horizontal and vertical links to the basic tile-based architecture. In further attempt to improve the throughput and bisection width of the network diagonal links were introduced in the tile-based architecture .The diagonal links are feasible on the chip as multiple layers of wiring are possible on chip [1–5]. The variants with these solutions are X-mesh [6] X torus, XX torus, MDMIN and honeycomb mesh [7]. On the introduction of honeycomb mesh the interesting property is that, use degree three for every node was identified and family of honeycombs was also introduced. As the above finding has an interesting property degree. Most of the direct topologies are studied on the four popular properties. The commonly used physical properties of direct interconnect networks are degree, diameter of the network, Degree of the nodes and the lenght of the edges used in the topology [8]. Degree of 2D honeycomb mesh will be three. Edge length will remain constant in case of 2D honeycomb mesh see in Fig. 1

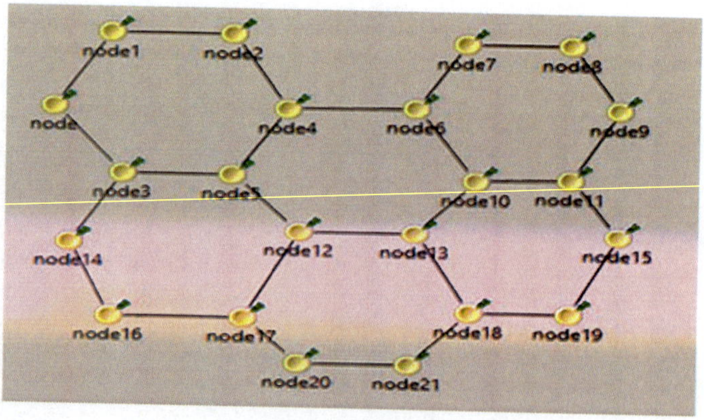

Fig. 1 Describes the 2D Honeycomb Mesh

2 Test Bed and Parameters Used

The following parameters have been used in the study of the Network on chips

2.1 Throughput

It is the most extreme rate at which the system can acknowledge the data [9, 10]

$$Th = (dr + dw)H \tag{2}$$

$$T_s = \frac{L}{W} \tag{3}$$

$$T = Th + T_s = (dr + dw)H + \frac{L}{W} \tag{4}$$

As shown in Eq. 2, where Th stands for header latency, dr is the router delay, dw stands for wire delay and H is the hop count. T_s described in the Eq. 3, basically propotion of the size of packet length to channel width. The total latency is basically the sum of header latency and Serialization latency. Our aim is to maximize the throughput [4, 11].

2.2 Latency

Latency of the network is basically accounts for the time consumption by the packets due to the various factors that are routing, queuing due to the huge flow of the packets trough certain channels, which can be avoided by diverting through different path which may be more hop counts. In our paper we have focused on latency of shortest path that may be congested as the number of packets incresses [12, 13].

2.3 Load Factor

Load Factor is basically proportion of current throughput of the topology to the maximum capacity of the networks. As the throughput is always less than 100% of bandwidth so is the load factor is always less than 1 and equal to 1 in the ideal case.

2.4 *Traffic Patterns*

It is the way toward blocking and analyzing messages therefore as to deduce data from patterns in communication, which can be performed even when the messages are encrypted. We are using different traffic pattern and they are as follows [4, 14, 15].

2.4.1 Bit Complement Traffic

The bitwise complement is basically a mathematical operation of complement of the packet starting address. However, there may be the cases where the number of nodes are not the power of 2. In order to generalize the definition for all number of nodes the relation is represented by the Eq. (5) [5].

$$D_{\mathrm{id}} = (N - 1) - S_{\mathrm{id}} \tag{5}$$

In the Eq. 5, N accounts for total number of nodes, S_{id} is the starting address and D is the destination node.

2.4.2 Uniform Distributed Traffic

Uniform traffic is basically maintains the uniformity in all the nodes. This type of traffic is very basic traffic. In this type of traffic the mathematics of uniform probability distribution is employed. In this distribution, there is a fixed probability P with which the event occurs in the experiment under consideration. To employ this idea, most of the programming languages have the random number generator which generates the traffic at the fixed rate [5]. Similarly, in OMNeT++ we can represent the relation as stated in Eq. 6. In Eq. 6 the N accounts for count of nodes in the topology.

$$\mathrm{int} d_{\mathrm{id}} = \mathrm{int\ unifrom}(1, N) - 1 \tag{6}$$

2.4.3 Tornado Traffic Pattern

The Tornado traffic actually used to tornado effect in the movement of the packets. To generate this effect the packet from source is send to the other half of the network both in the horizontal and vertical directions. As we are representing two components horizonal and vertical this leads to set of equations as stated by Eq. 7 [15].

$$\mathrm{Destination}_x = \mathrm{Source}_x + \left(0.5P + 1\right) \tag{7}$$

$$\text{Destination}_y = \text{Source}_y + \left(0.5Q + 1\right) \tag{8}$$

In Eq. 7, the P and Q represent the count of nodes in X and Y dimension respectively.

2.4.4 Neighbour Traffic

As the name suggest it is the communication between the neighbor. There can be 3 types of neighbors, that are horizontal, vertical and diagonal neighbor. In our study, we have considered diagonal neighbors. As the diagonal node will have the update in both the horizontal and vertical components the equation for the same has been represented in Eqs. 9 and 10 respectively.

$$\text{Destination}_x = \text{Source}_x + 1 \tag{9}$$

$$\text{Destination}_y = \text{Source}_y + 1 \tag{10}$$

2.5 *Power of Network*

In the some cases the throughput and latency could not provide the exact the picture of the performance. For getting the additive effect of the both parameter of the network, the power of network can be used for evaluation. Power of Network is directly proportional to the throughput and indirectly proportional to the latency of the network. The mathematical relation for the same is given by equation described below [16].

$$P_{\text{network}} = \frac{(\text{Throughtput})^{\alpha}}{\text{Latency}} \tag{11}$$

For the simplicity of the study the value of α is assumed to be one.

2.5.1 Omnet++ IDE- Software

Omnet++ is a General simulator that can be used for simulations of communication networks in various domains like wireless and ad hoc networks. It can be used for peer to peer networks and SAN (Storage Area Networks) simulations. In application area, it is used in distributed systems, multiprocessors and also in parallel computing. It is known to be open source and discrete event simulation tool. Its uses finds it in academic departments, in educational departments and by scientist in research departments for commercial institutions. It is basically C++ based simulator that attempts to fill between other simulation softwares like NS 2. This software is available in almost all platforms. Omnet++ is known to have framework representation. It does not directly provide simulation for data communication and computer networks rather it tools for writing simulations [17, 18].

3 Methodology

The simple mesh topology has already been designed earlier. With the advancement of the technologies more mesh topologies came into existence and were studied thoroughly resulting to that conclusion was made that the honeycomb mesh topology has better performance as compared to the square mesh topology. However, it was found out that 2D honeycomb mesh topology was less fault tolerant. Therefore, coming to the methodology part, we have worked on some characteristics of the mesh. The major emphasis being on efficiency, we have found out a method to increase efficiency that is increase power of the network. As we know that power of network is directly proportional to throughput, therefore increase in throughput means increase in power of the network resulting in increment of efficiency resulting in better performance than earlier. Similarly we also know that the power of the network is inversely proportional to latency, therefore we need to decrease latency to increase the throughput of the network thereby increasing the power of network to hold the high load and avoid congestion and thereby improving the performance of complete system on chip. Now when we got to know about the parameters to be studied, we started exploring the previously defined networks and the advancement that can be made in those networks to achieve our goal. Initially we studied the simple mesh topology of the network and found out some limitations. Further exploring these limitations in the depth, we moved further on to the square mesh 3D topology. Exploring the square mesh torus, we experimented on that and found out 2D honeycomb mesh giving better performance. The limitation being of less fault tolerance we started introducing new links and eventually came to proposed 3D honeycomb mesh topology. Working on proposed 3D honeycomb networks we compared the outcomes with the previous results and found out that proposed 3D honeycomb is giving better performance as compared to the rest. It had increased fault tolerance along with more efficiency.

In our proposed approach hypercube used as a 3-D variant of mesh topologyas described in Fig. 2. Modified version of honeycomb has been proposed. The proposed topology 3D honeycomb mesh has high throughput and low latency in comparison to the traditional honeycomb topology. Finally leading to the new results, we started experimenting the effects on different traffic patterns. Initially we tried one traffic pattern the result was positive. Then we moved onto the next traffic pattern and again had a positive result. After exploring two traffic patterns we wanted to

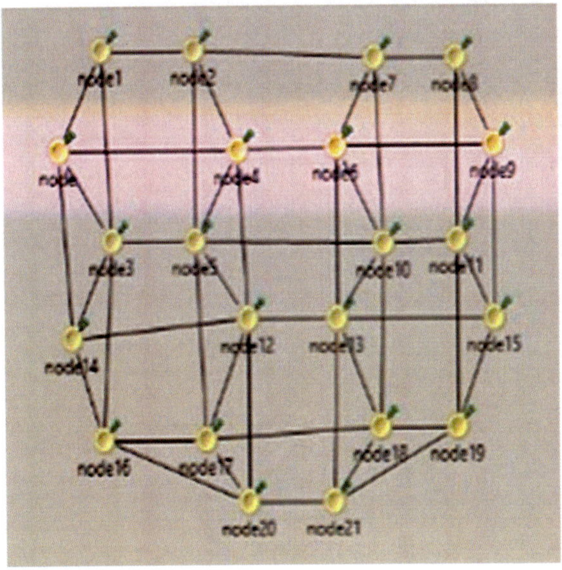

Fig. 2 Describes Proposed 3D Honeycomb

explore more and then we experimented on two more traffic patterns. In total we choose four traffic patterns and worked dedicatedly on these chosen traffic patterns.

Algorithm 1: Algorithm to Study the performace topology

Define the topology;
For each traffic pattern on different inter packet arrival delay, estimate the following;
Throughput;
Latency;
Power;
For each traffic pattern compare the values obtained from proposed 3D honeycomb with the values obtained from 2D honeycomb mesh topology.;
Combine all the findings in the form of graphs and tables for better comparison;
Calculate improvement from the findings;

The flowchart describing the complete process from the analysis and study of the performance is described by the diagram as Fig. 3

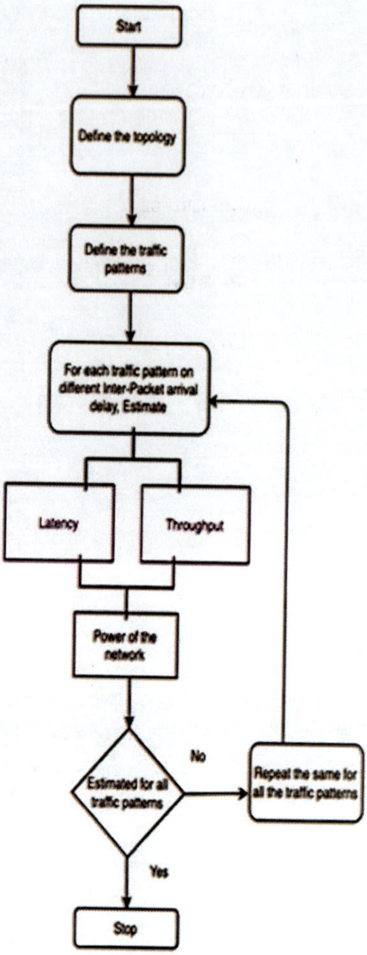

Fig. 3 Describes the flowchart of the proposed approach

4 Results and Discussions

The study of various result and discussion on different traffic it is given below:

4.1 Uniform Traffic

From Figs. 4a and b, it is observed that in the latency of 2D honeycomb mesh is more than proposed 3D honeycomb mesh and in Fig. 4b throughput of 2D honeycomb mesh is less than from proposed 3D honeycomb mesh. Even though the trend in the variation of latency is almost similar. Figure 4c is depicting the combined power of network for the two topologies on Uniform Distributed Traffic Pattern.

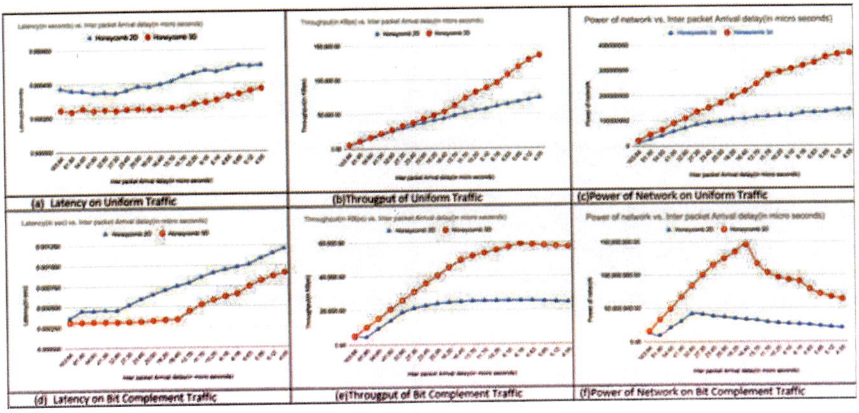

Fig. 4 Describes the perfomance comparison on uniform and bit complement traffic

4.2 *Bit Complement*

From Figs. 4d and e, it can be easily observed that latency of proposed 3D honeycomb mesh is less than 2D honeycomb mesh in Fig. 4d and e throughput of proposed3D honeycomb mesh is more than 2D honeycomb mesh. For example, on inter packet arrival delay 163.84 /micro s latency of 2D honeycomb mesh is 0.000357 s and latency of proposed 3D honeycomb mesh is 0.000306 s, therefore it's clearly observed that latency of 2D honeycomb mesh will always be more than latency of proposed 3D honeycomb mesh. Figure 4f is depicting the combined power of network for the two topologies on Bit Complement Traffic Pattern.

4.3 *Tornado Traffic*

From Figure 5a and b, it can be easily be observed that in Fig. 5a latency of 2D honeycomb mesh is more than proposed 3D honeycomb mesh and in Fig. 5b throughput of 2D honeycomb mesh is less than proposed 3D honeycomb mesh. For example, on inter packet arrival delay 81.90 /micros throughput of Proposed 3D honeycomb mesh is 10,088.17 kbps and throughput of 2D honeycomb mesh 9111.16 kbps,therefore it's clearly observed that throughput of proposed 3D honeycomb mesh will always be more than throughput of 2D honeycomb mesh. Figure 5c is depicting the combined power of network for the two topologies on Tornado Traffic Pattern.

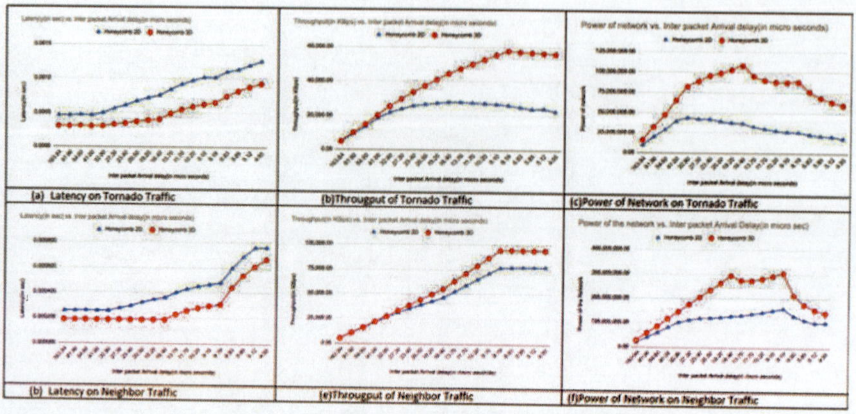

Fig. 5 Describes the perfomance comparison on tornado and neighbour traffic

4.4 *Neighbor Traffic*

From Fig. 5d and e it is observed that in neighbor traffic pattern latency of 2D honeycomb mesh will always be more than latency of proposed 3D honeycomb mesh on any of the inter packet arrival value and throughput of proposed 3D honeycomb mesh will always be more than 2D honeycomb mesh on any of the inter packet arrival value. Figure 5f is depicting the combined power of network for the two topologies on the Neighbour Traffic Pattern.

5 Conclusion

From the Pefromace observed by Figs. 4 and 5 on the different traffic patterns, it can be stated that Three Dimensional honeycomb mesh more effective in dealing the traffic in comparison to two dimensional honeycomb mesh.In all of the traffic patterns, the throughput of proposed 3D honeycomb mesh is more than 2D honeycomb mesh and latency of proposed 3D honeycomb mesh is less than 2D honeycomb mesh. We have also calculated the power of both 2D and 3D mesh topology and found that power of 3D mesh is more than that of 2D mesh. From the results, it is observed that the performance of the 3D honeycomb mesh has increased significantly.If the load factor is increases then the performance of the proposed 3D honeycomb is improved due to the extra links connected in the proposed approach.

References

1. Tavakkol, A., Moraveji, A., & Sarbazi-Azad, H. (2008). Mesh connected crossbars: A novel noc topology with scalable communication bandwidth. In *2008 IEEE International Symposium on Parallel and Distributed Processing with Applications* (pp. 319–326) IEEE.
2. Bouknight, W. J., Denenberg, S. A., McIntyre, D. E., Randall, J., Sameh, A. H., & Slotnick, D. L. (1972). The illiac iv system. *Proceedings of the IEEE, 60*(4), 369–388.
3. Wang, Y.-G., Du, H.-M., & Shen, X.-B. (2011). Topological properties and routing algorithm for semi-diagonal torus networks. *The Journal of China Universities of Posts and Telecommunications, 18*(5), 64–70.
4. Dally, W. J., & Towles, B. P. (2004). *Principles and practices of interconnection networks*. Elsevier.
5. Punhani, A., & Kumar, P., et al. (2014). A modified diagonal mesh interconnection network. In *2014 Annual IEEE India Conference (INDICON)* (pp. 1–6) IEEE.
6. Gulzari, U. A., Anjum, U. A., & Agha, S. (2015). Cross by pass-mesh architecture for on-chip communication. In *v2015 IEEE 9th International Symposium on Embedded Multicore/Many-core Systems-on-Chip* (pp. 267–274). IEEE.
7. Stojmenovic, I. (1997). Honeycomb networks: Topological properties and communication algorithms. *IEEE Transactions on parallel and distributed systems, 8*(10), 1036–1042.
8. Punhani, A., Kumar, P., et al. (2017). Optimal extra links placement in mesh interconnection network using improved environmental adaptation method. *Journal of Intelligent and Fuzzy Systems, 32*(5), 3285–3295.
9. Punhani, A., Kumar, P., & Nitin, N. (2017). Level based routing using dynamic programming for 2d mesh. *Cybernetics and information Technologies, 17*(2), 73–82.
10. Grot, B., Hestness, J., Keckler, S. W., & Mutlu, O. (2009). Express cube topologies for on-chip interconnects. In *2009 IEEE 15th International Symposium on High Performance Computer Architecture* (pp. 163–174) IEEE. 2009.
11. Punhani, A., Kumar, P., & Nitin, N. (2017). Routing for center concentrated mesh. *International Journal of Intelligent Engineering and Systems, 10*(1). https://doi.org/10.22266/ijies2017.0228.10.
12. Glass, C. J., & Ni, L. M. (1992). The turn model for adaptive routing. *ACM Sigarch Computer Architecture News, 20*(2), 278–287.
13. Punhani, A., Kumar, P., & Nitin, N. (2017). Three-dimensional topology based on modified diagonal mesh interconnection network. *Journal of Telecommunication Electronic and Computer Engineering (JTEC), 9*(3–6), 1–6.
14. Chauhan, D. S., et al. (2012). Comparative analysis of traffic patterns on k-ary n-tree using adaptive algorithms based on burton normal form. *The Journal of Supercomputing, 59*(2), 569–588.
15. Chauhan, A., Punhani, A., et al. (2015). Comparative analysis of traffic patterns on centre connected topologies based on burton normal form. In *TENCON 2015-2015 IEEE Region 10 Conference* (pp. 1–6). IEEE.
16. Punhani, A., Kumar, P., et al. (2017). A horizontal fat mesh interconnection network. In *2017 Tenth International Conference on Contemporary Computing (IC3)* (pp. 1–5) IEEE.
17. Varga, A. (1999) Using the omnet++ discrete event simulation system in education. *IEEE Transactions on Education, 42*(4), 11–pp.
18. Varga, A. (2010). Omnet++. In *Modeling and tools for network simulation* (pp. 35–59). Springer.

Pharmaceutical Supply Chain Management Blockchain

Neeraj Kumar Yadav, Harivans Pratap Singh, Rajat Verma, and Kush Kumar Singh

Abstract This paper gives a glimpse and a precise summary of this emerging technology that is blockchain and how we are going to utilize it to form a tamper-proof system to remove fake products/drugs in pharmaceutical industries. Study of various reports shows that fake products entering the pharma supply chain are a serious concern. By the use of blockchain across the whole supply chain, many problems of the distributors and manufacturer and consumer will be solved that includes—(i) robust system, (ii) improved transparency, (iii) immutable, (iv) authentic, (v) reliable, and (vi) secure. Overall, here, we are showing a raw outlet on how a blockchain transaction over a supply chain management system will be far better as compared to the conventional one.

Keywords Blockchain · Pharmaceutical supply chain · Smart contracts · Anti-counterfeit

1 Introduction

1.1 Blockchain

Blockchain is a distributed database. Its main USP has to be that it provides a tamper-proof system. It saves or stores the blocks that contains the transactions. They all are connected together in crypto-graphical method over a peer-to-peer network. With the

N. K. Yadav · H. P. Singh (✉) · R. Verma · K. K. Singh
ABES Engineering College, Ghaziabad, India
e-mail: harivans.singh@abes.ac.in

N. K. Yadav
e-mail: neeraj.16bcs1094@abes.ac.in

R. Verma
e-mail: rajat.16bcs1158@abes.ac.in

K. K. Singh
e-mail: kush.16bcs1168@abes.ac.in

S. Tiwari et al. (eds.), *Proceedings of International Conference on Big Data, Machine Learning and their Applications*, Lecture Notes in Networks and Systems 150,
https://doi.org/10.1007/978-981-15-8377-3_16

use of this build up design of blockchain, the users can have always the updated or as we say the latest copy of transaction every time a transaction takes place. Every time when a transaction occurs in this chain or matrix, it gets updated over whole system (network). This is possible because all the other participants of the network receives a copy of transaction. With this network in implementation, the role of the big player of any network or Web that was the central authority will be highly diminished.

1.2 Supply Chain

A supply chain is nothing but it shows how things are moving from one place to another by storing or we can say by creating checkpoints. The scope of the supply chain has shown a tremendous shift in terms of geographical locations in the past few decades. The role of this chain has changed in a dramatic way, and now the mishappening is very rare because it validates the transactions on various steps via several mechanisms. Here, the main entity in this whole scenario is buyer and seller. They are the ones who have to be satisfied in the whole chain. So if any entity in this whole process performs illegal activity, it becomes hard for everyone in the entire system.

1.3 Smart Contracts

Smart contract is inspired by the agreement phenomenon. It is a form of digital agreement that enhances the trust within a blockchain network. Since it is based on agreement, hence there is no need of mediator or as we call the third party. Smart contract generally contains code that ensures agreement between the nodes. This whole is executed over the network by its own. Smart contract is not separate from the transaction, and it is actually a part of it. Now, we are developing a new mechanism; hence, it should remove the hardship of the conventional one that includes speed and security like features. Every smart contract has a contract hash and a contract address, and this thing played a great role in the further process of getting back the contract in a very safe and tamper-proof manner.

2 Objective

The objective here is very glassy, we have a conventional method, and we are very much aware of its flaws. All the things that were performed on the traditional supply chain management are now going to be shifted on this new platform of blockchain. The feature provided by blockchain (data privacy and data accessibility) proves to be the best alternative for the current platform on which the supply chain is working.

This feature is going to be a boon in terms of time management and precision. The feature that blockchain is going to offer on existing supply chain is as follows

- It increases transparency of the system,
- Traceability,
- Authorization,
- Enhances the security,
- Immutable.

2.1 Problem Definition

In current scenario, there are certain problems:

Shipment visibility,
Slow process & prone paperwork,
Mutable and invalid source,
Lack of coordination.

2.2 Problem Formulation

Accuracy and precision in the whole chain at any checkpoints created.

- If tampering occurs at any stage, each entity should have real-time access.
- Visibility features should be there in the whole chain.
- Traceability means that we can trace back to the original source.
- All parties should be at a real-time connected to each other.

2.3 Motivation

The pharma industry is a very critical industry for the human kind, and one of the greatest threats we are facing nowadays in this industry is tampering with products. Though there are various kinds of tampering prevailing in society, this one is unavoidable because it is a great threat to human life. We are seeing the current supply chain management system of the pharma industry, but despite all of its advancement and security features there exists a great amount of loophole in this system, which we all witness from time to time. Now if you look at this blockchain technology, we can see that it provides a mechanism through which we can remove most of the loopholes persisting today in this network. Hence, we are focusing on shifting the conventional supply chain to this new blockchain platform.

The rise of this distributed network and how the various industries have shown positive response toward its acceptance and its applications over the year can be seen though Gartner Hype Cycle. Not only this, it also gives us a brief idea how

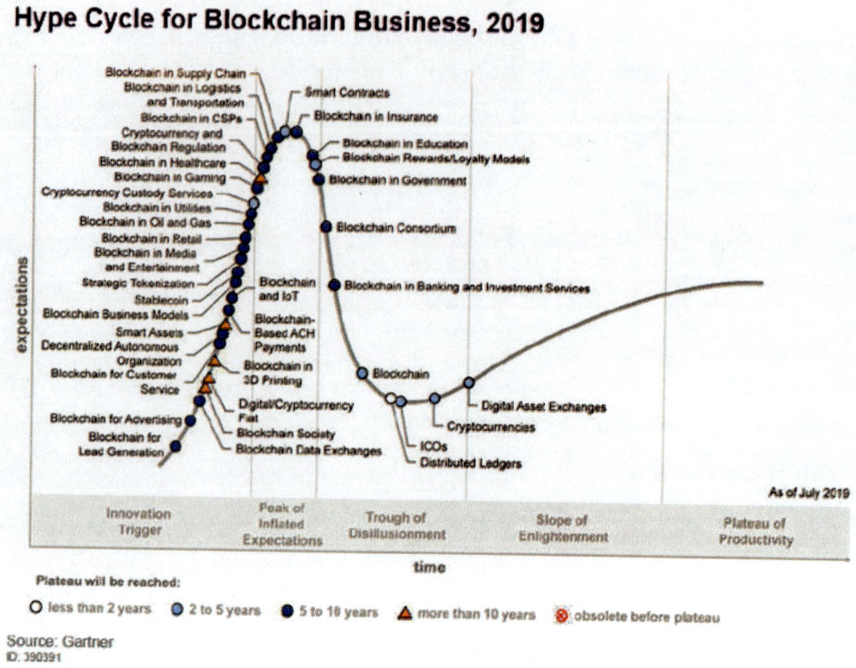

Fig. 1 Hype Cycle for Blockchain Technologies, 2019 [1]

this technology can be used to solve various real-life issues. This also gives us an idea that what future beholds for this amazing technology. We can get an idea of its evolution provides (Fig. 1).

3 Related Work

Blockchain technology has a dramatic beginning, and initially, it does not come in the knowledge as a great platform. There is a lot of work being done within a period of time on this. The initial phases were all about cryptocurrency and terms related to that, and we all became aware of bitcoin (a popular cryptocurrency). Things changed a lot for blockchain after the introduction of smart contracts. Now, blockchain became capable of various other things, and its scope increased a lot. Now, we started thinking of applying it in various fields/domains such as medical and supply chain. Benchoufi and Ravaud have introduced the use of blockchain for improving the clinical research quality [2]. They have discussed overall the use of blockchain in healthcare and medicine, but no explanation is given about the use of blockchain in the supply chain of drugs. Healthcare is a very critical field among all the files; hence, there is a great risk of attacks in this field [3]. M. Mettler has discussed the same phenomenon that is shifting the pharma supply chain to blockchain, but it lacks a lot on how it is going to be implemented. There are a whole lot of journals and papers published regarding it that one may consult [4].

4 Literature Review

So, it all started when Satoshi Nakamoto termed how we can use a digital system to perform peer-to-peer payment and not only this someone also lost a job here also that was the intermediate third party whose job was to validate our transaction. And there is a coincidence which related to this was that after the introduction of this technology in the above-mentioned paper the world witnessed the very first bitcoin client. Here, we are going to discuss only supply chain management and the flaws existing in the current supply chain management despite advancements like use of vouchers and all.

And how the new technology like blockchain and IOT can be proved as a boon in such scenarios.

There are various assumptions that if we see world annual report that was conducted a market research firm there they had done various predictions regarding its future growth rate in the coming years by 2022. They estimated it to be 6.3% CAGR globally. Yet, we are here facing old problems in this critical field like fake products and tampering in the chain of supply. As stated above, how fake products or tampering is not such a big issue in other industries, and if they are then it is limited only upto finance and economy. But here in the field of pharma industry, fake products can cost the life of an individual.

So as we are taking about its bright future, it is also notable that this whole forged drugs market values in range of 75 dollar to 200 dollars in a period of one year and not only this it is summed nearly 50 percentage of drugs used in some poor countries where people have less incomes. This whole forged drug market has a impact on all either it be across low-income countries or high-income countries like US and Europe. Although the statistics vary in these countries, here impact is of more concern. What is more dangerous is the fact that there exist some countries where its percentage is up to 20. This is really a very horrific fact and should be looked up immediately.

A report of 2006 states that forged drugs in India are at top, and there are two countries after that Thailand and China. This is a serious concern. After reviewing the situation in India, we can say that strong measures and changes are required in India to counter the situation. Introduction of new technology like blockchain in this corrupt conventional method is the best alternative available today.

5 Survey on Blockchain

5.1 *Bitcoin: A Peer-To-Peer Electronic Cash System by Satoshi Nakamoto*

Here, the whole focus is on the introduction of a new system of cash (electronic) that will remove the involvement of the central authority which means no third-party intervention. The sole purpose of that central authority can be seen as an intervention only, and it not only increases the cost but is also a great challenge to perform small transactions. This was not even the biggest challenge. The main challenge was double spending. Now after the removal of the third party, there is no problem of double spending [5]. There is a creation of chain in this process with the timestamp values. The privacy which is offered from this system is hard to obtain in the conventional banking system.

5.2 *It Increases Trust Among the Participants Because the Whole Network is Transparent in Terms of Functioning.*

Supply chain management system is a very complex system due to the involvement of various locations that are far distant. And not only this very critical operation is performed at those locations. Hence, it is obvious that trust issues will arise. The main reason that we are discussing blockchain here is the features it offer which are traceability and immutability. The end user or the mid user can trace back the product up to its very origin [6]. So the end user can easily trace his product, this provides him a lot of satisfaction, and also the authorization feature gives a sense of security that only authorized user have accessed the system.

5.3 *Blockchains Everywhere—A Use-Case of Blockchains in the Pharma Supply Chain by Thomas Bocek, Bruno B. Rodrigues, Tim Strasser, Burkhard Stiller*

This paper is different from others in various terms, and the main is implementation. Here, it has been shown in depth about the implementation part. They have also used the IOT feature in order to enhance the security feature. Also they have focused on smart contracts also. Their system comprises three things that includes the front end that can be implemented on any framework. And in the backend, they have an Ethereum network and also include the server. They have also used an IOT device.

The whole scenario is that they want to ensure the temperature of the drugs throughout the whole supply chain. So that end user can be satisfied with the state of the drugs [7].

6 Proposed Methodology

The aim of this is to achieve a system of supply chain management using blockchain that can remove the shortcomings of the existing supply chain management.

The proposed methodology is going to resolve the problem faced in the earlier supply chain management starting from the first step that is the drug design. There are going to be mainly five role such as designer who maintain drug design, manufacturer who manufacture drugs, distributor who distribute drug to retailers, and the end customer. Also there is a regulator who keep an eye on all such above role activity. (It approves drugs and the user role.)

Step 1. Drug designer designs the drug and uploads it on our blockchain network.

After the designer designs the drugs, he uploads it on the network, a unique hash code is generated, and we can perform the upcoming transactions using this hash code.

Step 2. Manufacturer purchases the drug design and that transaction is also recorded on the network.

Manufacturer performs various roles here firstly he uploads the environment in which drugs are created. After the manufacturing is done, he then updates the price of the product on the network while shipping the product.

Step 3. Distributor's role is to distribute the product to the retailer and upload the transaction on the network.

Step 4. Retailer takes the product from the distributor and updates the transaction on the network.

Step 5. End user can check his product or trace back his product to its very origin using the product ID.

6.1 Architecture Diagram of the Proposed System

See Fig. 2.

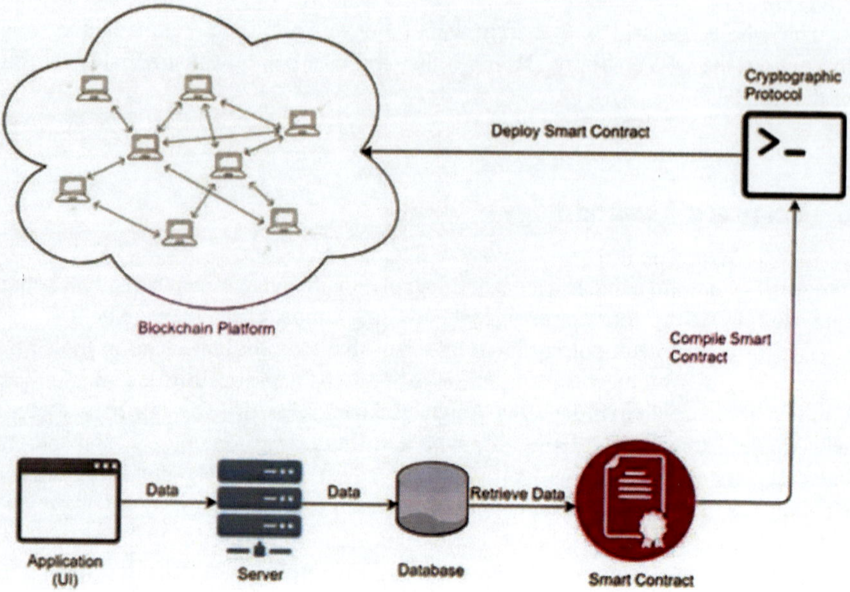

Fig. 2 Architecture diagram [8]

6.2 *Workflow Diagram of the Proposed System*

See Fig. 3.

7 Implementation

The functioning of the suggested system can be divided into two categories: (i) the entity involved into the supply chain and (ii) end user.

(i) **The participants involved into the supply chain network.**
The system interface used are generally the Ethereum blockchain; to make it easily accessible, we will use decentralized application (dapp); in general, we can say a Web site through which we can access our blockchain network; hence, the user will not face depth level difficulty as they can easily add the transaction and see the transaction over the network using proper user authorization. This transaction once added to the network are now immutable [10].

(ii) **End user**
After transferring it to blockchain platform, the goal which we had in the initial phase after the problem evaluation was of tampering which was due to involvement of various participants in the SCM, and now it is solved by the authorization

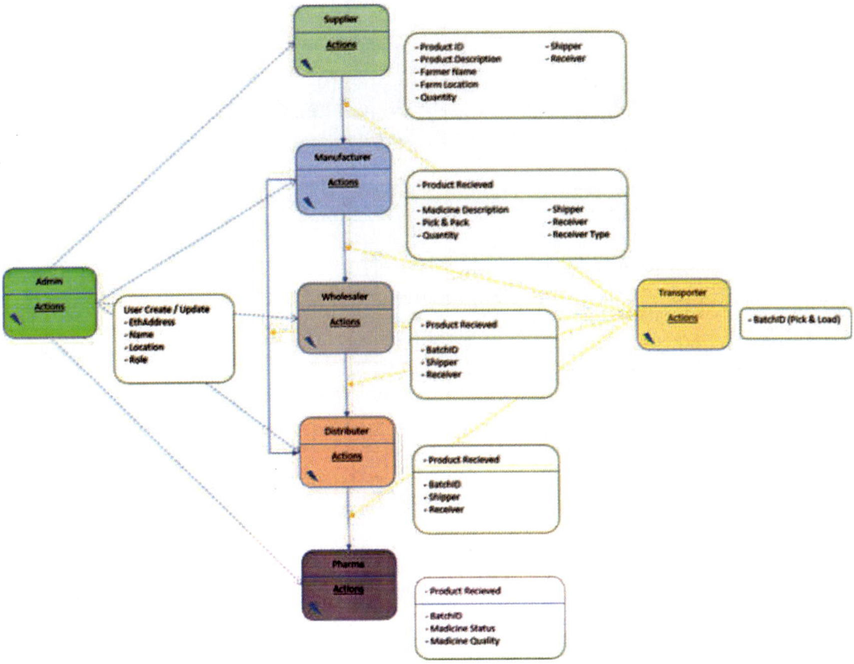

Fig. 3 Workflow diagram [9]

property. Due to the hash code generated, the end user is now able to track back the product to its origin, and he can be satisfied with his product [11].

8 Results and Discussion

Here, we discussed how critical pharma industry is for human life and how it is on constant risk of fake products/drugs. To overcome this shortcoming of the conventional supply chain management system of the pharma industry, we can use blockchain. Currently, it is the only platform available that can offer such a great security feature [12].

Blockchain seems to have very promising future due to the features that it offers. It can be seen that blockchain has the solution to many of the problems that IT industry is facing like security. And not only IT industry, various other industries such as pharma as we discussed here or banking sector are also facing issues like security and transparency. And as we see, many industries are already shifting their conventional platform to blockchain [13].

References

1. Gartner Hype Cycle for Blockchain Technologies, 2019.
2. Qi, X., Emmanuel, B. S., Kwame, O., Jianbin, G., Xiaojiang, D., & Mohsen, G. (2017). MeDShare: Trust-less medical data sharing. Among Cloud Service Providers via Blockchain. IEEE Access.
3. Asaph, A., Ariel, E., Thiago V., & Andrew, L. (2016). MedRec: Using blockchain for medical data access and permission management. In *2nd International Conference on Open and Big Data*, Cambridge, MA, 02139, USA.
4. Mettler, M. Blockchain technology in healthcare: The revolution starts here.
5. Nakamoto, S. (2008). Bitcoin: A peer-to-peer electronic cash system.
6. Hua, A. V., & Notland, J. S. Blockchain enabled trust & transparency in supply chains.
7. Bocek, T., Rodrigues, B. B., Strasser, T., & Stiller, B. Blockchains everywhere—A use-case of blockchains in the pharma supply-chain.
8. https://eudl.eu/pdf/10.4108/eai.11-4-2018.154550
9. https://connexiontechnologies.net/lustre-keria-solde.html
10. Casino, F., Dasaklis, T. K., & Patsakis, C. A systematic literature review of blockchain-based applications: Current status, classification and open issues.
11. Johansen, S. K. A comprehensive literature review on the blockchain as a technological enabler for innovation.
12. Wigley, B. The future is decentralised.
13. Peck, M. Reinforcing the links of the blockchain.

The Need of Smart Guidance Systems for Blind People in the World

Maneesh Vishwakarma, Harivans Pratap Singh, Nitin Kumar, and Manan Arora

Abstract The number of people with visual disability has reached an alarming number of 285 million. Out of these 285 million, 39 million people are facing complete blindness and 246 million suffer from low vision. This low vision can be due to certain reasons. One of the factors of low vision can be the age of the individual (World Health Organization in Global data on visual impairments 2010. World Health Organization, Geneva (2012), [1]). Out of 39 million, 15% of the people are below the age of 50. A better system is required to help all the people to let them lead a better life with better assistance. Despite of the invention of recent assisting systems, we still lack in helping blind people with assisting devices so that they can explore the world more easily. The devices which are currently in the market either have a short range and also do not have enough features to help blind people in every situation. We aim to develop a system which will help blind people by providing assistance against obstacles in their way and alerting them about the obstacles so as they can move easily without the use of sticks.

Keywords BlindMate · Metrobus · ARIANNA app · EyeMate · AECA · Algorithm · NavCog · RFID

M. Vishwakarma (✉) · H. P. Singh (✉) · N. Kumar · M. Arora
ABES Engineering College, Ghaziabad, India
e-mail: maneesh.16bcs1153@abes.ac.in

H. P. Singh
e-mail: harivans.singh@abes.ac.in

N. Kumar
e-mail: nitin.16bcs1065@abes.ac.in

M. Arora
e-mail: manan.16bcs1034@abes.ac.in

S. Tiwari et al. (eds.), *Proceedings of International Conference on Big Data, Machine Learning and their Applications*, Lecture Notes in Networks and Systems 150,
https://doi.org/10.1007/978-981-15-8377-3_17

1 Introduction

We are advancing in every technology in the world with the advanced requirements and resources. New technologies are being developed everyday and new inventions are been performed in order to provide the ease of living to all the people in the world. In this era of discoveries and inventions, some emphasis needs to be laid for the blind people to ease their lives more. They are unable to explore the world like others and face problems with daily activities such as going from one place to other, recognizing objects and the main problem to them is the obstacles they encounter while moving on the road. This is the time when there must be some development with the instruments or gadgets available for the blind people. There are smart sticks and many other gadgets in the world but they have short range to monitor. The advanced gadgets must be induced with long range monitoring and must provide information to the blind person about the obstacles he/she must have to dodge in order to walk safely and move from one place to another.

2 Background Details and Related Work

Sandra Mau et al. [2] proposed a method of using RFID tags. These tags would be carried by the user as well as would be implanted on the building for travel assistance. This project resulted in the development of low-cost travel assistance for the blind called BlindMate.

Dunai et al. [3] presented a new prototype for being used as a travel aid for the blind people. The prototype uses stereo cameras and a computer for calculation. The primary aim of system is to detect objects from surrounding environment. After detection, the result is converted to acoustic signals. The experiments provided satisfactory results and can be used by blind people to navigate safely.

In an article by Simon Hill [4], he discussed about various devices which were invented for the help of blind people. These devices invented had the very purpose of letting blind people leading an independent yet secure life. These devices include Assisted Vision Smart Glasses, AI glasses, ARIANNA app for vision support and path detection and other devices such as Finger Reader and Braille e-book reader to let them read and lead their life easily.

Siddiqur Rahman Tanveer et al. [5] implemented a system as an approach for navigating blind people and tracking by flexible architecture that could be used for blind mobility. An android application can be deployed using this assistive device with Eyemate can help them moving without help and is equiped with emergency voice call help feature. Also, blind person's live location can also be tracked with the help of BlindTracker application.

Table 1 Brief introduction to technologies being used currently for blind people

Technology	Primary focus
Blind Aid [2]	An cost-efficient assistance project which tells the path through RFID tags
Device for Metrobus	This project helps visually impaired people get oriented and aligned in a metrobus system
ARIANNA app [5]	It solves the problem for indoor environment navigation
AI Glasses [5]	This combines AI, ultrasound techniques, and computational geometry to create useful aid for visually impaired
EyeMate [6]	This helps blind people to travel without aid. This can also be used to track blind people
NavCog [7]	It stands for Navigational Cognitive. It is a smartphone-based mobility aid to help visually impaired people
NavCog [3]	It is an advanced version of NavCog and it helps blind people by providing indoor navigation support

Dragan Ahmetovic et al. [6] proposed a smartphone mobility aid which is designed to help visually impaired people. This aid is much helpful in navigation in completely unfamiliar places. To balance accuracy, a localization algorithm was designed and implemented. The evaluation conducted results in the robustness of the localization algorithm and approved for its use in unfamiliar conditions and places.

Elmannai [7] evaluated a system which showed 100% satisfactorily results. The features are the backbone for designing a device to provide services to blind people. The evaluation performed by them led the way to design more devices that ensures safety and independent mobility to visually impaired people.

Cheng et al. [8] implemented the AECA algorithm. They used an approach on a wearable device to help visually impaired people cross roads. The algorithm proposed in this paper for detection shows better results as compared to conventional algorithms in many different scenarios with the result being accurate by more than 80% on dataset.

Kaur et al. [9] developed a design for the electronic travel aid kit thus reducing the cost of assisting devices and making it more accurate and efficient. Since all the existing algorithms and approaches were not delivering best results. Also, the pre-existing algorithms were unable to detect hanging articles in the frame.

The description of the technologies and projects being used in the market to help blind people currently are mentioned in Table 1.

3 Problem Statement

Despite the development of invention of a large number of algorithms, applications and systems, the blind people are still far away from such featured support systems that can help them lead a safe and better life. We, in this situation, are trying to develop a system which would be a personal assistant for blind people. The system will perform obstacle detection and will alert the blind person in case of presence of any obstacle. We aim to make the output audible by conveying the output in the ears of blind people with the help of earphones or headphones to make them more attentive on road and help them to explore this world without aiding sticks. In indoor facility, we aim to have the high accuracy and precision in order to secure them from any obstacle (Fig. 1).

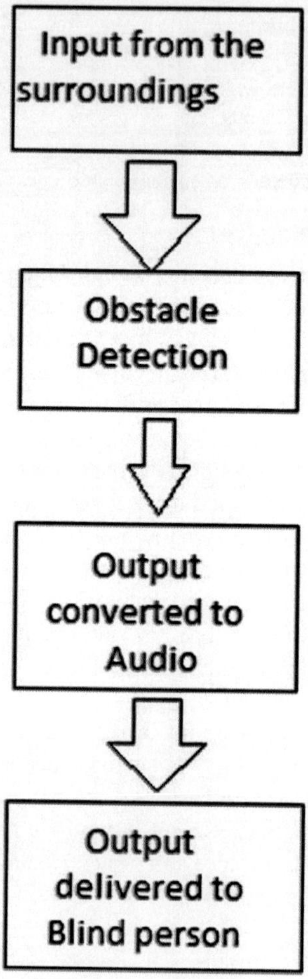

Fig. 1 Proposed flow of data

4 Conclusions

The current deployed systems in the market for the help of blind people are currently not able to cover every aspect or every situation everyone faces daily. The world still requires a strong solution for blind people so that they can also pace up in the development of the world. Machine learning and artificial intelligence are two best technologies that can be used to obtain the best and most accurate solution for the blind people to assist them in crucial conditions and to have a more impacting support as compared to aiding sticks and current deployed solutions for the blind people.

References

1. World Health Organization. (2012). *Global data on visual impairments 2010*. Geneva: World Health Organization.
2. Mau, S., et al. (2008). *Blindaid: An electronic travel aid for the blind.* Robotics Institute Report, CMU-RI-TR-07-39.
3. Dunai, L., et al. (2010). Real-time assistance prototype—A new navigation aid for blind people. In *IECON 2010–36th Annual Conference on IEEE Industrial Electronics Society*. IEEE.
4. Hill, S. (2014, June 20). *5 amazing gadgets that are helping the blind see.* Digital Trends.
5. Tanveer, W., Siddiqur Rahman, Md., Hashem, M. M. A., & Kowsar Hossain, Md. (2015). Android assistant EyeMate for blind and blind tracker. In *2015 18th International Conference on Computer and Information Technology (ICCIT)*. IEEE.
6. Ahmetovic, D., et al. (2016). NavCog: a navigational cognitive assistant for the blind. In *Proceedings of the 18th International Conference on Human-Computer Interaction with Mobile Devices and Services*. ACM.
7. Elmannai, W., & Elleithy, K. (2017). Sensor-based assistive devices for visually-impaired people: Current status, challenges, and future directions. *Sensors, 17*(3), 565.
8. Cheng, R., et al. Crosswalk navigation for people with visual impairments on a wearable device. *Journal of Electronic Imaging, 26*(5), 053025.
9. Kaur, N., et al. (2017). Device for blind people navigation. *International Journal of New Technology and Research, 3*(4).

Behavioral Cloning for Self-driving Cars Using Deep Learning

Rajesh Tripathi, Sakshi Vyas, and Amit Tewari

Abstract We are presenting a model for self-driving car simulator provided by Udacity as an open-source simulator. We will then use the simulator to create our own training data for our model which will be driving a car through training mode on its track in the simulator. We are taking images at each instance of the drive. These images are used as a training dataset, and the labels are steering angle for each specific image at that instance. We will then input all those images to our Nvidia's convolutional neural network model and allow it to learn how to drive autonomously by learning from our behavior as the manual driver. Our main variable is the steering angle which our model learns to adjust at any given instance. Now, as our model is perfectly trained, we use autonomous mode to find the performance of our model by driving the car autonomously.

Keywords Behavioral cloning · Nvidia's CNN · Data augmentation · Continuous regression

1 Introduction

Self-driving cars use cloning of behavior by taking various aspects of manual training of a driver and try to make that training more efficient with the help of tuning of hyperparameters. Use of a good convolutional neural network [5] helped us to learn all the low-level features of our training set. Our center of focus is to clone the behavior of manual driving car by making use of an open-source simulator to extract the training dataset comprising of images as the features and steering angles as the labels. Since we do not have had enough data to train, we will have to augment this

R. Tripathi · S. Vyas (✉) · A. Tewari
Motilal Nehru National Institute of Technology, Allahabad, India
e-mail: sakshi2401vy@gmail.com

R. Tripathi
e-mail: rajeshtcsed@mnnit.ac.in

A. Tewari
e-mail: tiwariamit6250@gmail.com

S. Tiwari et al. (eds.), *Proceedings of International Conference on Big Data, Machine Learning and their Applications*, Lecture Notes in Networks and Systems 150,
https://doi.org/10.1007/978-981-15-8377-3_18

data by using different data augmentation [8] techniques, and then, we finally use Nvidia's CNN to train our model, and then, we are using autonomous mode to find the performance of our model. As we have already split our data into training dataset and validation dataset, then we finally train our model on training data, tune all our hyperparameters on the validation set [12] and then evaluate the final performance on test data. We are taking a sample drive of three laps with the left turn and three laps with the right turn for getting rid of skewness. Based on the training set, we are getting the training loss of 4.17% and validation loss of 3.02%. We are using the batch generator by taking help of keras data augmentation to train our model using fit_generator [11] So that we get our improved model by cutting off some of the dropout layers [2]. Through this, we are getting an instance steering angle of 0.34218859, the throttle of 0.09456000 and the speed of 9.0544.

2 Preprocessing

2.1 *Sampling*

As our data is of many variations, we need to have some preprocessing done to get the relevant data to train our model [10]. Initially, as our data is more centric toward the center, but the problem is if 90% of our data is like this, then we are facing problems in taking sharp turns. So, to get rid of this problem, we cut off the samples (steering angles) whose frequency is greater than 200 for getting the data without any bias, but the cutoff [7] is not direct as we can lose the information at the end of the track, so we are shuffling the data and then cutting off the samples. Now, as we are having the samples, our next step is to get the list of images and the steering angles [9] corresponding to those images, but to input all these images and the steering angle into our model, we need to convert into numpy array using np.asarray() (Fig. 1).

2.2 *Train-Test Split*

Now, in this step, we need to split the data that we collected from our simulator into training and validation dataset, and each image for each set of data corresponds to a specific steering angle. So, a general 20% split is best for splitting the data before going to the next step as it is very important to check whether the split data follows the same trend or not [6]. The histogram in Fig. 2 for training data and for testing data validates the same trend for steering angles which is perfectly balanced.

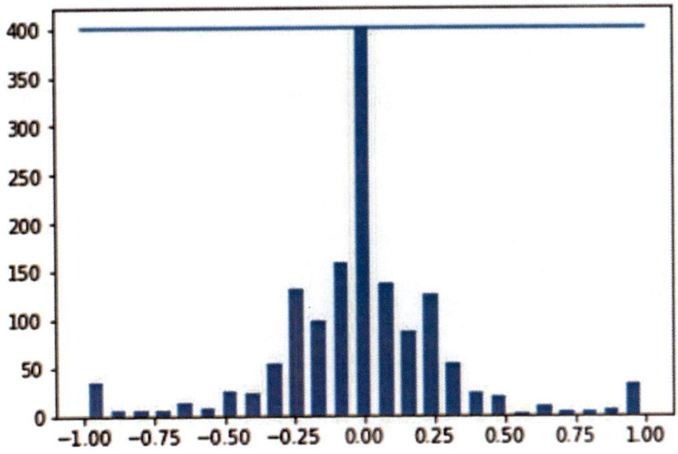

Fig. 1 Sampling

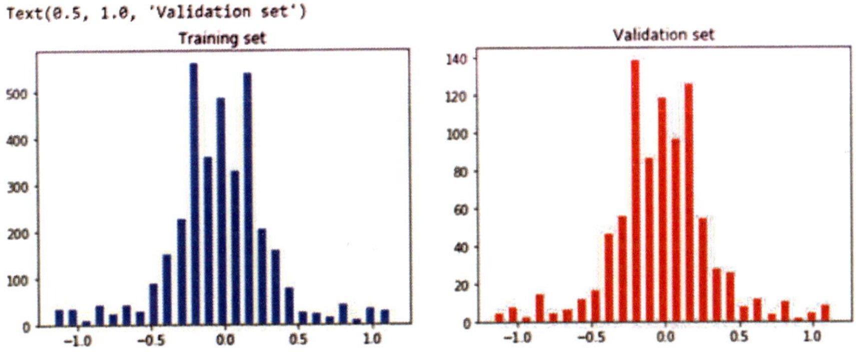

Fig. 2 Training and validation set

2.3 *Image Processing*

In this step first, we need to convert RGB channel to the YUV (luminance, chrominance) channel, and usage of YUV channel is taken as they can reduce the resolution of the U and V channels while keeping Y at full resolution because luminance is more important than the color. If you can reduce the resolution of U and V in a way that is compatible with convolutional nets [11], your net should be half the size and therefore twice as fast, without much loss of accuracy. Now, we apply the Gaussian blur using Gaussian kernel convolution [2] to the image for smoothing the image and reducing the noise of the image such that it helps our model to extract only important features. Now, we need to resize the preprocessed image to convert it into the same dimension as Nvidia's model wants it to be, and as a final step, we need to normalize the image by dividing it by 255 (Fig. 3).

Fig. 3 Original versus preprocessed image

3 Defining the Model

As our dataset of dimension (200,66) is more complex as compared to other datasets like mnist of dimension (28,28), traffic signs (32,32) are used for image classification. Therefore, we used a more capable and complex structure like Nvidia's model which gives us the steering angle which is more of regression type problem.

3.1 Nvidia's Model

First, we have an input plane of size (66,200,3) with normalization as we have already normalized the input and we skipped this step. Now, input this into an convolutional layer, and this layer uses 24 filters along with a kernel of 5×5, with a stride of 2×2 as now we have a larger image to process, and we are taking relu as the activation function. Secondly, the layer is also a convolutional layer with 36 filters with a kernel size of 5×5, with a stride of 2×2 and relu as the activation function [2].

The third layer is also a convolutional layer with 48 filters, with a kernel size of 5×5, with a stride of 2×2 and relu as the activation function. The fourth layer is also a convolutional layer with 64 filters with kernel size of 3×3 and relu as the activation function. The fifth layer is also a convolutional layer with 64 filters with kernel size of 3×3 and relu as the activation function. The next layer is the flattening layer which takes input the output of the convolutional layer and flattens it into a 1-D array which feeds up into fully connected layer with 1152 pixels. Now, we have three dense layers with 100, 50 and 10 nodes, respectively, and the final output layer has 1 node as output which predicts the steering angle of the self-driving car. At the compilation, we use error metric [1] as mean squared error (MSE) and Adam as optimizer with learning rate as pow(10, −4) as it is a regression problem. And for the better performance, we add two dropout layers such that randomly, 50% of the nodes are off for getting the better performance (Fig. 4).

Layer (type)	Output Shape	Param #
conv2d_1 (Conv2D)	(None, 31, 98, 24)	1824
conv2d_2 (Conv2D)	(None, 14, 47, 36)	21636
conv2d_3 (Conv2D)	(None, 5, 22, 48)	43248
conv2d_4 (Conv2D)	(None, 3, 20, 64)	27712
conv2d_5 (Conv2D)	(None, 1, 18, 64)	36928
flatten_1 (Flatten)	(None, 1152)	0
dense_1 (Dense)	(None, 100)	115300
dense_2 (Dense)	(None, 50)	5050
dense_3 (Dense)	(None, 10)	510
dense_4 (Dense)	(None, 1)	11

Total params: 252,219
Trainable params: 252,219
Non-trainable params: 0

Fig. 4 Nvidia's model

3.2 *Hyperparameters for Nvidia's Mode*

We are using Adam as the optimizer with a learning rate of (le-4), with 150 epochs and batch size of 100 with shuffling and relu as the activation function.

4 Problems with the Mode

As the epochs are continuously running but the training and the validation loss are not converging and we can see that our validation loss is quite consistent which is an indication of the poorly trained model [12], there are some problems which can be problematic in driving the car appropriately.

4.1 Activation Function

The gradient of the relu activation function in the positive region is 1, but in the negative region is zero; it means that if a node gets an input as negative number, it will return a 0; however, since the gradient at this point is 0, so the weight of this node will never change, as back-propagation uses the gradient to change the weights of the nodes, so no learning happens [4]; if enough relu dies, then the loss of model remains stagnant and never decreases.

4.2 Overfitting

As we can see that our model is overfitted by seeing the validation loss and the training loss which makes our model not able to generalize on the new data due to memorizing the pattern. As we can see in the track that when our car is approaching the river on the side of the road in the track, it is not able to navigate from there, and our car crashes. So there is a predominant factor which is lack of training data.

4.3 Small Dataset

Due to small dataset, our model is not able to generalize well on the new track and this causes a big problem.

5 Improvement in the Model

As we saw above that there are some problems that take place with this model, so in this section, we will focus on the improvements regarding the model.

5.1 Activation Function

So, we are using elu(leaky-relu) as an activation function [8], and it is similar to behavior in the positive region, but it has the nonzero value in the negative region, and it means that elu function has always chance to recover and fix its weight and parameter to decrease the error and rectify died relu problem [4].

Fig. 5 Original versus zoomed image

5.2 *Overfitting*

We can eliminate this problem by using the dropout layers [5] in between the dense layers of the fully connected dense layers. It randomly cuts off 50% nodes in each epoch such that it helps to generalize the model instead of memorizing the pattern.

5.3 *Data Augmentation*

As our model has very less amount of training images, we need different data augmentation techniques for generating new images from our existing dataset.

It is done by applying different transformations [3] to our current set of images which results in a brand new set of images. This helps our model to learn efficiently as now our model has more data to learn with it [8]. So, for this model, I used my own customized data generator consisting of different data augmentation techniques using imgaug library

5.3.1 Zooming

It helps our model to get a closer look at some of the features in the image, which improves the performance of our model and helps our model to generalize (Fig. 5).

5.3.2 Image Panning

It is the horizontal or vertical translation of the image, which helps our model to learn the features on the corner horizontally or vertically, which makes the training set more robust. It is shown in Fig. 6.

Fig. 6 Original versus panned image

Fig. 7 Original versus altered image

5.3.3 Altered Brightness

It essentially multiplies all the pixel intensities inside the image with a specific value; thus, any pixel intensity multiplied by a value less than 1 will become darker. As research says that convolutional networks react better to a higher fraction of darker images. It is shown in Fig. 7.

5.3.4 Flipping to Balance Data

This data augmentation technique helps to balance our dataset as our dataset has skewness toward left turn flipping the dataset figure it out the skewness and better generalize the dataset. We are not only flipping the image [8]; we also need to flip the steering angle by negating the steering angle as well. It is shown in Fig. 8.

6 Connection to the Simulator

For running our model, we need a script to set up a bidirectional client–server communication which makes a persistent connection to connect our model [7]. This code is used to control the simulator based on speed, throttle, breaks, etc. This is not related

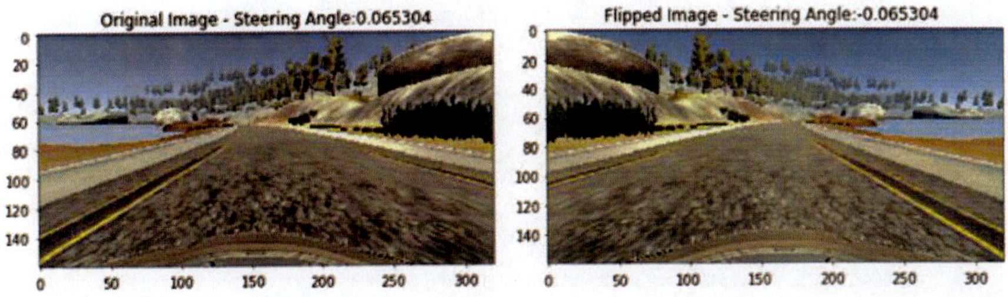

Fig. 8 Original versus flipped image

to deep learning anyway, but we need this step to run our model on the car and get the output accordingly.

7 Running the Simulator

As soon as the connection is established, we are setting the initial speed and throttle values and emitting it to the simulator such that the car starts off as stationary and facing forward, but then, the simulator will send us back the data which contains the current image of the frame where the car is presently located in the track, and based on the image, we want to run it through our model, and the model will extract the features from the image and predict the steering angle which we send back to the simulation and we keep doing this for entire simulation such that the car starts driving on its own in Fig. 9.

8 Experiments

For better results, we opted for steering angle. We started comparing with our results while considering different hyperparameters and choosing relu as our activation function. We first collected the dataset by driving three laps either left or right. But just collecting the data through either driving left or right was not enough to drive the car for sharp turns in autonomous mode. So, we tried to cut off the samples to balance the dataset and reduce the losses. We incurred 0.0300 as the training loss and 0.040 as a validation loss shown in Fig. 10. Since dataset was small and validation loss is more, hence learning becomes poor. So, to improvise the results, increase our leaning rate and minimize the losses, we used Nvidia's model in which we first preprocessed the dataset and also augmented it through different techniques discussed previously. Along with this, we used appropriate number of hyperparameters and elu as activation function and then ran our model with less epochs to make it less computive, hence faster. And thus, we were able to optimize the results as training loss of 0.0417 and validation loss of 0.0302 shown in Fig. 11. Here, reduced validation loss proves better

Fig. 9 Simulation in autonomous environment

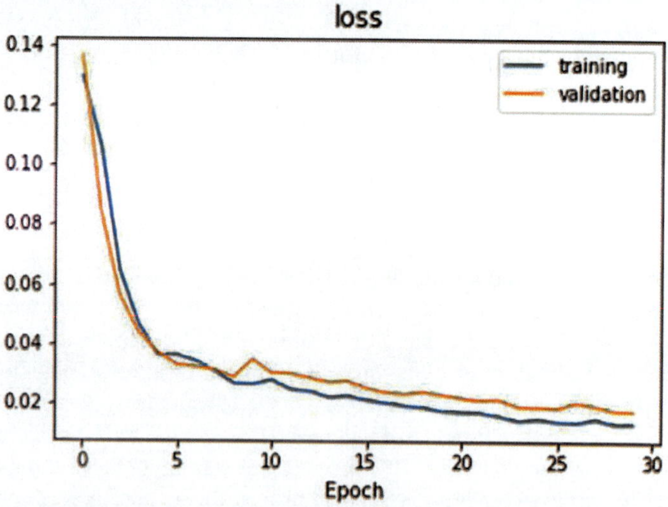

Fig. 10 Training and validation loss on small dataset

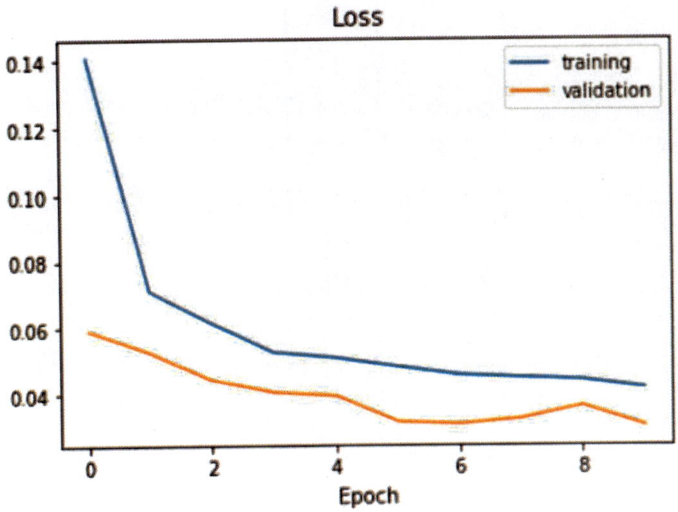

Fig. 11 Training and validation loss

learning of our optimized model.

9 Results

We are having the input to the model as an image; so after enough preprocessing, it is fed up into the model for predicting the loss, the steering angle, throttle and the speed for the car to move. We are using in total of 252,219 hyperparameters which are trained on the Nvidia's model [1] [2] and outputs a training loss of 0.0417 and validation loss of 0.0302 which is perfect to clone the behavior of manual driving which is shown in Fig. 11; at an instance, we are getting steering angle, throttle and the speed for the car as we shown in the Fig. 12.

10 Discussion

We are building a very powerful model in this paper which clones the behavior of the manual driving based on the steering [1], but if we want to clone the exact behavior, we have to think about external factors like speed control and critical decisions based on environment [3]. As self-driving cars [9] are the future of artificial intelligence, we have to be pretty cautious as this is a life-threatening invention if our model is not perfectly trained, and not only model is the factor, proper sensor communication and surrounding information [10] are also a major factor.

Fig. 12 Final output

11 Conclusion

This paper is dedicated to one of the major aspects of the cloning [6] of the self-driving cars which is steering angle. We are building a powerful model which is connected to the simulator with the help of bi-directional Web sockets that takes the images from the simulator and then process it by extracting its features and finding its instantaneous steering angle and giving it to the simulator and driving accordingly.

References

1. Al-Qizwini, M., Barjasteh, I., Al-Qassab, H., & Radha, H. (2017). Deep learning algorithm for autonomous driving usinggooglenet. In *2017 IEEE Intelligent Vehicles Symposium (IV)* (pp. 89–96). IEEE, 2017.
2. Bernauer, J. (2017). Nvidia deep learning tutorial. In *2017 IEEE International Parallel and Distributed Processing Symposium (IPDPS)* (p. 491). IEEE.
3. Bojarski, M., Del Testa, D., Dworakowski, D., Firner, B., Flepp, B., Goyal, P., Jackel, L. D., Monfort, M., Muller, U., Zhang, J., et al. (2016). End to end learning for self-driving cars. arXiv preprint arXiv:1604.07316, 2016.
4. Ding, B., Qian, H., & Zhou, J. (2018). Activation functions and theircharacteristics in deep neural networks. In *2018 Chinese Control and Decision Conference (CCDC)* (pp. 1836–1841). IEEE.
5. Gavrilov, A., Jordache, A., Vasdani, M., & Deng. J. (2018). Con-volutional neural networks: Estimating relations in the ising model on overfitting. In *2018 IEEE 17th International Conference on CognitiveInformatics & Cognitive Computing (ICCI*CC)* (pp. 154–158). IEEE.
6. Huval, B., Wang, T., Tandon, S., Kiske, J., Song, W., Pazhayampallil, J., Andriluka, M., Rajpurkar, P., Migimatsu, T., Cheng-Yue, R., et al. (2015). An empirical evaluation of deep learning on highway driving. arXiv preprint arXiv:1504.01716.

7. Kisačanin, B. (2017). Deep learning for autonomous vehicles. In *2017 IEEE 47th International Symposium on Multiple-Valued Logic (ISMVL)* (p. 142). IEEE.
8. Mikołajczyk, A., & Grochowski, M. (2018). Data augmentation for improving deep learning in image classification problem. In *2018 International Interdisciplinary Ph.D. Workshop (IIPhDW)* (pp. 117–122). IEEE.
9. Rao, Q., & Frtunikj, J. (2018). Deep learning for self-driving cars: Chances and challenges. In *Proceedings of the 1st International Workshop on Software Engineering for AI in Autonomous Systems* (pp. 35–38).
10. Schoettle, B., & Sivak, M. (2014). A survey of public opinion about autonomous and self-driving vehicles in the US, the UK, and Australia, 2014.
11. Shin, H.-C., Roth, H. R., Gao, M., Lu, L., Xu, Z., Nogues, I., Yao, J., Mollura, D., & Summers, R. M. (2016). Deep convolutional neural networks for computer-aided detection: Cnn architectures, dataset characteristics and transfer learning. *IEEE Transactions on Medical Imaging, 35*(5), 1285–1298.
12. Woo, J.-h., Song, J.-Y., & Choi, Y.-J. (2019). Performance enhancement of deep neural network using feature selection and preprocessing for intrusion detection. In *2019 International Conference on Artificial Intelligence in Information and Communication (ICAIIC)* (pp. 415–417). IEEE.

Performance Analysis of Chatbots and Brief Study About Technical Aspects

Rajesh Tripathi and Vishnu Kumar Sharma

Abstract Machines using Artificial intelligence is a very challenging task. It contains the formulation of machines by which it can simulate intelligence. In this paper, we will discuss about implementation and design of the chatbot using machine learning which is a part of artificial intelligence. Nowadays, chatbot using machine learning is a hot topic and widely popular, and we need an application for computer communication to catching the speed of natural language. There is statement that is "can machine think" which gives a idea that there can be intelligent application that respond like human, and this application we can name as chatbot [17]. Smart assistants such as cortana, siri, and text-based search engines are recently on boom in today's generation. For analyzing, text techniques of Python like NLTK can be applied in natural language processing and for getting the convenient human like responses by designing an engine for finding the intelligent responses.

Keywords Natural language processing · AIML · Chatbot · Loebner price · Pattern matching

1 Introduction

Chatbot is a computer program that can talk to humans in natural language. Intelligence in machines is the part of today's era. In the AI with the advances, machines have begun to masquerade as different type of human in today's trails. An exemplary example of intelligent machines and conversational entities of artificial intelligence is also known as chatbot. A computer program, which is chatbot, can communicate like natural conversation of humans with other people. Human to computer interaction has many approaches. Among these approaches, one popular approach via natural

R. Tripathi (✉) · V. K. Sharma
Motilal Nehru National Institute of Technology, Allahabad, India
e-mail: rajeshtcsed@mnnit.ac.in

V. K. Sharma
e-mail: vishnu.ks.mnnit@gmail.com

S. Tiwari et al. (eds.), *Proceedings of International Conference on Big Data, Machine Learning and their Applications*, Lecture Notes in Networks and Systems 150,
https://doi.org/10.1007/978-981-15-8377-3_19

language processing (NLP) has again many goals and sub-approaches. Because of success of virtual assistance like Alexa, Cortana, Siri, the popularity of chatbots is boosting. The purpose and the idea of chatbot is to achieving a conversation of natural language of human clients with computers, and it should pretend computer-like human. Understanding the responses of human inputs and maintaining the context is the enormous challenge that existing smart assistance have. The functionality according to pattern matching is finding a scripted response matches with the input that is still used by most of the chatbots [9]. The functionality which used by most of the chatbot cannot result in a satisfactory conversation.

Due to various lackings of scripted responses, researchers and developers keep working on adding new functionalities to the existing ways of working of chatbots, especially various new ideas and outlooks and retaining facts from conversation. These modifications have added to competitive advantage to chatbots. In the growing world of digitization, it has become a challenging task for users to outburst the required information quickly in an easier way.

With the view to increase the efficiency of user interaction with any system, cooperation of human and artificial intelligence is required. Chatbot can act as question answer system where experts provide information for solicitation of user.

2 Design of Chatbot

A chatbot is a communication appealing computer program. Its work is conversation with human. The conversation with chatbot is quite easy. Its work is to answer the questions raised by the user. The design of chatbot is exhibited with the help of following diagram [6] (Fig. 1).

Its designing requires recognition of its various components—It is divided into three parts: responder, classifier, and graphmaster (shown in Fig. 2) which is explained as follows:

1. **Responder**: It acts as an interface between chatbot's main routine and the user. Its function is to transfer the data from user to the computer and to monitor input and output.
2. **Classifier**: It is an element between responder and graphmaster. Its function is to modify and normalize the inputs, dividing the inputs into logical parts, and transfer the modified sentences into the graphmaster. Then, it deals with the output produced by the graphmaster and handles the instructions of the database system.
3. **Graphmaster**: This is the component for matching sequence of data. It constitutes brain of chatbot and stores the pattern matching algorithms.

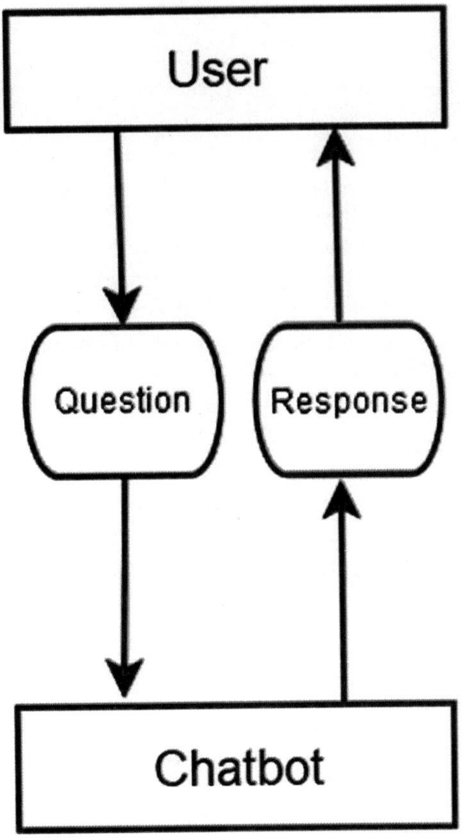

Fig. 1 Diagram of chatbot design [6]

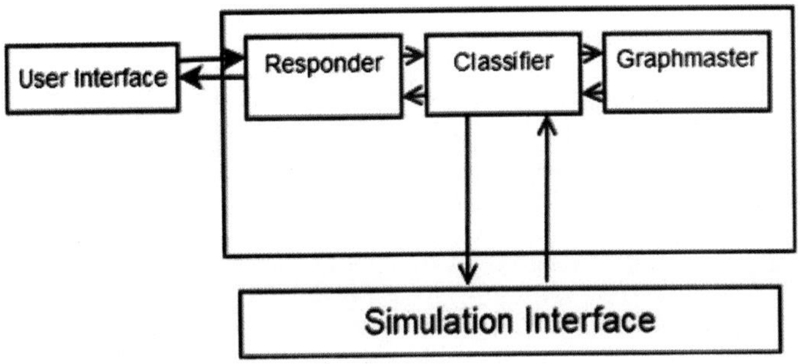

Fig. 2 Diagram of chatbot components [1]

3 Loebner Prize

The Loebner prize is an yearly competition in artificial intelligence that gives recognition to the most human-like computer program by awarding prize (for conversational agents or chatbots). In this competition, they conduct a test which is known as Turing Test, where they simultaneously hold textual conversations with the computer program and on the basis of these responses they give ranks.

This competition was organized for the first time in 1990 by Hugh Loebner in collaboration with the Cambridge Center for Behavioral Studies, Massachusetts, United States. Since 2014, it has been conducted by the AISB at Bletchly Park. $2000 were granted for the most human-like program. It was $3000 in 2005, $2250 in 2006, and $3000 in 2008. The disputation is that whether this competition is making artificial intelligence grow or is acting as hindrance in the process of development of artificial intelligence [4]. This controversy has arisen because this competition is obliging chatbots to act like human which causes chatbots to simply assume that they are thinking beyond actual intelligence.

Here, we will discuss about the technologies used in chatbots. Artificial Linguistic Internet Computer Entity chatbot uses artificial intelligence markup language which we will discuss later; Ella chatbot uses pattern matching and parse normalization, Jabberwock chatbot uses pattern matching, parser, and context-free grammar; George and Joan chatbot uses large database of people's responses; UltraHAL chatbot uses combination of visual basic code and pattern matching scripts; Elbot chatbot makes use of commercial natural language interface system which supports a person-to-machine communication; Do-Much-More chatbot is programmed in C and C++, and 75% of code is written in C++ and also uses large database; Suzette and Rosette chatbot uses AIML successor, concepts, triples, and variable; Chip Vivant chatbot used technology is not publicly disclosed; Mitsuku chatbot uses AIML and is a pandorabot; Rose chatbot uses chatscript [14] (Table 1).

4 Technical Working and Algorithm

4.1 Markov Chain Model

Markov chain model is a concept which is used in the speech identification, identification of text path recognition, and in many other artificial intelligence tools in some other forms. This model is based on the principle of 'Memorylessness' or simply we can say that the subsequent state is based on the preceding state and not the chain of states as shown in Fig. 3.

Figure 3 shows twofold Markov chain, with the two states mentioned as E and A. Each number shows the probability of the Markov chain transforming from one state to another in accordance with directions shown by an arrow. For example,

Table 1 List of Loebner prize winners chatbots [3]

Year	Winner chatbot	Score(/100)
2000	A.L.I.C.E.	76
2001	A.L.I.C.E.	41
2002	Ella	89
2003	Jabberwock	82.2
2004	A.L.I.C.E.	56
2005	George	64.21
2006	Joan	29
2007	UlrtraHAL	71
2008	Elbot	83.05
2009	Do-Much-More	66.9
2010	Suzette	78
2011	Rosette	18.01
2012	Chip Vivant	70
2013	Mitsuku	90
2014	Rose	89.17
2015	Mitsuku	83.33
2016	Mitsuku	90
2017	Mitsuku	27
2018	Mitsuku	33
2019	Mitsuku	24

Fig. 3 Two-state Markov chain model [20]

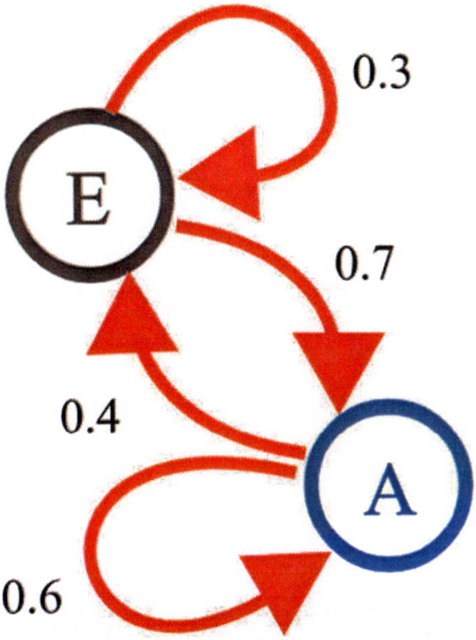

the probability of Markov state changing into state E from state A is 0.4, while its probability of being into state A is 0.6.

In chatbots, the Markov chain models were used to give more authentic and valid responses [20].

4.2 Pattern Matching

This technique is widely used in chatbots. Diversifications in some pattern matching algorithm are found in every existing chatbot system. They can vary in their complexity, but the basic purpose and idea is same.

Earlier chatbots were based on simplest and easiest patterns such as ELIZA and PC Therapist.

Chatbot's response on the question asked by human:

- **Human**: Who is the master of Greek comedy?
- **Chatbot**: Aristophanes was the master of Greek comedy.

Example of pattern matching:

```
<aiml version = "1.0.1" encoding = "UTF-8"?>
<category>
<pattern>WHO IS ABRAHAM LINCOLN</pattern>
<template>Abraham lincoln was the US President</template>
</category>

<category>
<pattern>DO YOU KNOW WHO * IS</pattern>
<template>
<srai>WHO IS</star></srai>
</template>
</category>
</aiml>
```

4.3 Parsing

Parsing is also known as breaking data block into smaller chunks by set of rules, so that it can be more easily interpreted, managed, or transmitted by a machine (computer). Parsing is a technique which converts original text into set of words. Mostly, it is used to determine grammatical structure.

The parsers that were used earlier were simple that used to work on predetermined keywords in approved order. Example of such parsing is as follows: "please take the gold" and "can you get the gold," both the sentences can be phrased as "take gold." So this way, the chatbots can cover multiple input sentence with a limited set of patterns.

The more complicated parsers used in latter chatbots do the complete grammatical parsing of the natural language sentences [4].

4.4 Semantic Network

To find solution of any problem and take decisions, we have to approach information that is already stored in our brain. How is that information stored in our brain? (Figs. 4 and 5)

There are so many ways, but the one way that we are going to study is semantic network approach, which contends that concepts are stored in your mind in terms of interlinked ideas. You can think of it as how information is stored in computer. You have different nodules, and those nodules are concepts that are connected by links. The links are shorter for closely related ideas, whereas larger for less related ideas as it depends on how connected the nodules are [19].

It is a knowledge narration technique used for showing information in logical and mathematical manner. So it is also known as propositional network. Mathematically, a semantic network is regarded as labeled directed graph. Some chatbot systems is a set of sequential and analogous interconnected concepts [15].

These concepts can have common talk names that can be used directly in chatbots to find out synonyms, antonyms, homophones, homonyms, and other relations between the concepts [4] (Fig. 6).

4.5 Artificial Intelligence Markup Language (AIML)

AIML is a markup language based on XML which is used to create AI applications. AIML is used to create human interface systems that is simple to operate and program, easy to understand, and maintainable. It was developed by Richard Wallace and he used it to build a bot called Artificial Linguistics Internet Computer Entity (A.L.I.C.E.) which has been awarded with several artificial intelligence awards (Loebner Prize).

Figure 7 at the top and bottom of every AIML file you have the AIML tag. The AIML tag just lets the bot to know that the file that its reading is in fact AIML file. Next, we have the category tag as the name implies categorizes the different inputs and outputs that the box sends and receives. Next one is pattern tag, the input received from the user is called pattern tag. After the pattern tag is the template tag which is the output that we are going to get from the bot.

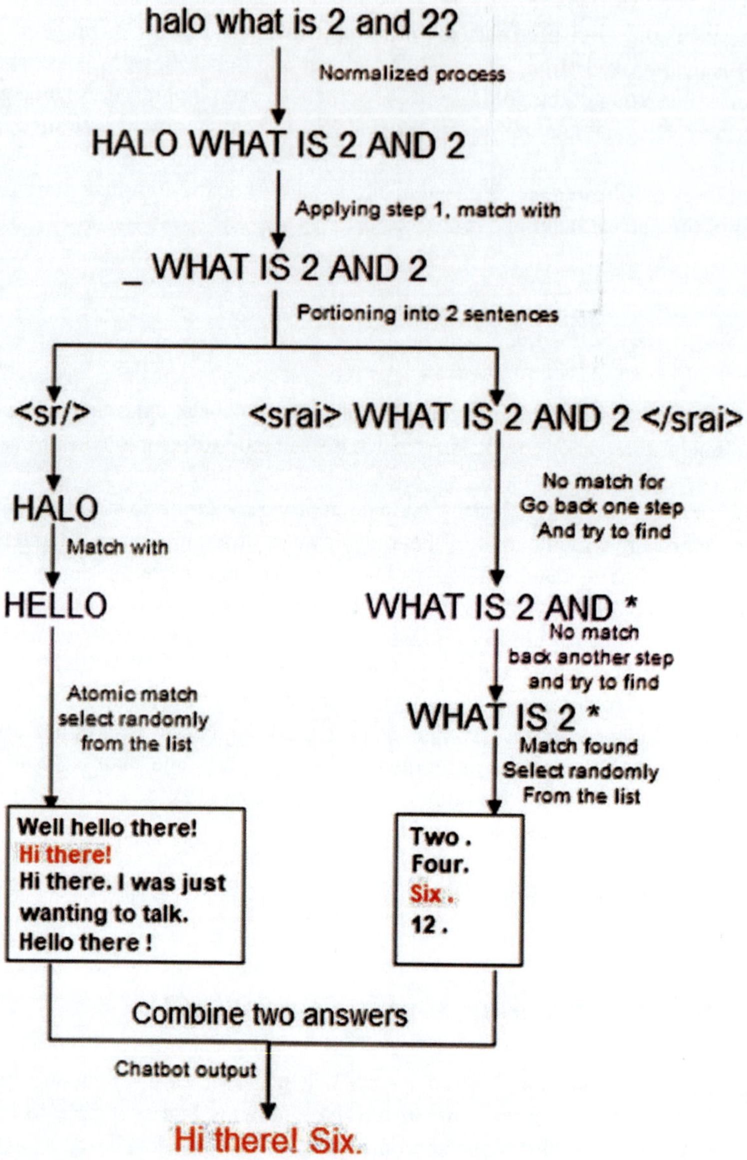

Fig. 4 Tree structure of pattern matching

Fig. 5 Example of a parse tree [4]

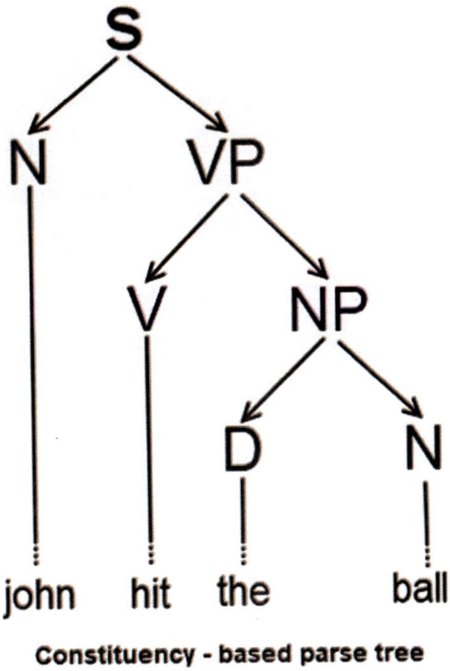

In Figure 8, in AIML sometimes, it is mandatory to use a wild-card character. This is in case you are unsure of how are usually about to say a particular pattern, so the wild-card characters used in AIML are underscore (_) and asterisk (*). Underscore is placed in front, and asterisk is placed at the back.

AIML can recursively call itself, and it can yield input to itself with the help of srai tag and content of * by using star tag [4].

In Figure 9, if you are looking to make bot seem more human, it is beneficial to use the random and the LI tags. The random and LI tags allow you to use different responses for the same pattern. The random tag lets your bot know that you are using different responses, and the LI tag lists the different responses. Random is used with the LI tags to give the different responses for the same input on the random basis.

As we have discussed, some of the tags which is used in AIML to creating a chatbot, but AIML is a vast language, and it contains the different tags for different purpose.

4.6 ChatScript

ChatScript is giving cut throat competition to AIML language and aims to succeed it by working on better syntax which makes it easier to maintain and program. It fixes

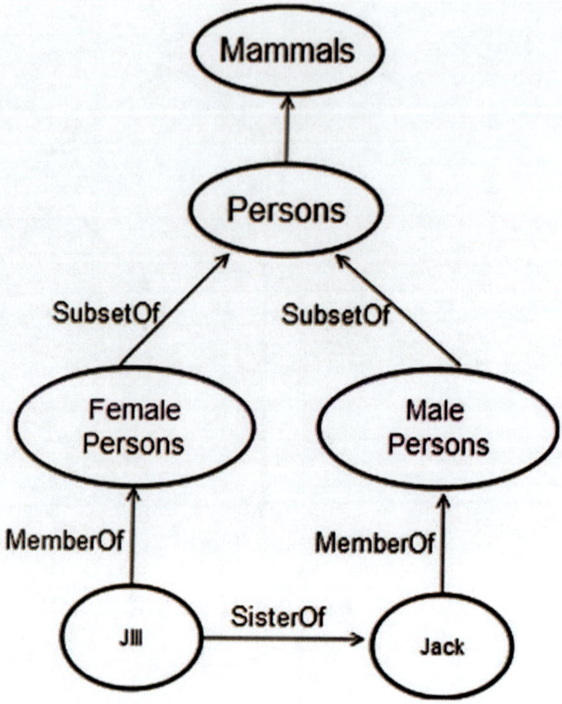

Fig. 6 Semantic network

```
<aiml>
<category>
<pattern> Hello World! </pattern>
<template>The world say hello back! </template>
</category>
</aiml>
```

Fig. 7 AIML pattern

the zero word matching problems and adds new functionalities such as logical and variables, fact triples, concepts, and functions. With the help of these, functionalities complete need for ontologies in script [4].

```
<aiml>
<category>
<pattern> _IS* </pattern>
<template> I Completely Agree </template>
</category>
</aiml>
```

Fig. 8 Wild-card character

```
<aiml>
<category>
<pattern> _IS* </pattern>
<template>
<random>
<li> I completely agree. </li>
<li> I concur. </li>
<li> You are preaching to the choir my friend. </li>
</random>
</template>
</category>
</aiml>
```

Fig. 9 Random and LI tag [4]

5 A.L.I.C.E. Chatbot Architecture

The Artificial Linguistic Internet Computer Entity is a chatbot with which you can chat. It stores its knowledge in AIML files.

5.1 AIML Files

These files initiate with an "aiml" tag which shows the version of AIML being used and accommodate AIML elements which carries data objects known as AIML objects. These objects are composed of topics and categories, which can contain either parsed or unparsed data.

AIML format:

```
< aiml version="1.0" >
< topic name=" the topic" >
<category>
<pattern>PATTERN</pattern>
<that>THAT</that>
<template>Template</template>
</category>
....
....
....
</topic>
</aiml>
```

- Each element has opening and closing tags.
- That tag allows to recall previous templates that already used and that tags come before the template not in the template [16].

5.2 *Types of Categories*

There are three kinds of categories:

1. **Atomic Categories**: In these categories, pattern does not have wildcards "*" or "_".
2. **Default Categories**: These category patterns carry wildcards "_" or "*." When chat-box searches for relevant matching, these patterns result from a reduction process. These wildcards instead of varying in alphabetical sequence can match any input. According to preceding category, if chat-box is not able to find the foregoing atomic pattern, then it will find out the successive default pattern [16].
3. **Recursive Categories**: It is not a pattern, and it is a property of template. Using the srai tags and sr tags, the pattern matcher recursively called via the template calls which refers to simply symbolic reduction and recursive artificial intelligence [16].

6 Natural Language Processing (NLP)

When people say natural language, they usually mean human language does not mean for example language of some animal. Natural language processing is sub-field of linguistics, computer science, information engineering, and artificial intelligence concerned with the interaction between computers and natural languages, which tells you how to program computers to process and analyze large amounts of natural language data. Language processing and NLP serve a lot of use cases when you are dealing with text or unstructured text data [12, 18].

Natural language processing is used as the manipulation of human language, like text, by software and also to convert the speech into text. Sometimes it is considered to be as the field of computer science and artificial intelligence used to find linguistic information from the respective data. It helps computers or devices to extract meaning from human language input [2]. Now, the various economies of the world are interlinked and connected to each other due to the technological advancement, now information is exchanged in seconds, and this has led to a number of challenges in analyzing large amount of data in a form of text. Identifying approximately 6500 languages and dialects followed across the globe and applying quantitative analysis on large amount of data and handling vagueness while analyzing and extracting information is the main purpose of natural language processing [10, 13].

Natural language processing algorithms cannot understand text like humans do but find patterns by converting text into matrices. While doing this analysis, words that do not offer much value like "A," "Is," "Are," "The," etc., are removed very first time. These words are known as stop words. After that the algorithm divides the sentences into group of words and counts that how many times a particular group of words appears in a document and how many documents have that group of words out of all the documents analyzed. Without understanding anything about the text, the algorithm can tell that how many times a particular group of words or phrase appears in a document and how many documents contain that particular group of words out of all the documents analyzed, so tokens that appear lots of time in lots of documents may not mean much, but token that appear frequently in only a few documents tells us that something is going on [8, 12].

6.1 *Natural Language Text Processing Systems*

Modification of text for extraction of knowledge, for automatic abstracting and indexing, or for getting the text in a required format is the main area of research in NLP. It helps in analyzing large bodies of textual information and finding or extracting knowledge from that for a particular purpose. Automatic text processing system converts some form of text input into some other form of text output [10].

The main task of natural language text processing system is to translate potentially vague natural language doubts or texts into explicit internal representations on which pairing and repossession can be done. A natural language text processing system may start with structural analysis. To get the structural variants or alternatives of the words involved, derivation of both the queries and documents is done. The rhetorical and linguistic or semantic processing involves utilization of lexicons for analyzing and determining the nature the words, parts-of speech, words and phrases, and parsing the sentences [8].

6.2 Applications of Natural Language Processing

Some of most common applications for NLP in today's era:

1. **Spam Filters**: Email has two categories one is spam (unsolicited) mails and other one is ham (solicited) mail. Email spam is also known as junk mail, and it is unsolicited information sent in message by email. Spam filters scan text in all emails and understand meaning of text and detect unsolicited, unwanted, and virus-infected email and stop it from getting into email inboxes.
2. **Algorithmic Trading**: Trading is of two types, one is algorithmic trading (no emotions present), and another is discretionary trading (Emotions are added) or manual trading. Algorithmic trading is as good as your code, and it can make you master in stock market without doing anything. By adding natural language processing to the algorithmic trading to make use of reading news stories concerning companies and stocks and based on this analysis, determine if you should buy, sell, or hold onto certain stocks [13].
3. **Answering Questions**: This application of natural language processing is mostly used in chatbot to make them more effective and help it to generate relevant response.
4. **Summarizing Data**: Web is a collection of huge amount of data, and some of the long documents or articles contain the large amount of data. To understanding the meaning of that data, most common technique is natural language processing. After understanding the context of the data, create a summary of data, so humans can understand it quickly [7].

7 Conclusion

This paper investigated chatbot history, technical aspects of chatbot, algorithms used in chatbot, Loebner prize winners, and natural language processing. The basic purpose to create the chatbot is to stimulate conversation with a user (human) in natural language processing. Its purpose is to do conversation with human clients in natural human language. Chatbots cannot understand human context, it only responds on the basis of rules, and due to these limitations, developers and researchers have decided to add new functions to the already existing chatbots way that how chatbots works which was basically based on the artificial intelligence. In this paper, we have given the brief introduction about the Markov chain model which is used for speech recognition, text identifier, and in the other artificial intelligence applications, and other supporting algorithms are pattern matching algorithms, parsing technique, semantic network technique, artificial intelligence markup language (AIML), chatscript, and the major thing is natural language processing (NLP) which help to manipulate the text. AIML was used in Artificial Linguistics Internet Computer Entity (A.L.I.C.E.) chatbot.

References

1. Abdul-Kader, S. A., & Woods, J. (2015). Survey on chatbot design techniques in speech conversation systems. *International Journal of Advanced Computer Science and Applications, 6*, 7.
2. AbuShawar, B., & Atwell, E. (2015). ALICE chatbot: Trials and outputs. *Computación y Sistemas, 19*(4), 625–632.
3. Al Rifaie, M. M. (2018). https://www.aisb.org.uk, Nov. 2018. [Online]. Available: https://www.aisb.org.uk/events/loebner-prize.
4. Bradeško, L., & Mladenic, D. (2012). A survey of chatbot systems through a loebner prize competition. *Proceedings of Slovenian Language Technologies Society Eighth Conference of Language Technologies*, 34–37.
5. Collobert, R., et al. (2011). Natural language processing (almost) from scratch. *Journal of Machine Learning Research, 12*, 2493–2537.
6. Dahiya, M. (2017). A tool of conversation: Chatbot. *International Journal of Computer Sciences and Engineering, 5*(5), 158–161.
7. Fryer, L., & Carpenter, R. (2006). Bots as language learning tools. *Language Learning & Technology, 10*(3), 8–14.
8. GGChowdhury. (2003). Natural language processing. *Annual Review of Information Science and Technology, 37*(1), 51–89.
9. Khanna, A., et al. (2015). A study of today's ai through chatbots and rediscovery of machine intelligence. *International Journal of u-and e-Service, Science and Technology, 8*(7), 277–284.
10. Klüwer, T. (2011). *From chatbots to dialog systems* (pp. 1–22). Conversational agents and natural language interaction: Techniques and Effective Practices. IGI Global.
11. Liddy, E. D. (2001). Natural Panguage Processing.
12. McNeal, M. L., & Newyear, D. (2013). Introducing chatbots in libraries. *Library technology reports, 49*(8), 5–10.
13. Neo speech "nlp," Nov. 2018. [Online]. Available: https://blog.neospeech.com/what-is-natural-language-processing/.
14. Rahman, A. M., Al Mamun, A., & Islam, A. (2017). Programming challenges of chatbot: Current and future prospective. In *2017 IEEE Region 10 Humanitarian Technology Conference (R10-HTC)*. IEEE.
15. Robin, "semantic," Nov. 2018. [Online]. Available: http://intelligence.worldofcomputing.net/knowledge-representation/semantic-nets.html.
16. Shawar, B. A., & Atwell, E. (2002). *A comparison between Alice and Elizabeth chatbot systems* (p. 19). School of Computing Research Report: University of Leeds.
17. Shawar, B. A., & Atwell, E. (2007). Different measurement metrics to evaluate a chatbot system. In *Proceedings of the Workshop on Bridging the Gap: Academic and Industrial Research in Dialog Technologies*.
18. Shawar, B. A., & Atwell, E. (2015). Using dialogue corpora to train a chatbot. In *Proceedings of the Corpus Linguistics 2003 Conference*.

19. Sowa, J. F. (1987). Semantic networks.
20. Wikipedia, "markov chain," Nov. 2018. [Online]. Available: https://en.wikipedia.org/wiki/Markov_chain.

Evolving Evidence Gathering Process: Cloud Forensics

Pranjal Srivastava and Arjun Choudhary

Abstract Identification of data which is stored in the cloud is complicated than what we expect and collection of evidence which is robust, tamper-proof and stands in the court of law is one of the most challenging aspects of the whole procedure. The data for which the whole process is being executed can be reserved on the data-centers, can be placed anywhere across the globe. Feasibility of accessing the data on the physical drive is much less when we see the scenario of cloud storage and processing. Case studies, guidelines and advisories till date majorly describes those traditional processes i.e maintaining chain of custody, search and seizure, etc. But the principle concern are gathering the data, integrity and reliability of the data that is to be analyzed after seizure and Formation of essential requirements like, Cloud Infrastructure, instances, services to run upon and most important—Client, for testing the procedure. Infrastructure to be used, compatibility issues, guidelines and advisories have been finalized to achieve the primary aspects that is: Assess the nobility and dependability of the extracted information that is not compromised, tampered.

Keywords Cloud forensics investigation · Virtualization · CSP · SaaS · IaaS · Evidence

1 Introduction

The procedure which involves utilizing indicated strategies, systems and devices to recognize, separate and investigate information found in advanced media that can be

P. Srivastava · A. Choudhary (✉)
Center for Cyber Security, Sardar Patel University of Police, Security and Criminal Justice, Jodhpur, India
e-mail: a.choudhary@policeuniversity.ac.in
URL: https://policeuniversity.ac.in/net_team.php

P. Srivastava
e-mail: spu18cs06@policeuniversity.ac.in

S. Tiwari et al. (eds.), *Proceedings of International Conference on Big Data, Machine Learning and their Applications*, Lecture Notes in Networks and Systems 150,
https://doi.org/10.1007/978-981-15-8377-3_20

exhibited as dependable proof in a courtroom has now found a niche in the world of forensics and known as **Digital Forensics.** The way toward separating information may shift contingent upon the gadget or information type being prepared.

For instance, getting and breaking down information from a customary PC hard-plate drive requires an alternate procedure at that point acquiring and examining information over a working system that has diverse advancements including proof isolation in disseminated conditions. Despite the procedure, explicit scientific strategies must be fastidiously followed to acquire and safeguard reasonable computerized proof.

The development of cloud innovation has constrained a re-assessment of regular digital forensics strategies, systems and analytical tools utilized by advanced experts to address cloud security challenges in a consistently changing and imaginative advanced world. Because of the data warehousing, inland as well as offshore, customarily prepared investigators are looked with specialized and lawful difficulties where customary strategies don't have any significant procedures.

- Examine integrity, trustworthiness of facts and figures gathered in the course of cloud forensics.
- Evaluate along with adequate strategies, apparatuses for examination of data.

2 Standards

A risk is added in the equation whenever control is given away. So Mitigation of corporate risks and opportunities take serious efforts by the higher-end management when rendering services to the Cloud Service Provider. The Industrial safety protocols often act as a key process in threat reduction and risk management which includes smooth execution of the process and entails expected output. But, these industrial protocols are sub-par in the emerging boom of cloud services which are mainly concerned with domains of Privacy, Security, and compatibility. In the absence of these security codes, the consumer of the services will always be in the presumption of the risks, and its effective management. The slacking in cloud standards adversely impacted the management of cloud computing which further affected the implementation of it. A unilateral strategy for cloud requirements may push the barriers and lead to transparent process subscribers so that basic amenities can be easily assessed and in case of conflict the services can be partially or completely be transferred conveniently from one provider to another [1]. However, the inconsistencies in the transition and enactment can be categorized there below:

- Technological Aspects: Reliability, Data, and Integrity.
- Organizational Aspects: Risk Management and Governance.
- Syntactical and Semantically: Portability and Interoperability.

3 Virtualization

Concerning computation oriented network domains, virtualization is defined as working of numerous operating system on an individual system. Whilst maximum computers solely operate a single machine, Virtualization permits the system to function numerous working devices at the identical time. The virtualization is the consideration of computer assets like Processing unit, Disks, Network, and other utilities from packages and provide to end-users utilizing those services. The concept aid democratization which is initiated as a result of Infrastructure abstraction, gives the capability to users for utilizing the utilities and storage as a pooled service, irrespective of authorization aspects as per the protocols settled by the organizations.

The host system capable of managing guest ones without any conflict, process scheduling has to be fluent when it comes to delivering services with the help of distributed and discrete systems in that environment. Establishment of a server for a [2] specific assignment is mainly done by administrators because compatibility between most of the tasks becomes complicated [3, 4]. Since a single process or aptly putting a single task through a server causes less difficulty in discovering troubles at the accurate time and also to align the network which poses a limitation.

For instance, this kind of configuration under-utilizes the whole skill of the processing units of taking in-power. But with the expansion of the network, the complexity increases which makes the server to occupy more space. The statistics pointing to a result that is overcrowded with servers that take derive more power and emanates a huge amount of heat. These worries can be properly addressed by virtualization of servers along with using particularly owned software programs which results in transitioning of lone gadgets to a couple of systems placed at data-centers. Every digital server behaves as a special device adept at enhancing its processing unit. Cloud virtualization relies upon on isolating consumer functions from the fundamental framework. The host working framework gives a deliberation layer to executing a virtual temporary working framework. An important division of virtualization is the hypervisor also called "Virtual Machine Manager (VMM)". The Virtual Machine Manager is an application that enables quite several working frameworks to share a solitary device. Each working framework looks occupy the processor, memory, and implicit belongings of the host for its self. In any case, Virtual Machine Manager genuinely has the control remote of the processor and the assets of the host, apportioning that is expected of every working framework whilst ensuring that the guests' working frameworks can't upset one another.

3.1 Requirements of Virtualization

The visitor working frameworks know that they are executing on a Virtual Machine taking into account close local execution. Para-virtualization is likewise the hypervisor virtualization model and disposes, part of the catching and-copying overhead connected with programming performed virtualization. It necessitates that the foreign working framework be readjusted before the virtual processing system.

Virtualization is the essential model that we can see in XEN hypervisor, which utilizes Operating System Linux to maintain its procedural condition, known as Domain0.

Xen can likewise exploit tools virtualization to run unmodified editions of working frameworks over its hypervisor [5].

3.2 *Issues Related to Virtualization*

The benefits of the usage of virtualization do not exist without its own set of challenges, hurdles, and risks. Procuring security for virtualized surroundings becomes greatly complicated from a normal network mainly when more than one VM is settled upon a sole working system. Virtualized devices are flexible and ever-evolving which makes security little callous to define. If a VM ever gets compromised, it can lead to various tenants being at risk and if no intent supervision is done, it can render the complete virtualized surroundings to cyber attacks.

4 Digital Evidence in a Virtualized Environment

In Forensics examinations, proof is appropriated over quite several gadgets. For example—hard disks, settled systems either servers or communication setups, investigation of setup like this enables the forensics person to get better data to appreciate the speculators' exercises. A bit of leeway of disbursed computing is that more than a few functions and servers crosswise over geopolitical areas can connect consistently to give the administrations and interface that customer utilizes [6]. Distributed systems with computation capabilities are autonomous of tools and working framework profiles. What's more, the end-client expects information to be prepared through specific purposes or PCs but conveyed as although it starts from a solitary supply PC making distinguishing proof and assembling of proof. As an outcome, virtualization programming can alters the Cloud Forensics evidence. Infrastructure information lives inside a digital event and shutting the occurrence with the end aim of a scientific inquiry might compel more digital ongoing cases to be abandoned. Virtualization with the aid of configuration count on the gear the executive's obligations of the processing system and in addition to this ever-increasing range of online utilities that extract the executive services from the OS. After long, a dispensable OS shall be made that would be making use of a combination of Virtual Machine Manager capacities with that work in a solitary session and then disassemble when shut down.

4.1 *When to Virtualize*

According to US Research Company "*Forrester Research*", seventy percent of the consumers transitioned to virtualization at a point to elevate an important hardware refresh to avoid the price of enhancing greater numbers of physical hardware. Fifty-two percentage virtualized their surroundings in time for most important functional machine migration. For instance: updating from Windows XP to Windows 7. Fifty-One percentage of users virtualized their centralized systems at the time for renewal of the license of an essential application, for example: Oracle or SAP. Consolidated servers can help in making these sizable savings. Companies have been making move to virtualization in form of unplanned system outage which serve as other immediate measures for them [7].

Under "US Software and Solutions Company CA Technologies" extensive research that provided reports about the spontaneous disruptions, downtime that was deemed responsible for hefty billion dollars being misplaced in the revenue of organizations associated to North America and Europe.

5 Existing Studies

The customer can become a tool to be up aware of the existential services related to the cloud, besides acquiring facts saved as a means by the client. Hence it is necessary to carry out the client's forensics analysis carefully before evaluating the environment of the server. While conducting a traditional as well as distributed system forensics, expert must observe in the same environment the following things:

1. In Digital forensics one should be concerned with the identification of sources of proof looking through the perspective regarding the primary and secondary evidence, these proofs can range from end-point devices, operative or communicative which can be commonly found in a possession of a potential suspect. However, the file systems already exist in the central systems, in cases of Distributed File System which is employed in Cloud Computing Environment. Conservation is essential for maintaining the authenticity of forensic inquiry and investigations, and set up perfect techniques are required for the maintenance of the former irrespective of the source of the evidence.
2. The authentic seizure of evidence, there are numerous methodologies for statistics of evidence collection which is applicable in multiple cloud paradigm for assessment and deployment. Whist Infrastructure as a Service can lead an end-result: Series relating to virtual disk memory Software as a Service can lead to an export in distinction to cloud-applications while a large series of FS supporting deployment of cloud infrastructure supposed to be deeply involved. Considering the case in which the hosting of the File System lies outside the jurisdiction of investigating law agency suitable venues must be obtained legally to grant authorization in the File System remotely.

3. It is concerned with the observation, interpretation, and deduction of the forensic data. A critical examination is fundamental to gain a total insight regarding the running factors from the extracted file system. This evaluation is necessary as it leads to the evidence reconstruction.
4. The legal demonstration of the proof collection which is comparable to the skeletal framework of International advisory bodies. Keeping a general view, records ought to consist of facts that shall not put barriers to access or examine the extracted data, process logs etc, as not to get any further inferences for wrong preemption.

6 Research Questions and Hypothesis

A lot of the research has been assessed to explore the subject that incorporated the different cloud types and investigations, industry standards, information security, information protection, administration and cloud lawful structures (Table 1).

The way toward directing an ordinary CF examination was additionally checked parallel with explicit advanced measurable procedures in a "Cloud". Also, research recognizes instances of conceivable sources of proof, apparatuses alongside procedures required for assembling these proof inside situations like cloud infrastructure, for instance, logging records, arrange observing instruments and CF investigation devices (Table 2).

The improvement depends on the audit of comparable research. Acquiring evidence in a cloud, potential accompanies that will scrutinize the trustworthiness and unwavering quality of the search and seizure and whether the evidence is tested and maintained in the courtroom. For instance, authority's chain of custody, processing the evidence and its time undertaking, wellsprings for proof with examiner capa-

Table 1 Representation of hypothesis

Question arises	While conducting cloud forensics investigation, how reliable can be the evidence and also maintain the integrity
Hypothesis	Current supporting infrastructure are capable of current investigation

Table 2 Questions evolved during research

Questions evolved	
1	Supporting requirements i.e Hardware and Software for evidence gathering and CF
2	Analyzing the effectiveness of the proposed system
3	Relationship between what is to monitor and justifying the evidence
4	Procedures, guidelines and tools that be used to gather evidence and also stands in the court of law

bilities curtail to the regions bound to prove social occasion development that is quite testing while leading CF. Furthermore, many questions evolved while going through the whole process and somehow related to the main problem statement and concurrent phases.

Now, with respect to the questions, additional hypothesis have been framed to clear the vision and go ahead to explore with further phases. To respond to the proposed problem statement, the validation of the chosen hypothesis and behavior in a systematic manner, a data demographic needs to be elaborated. The flow of phases as described below provides well-articulated information, demographic outlining—essential lookup queries, alternative view, and reference of the concerned literature (Table 3).

Every testing phase additionally associating to the related facts series which have been accomplished from testing (Fig. 1) [8].

Table 3 Further to know

Additional research: to be found during proof of concept	
1	For traffic monitoring i.e network analysis techniques or hardware|software to be used
2	Memory management to determine and reconstruct the same environment in which the incident took place
3	Evidence should be determined, without disrupting the capacity of the system

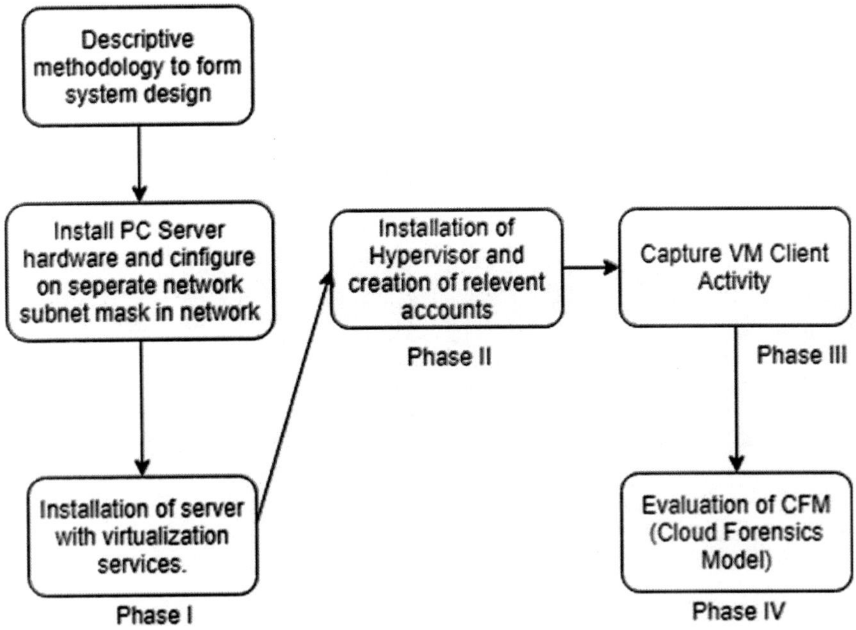

Fig. 1 Proposed research model

Conclusively, the insights obtained after thorough literature reviews hypothesis testing and its related phases along with the records shall prove to be of substance for finding out the reliability of the tested hypothesis as a proven fact.

6.1 *PHASE I*

As we can see, the distribution of the whole process of recreation of attack and measuring its metrics includes multiple phases.

Phase 1 is the installation of a server with virtualization services. So, Infrastructure as a Service (IAAS) is created with the help of the XCP-NG virtualization platform which is capable of inheriting a distributed ecosystem.

As we know, XenServer, Support for this product is the combination of individuals and companies which is capable of delivering these extraordinary products with no limit. Everything related to this product can be found on GitHub.

Mentioned below are the screenshots that give a glimpse of the installation process (Figs. 2, 3, 4 and 5).

Source: https://docs.citrix.com/en-us/xencenter/current-release/.

6.2 *PHASE II*

This phase includes the overall management for the creation of the base for relevant Repository related to virtualizer.

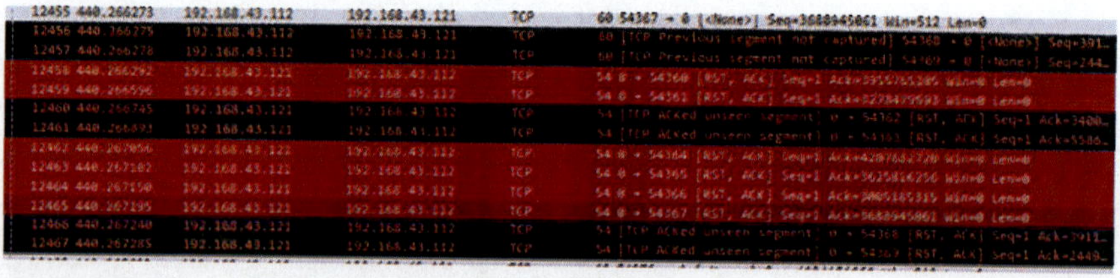

Fig. 2 Client system—attack logs

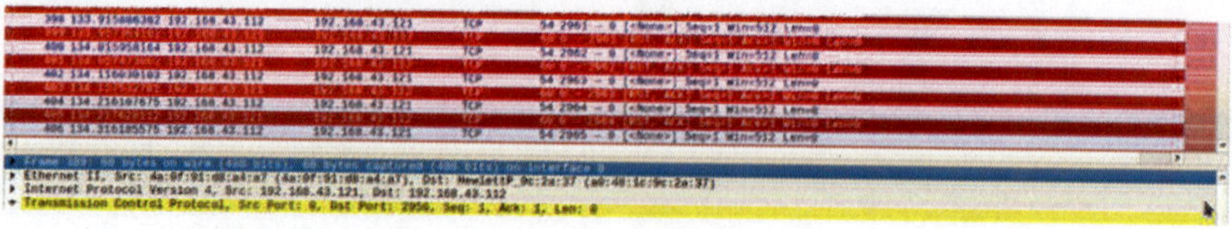

Fig. 3 Attacker system—log

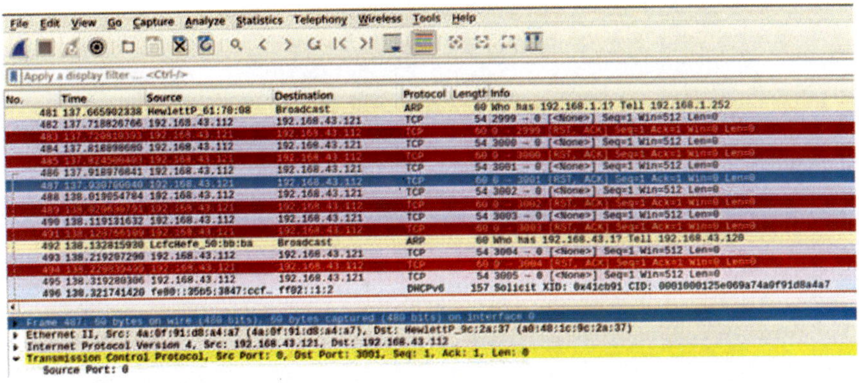

Fig. 4 Attacker system—log

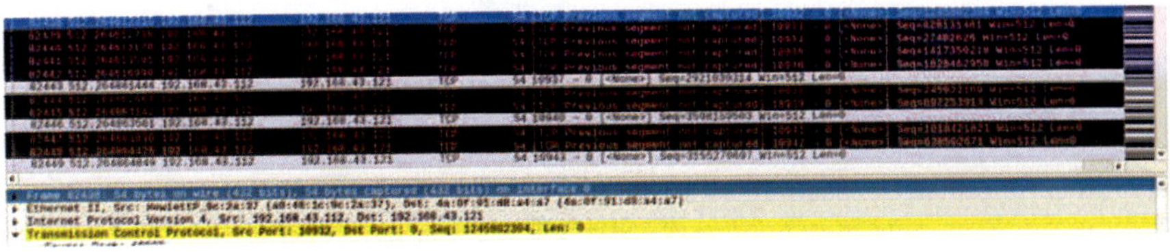

Fig. 5 Attacker system—log

Storage in XCP-ng Storage capacity in SCP-ng is one of the vital and huge problems in itself. This includes [9]

1. SR: Storage Repository
2. VDI: virtual Disk
3. ISO SR : Special SR only for ISO Files (Read Only) (Fig. 6).

Storage Types There is two types of storage available:

1. Thin provisioned.
2. Thick provisioned.

Here in this research, ISO SR will be in use [10]. Figure 6 uploading an ISO is always a challenge, in XCP-ng we need to create a dedicated “Place/ Space” for specific ISO SR.

Virtual Machine Creation in XCP-NG Inside XCP-ng Center pick “Add New Server” and enter the IP of the hypervisors management interface.

Now you can access the hypervisor settings from the XCP-ng Center.

For you to be able to create a VM, XCP-ng needs a network share where it can access ISO files so that you can install an OS on your VM. Create it by:

1. Share a folder on your Windows PC or use local NAS if you have one.
2. Copy or download the ISO of the OS you want to install and put it in the shared folder.

Type of Storage Repository	Name	Thin Provisioned	Thick Provisioned
file based	local Ext3/4	X	
	NFS	X	
	File	X	
	XOSAN	X	
block based	local LVM		X
	iSCSI		X
	HBA		X

Fig. 6 Storage repository

3. In XCP-ng select "New Storage" and select ISO library and "Windows File Sharing". Enter the location of your Windows share (easiest by using IP address and share name) and the login/password needed. If everything worked you should be able to see "SMB ISO library" in XCP-ng Center under your hypervisor. Inside that, you'll see the ISO files stored on your shared Windows folder.

Create a VM by clicking "New VM". XCP-ng uses templates to set some basic minimum for the OS you intend to install. Pick the OS you are installing or the closest one to it. If you are installing something that is not there at all, you can pick "Other install media".

Select your ISO file when it asks for installation media. Pick how many cores and memory you need. And add a virtual disk or change the size of the one it suggests if needed. Then finally network settings and you are done.

XCP-ng will create the VM and automatically boot it up. In XCP-ng Center you can see the VM under your hypervisor in the left pane. If you click the "Console" tab you can see the screen and interact with the VM. Go through with the installation of the OS here.

6.3 PHASE III

As we can see, During Implementation we got four major points:

1. XCP-NG module
   Here, Two major modules are worth identifying.
   (a) Hyper-Calls—This is to verify the logs created while an attack on the hypervisor is going on.
   (b) Running VM Calls—To Cross- verify the logs at the same time.
2. Instances Evaluation
   Here, a standalone system and Virtual machine logs made available to verify the proof of the attack.
3. Detection and Monitoring
   (a) To examine the flow of packets—Multiple tools like Wireshark are used to achieve this goal.
   (b) Verification—More than one tool, logs, and presence of footprints verifies the occurrence of the activity.
   (c) Obviously, Supporting Evidence and data for the same.
4. Attack: Logs from attacker system—to re-verify the scenario.

So, Some screenshots as proof of attack and monitoring are also been attached (Figs. 7 and 8).

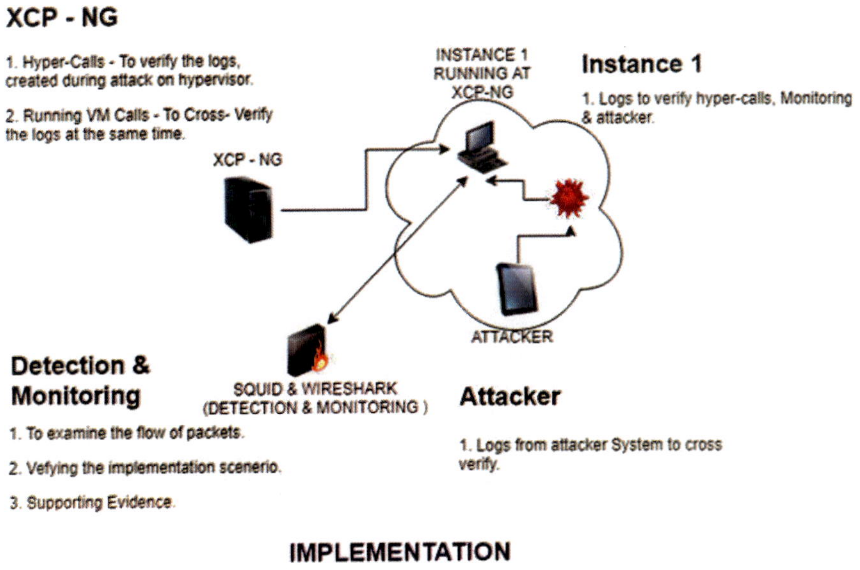

Fig. 7 Implementation

Fig. 8 Attacker system—log: flooding

6.4 PHASE IV

As specified in Section VIII, research Question and Hypothesis, Phase 4 of the research will determine the Evaluation of Cloud Forensics Model and its mapping.

But before that, throughout this research essential key to evaluate and make base of evidence evolving process is LOGS.

So, we'll discuss the various logs involved in the process and essential for investigation.

There are many type of Logs present, some of them are mentioned below :

1. General Logs
   Location : /var/log/daemon.log
   Description: Every task handled by XCP-ng logged here. Like Storage Related, Migration Related, Network etc.
2. XCP-ng API(XAPI) Logs
   Location: /var/log/XenSource.log
   Description: Here you can find everything related to Toolstack for API and its output.
3. Storage Logs
   Location: /var/log/SMlog
   Description: In this file, you can find everything about storage, like modification, new alterations, addition and removal of storage. This will help to establish the relation between self-made storage or malicious ones.
4. Installation Related
   Location: /var/log/installer/ although main file is /var/log/installer/install-log
   Description: You can also find the very helpful debug information to identify some critical faults.
5. Kernel Message
   Location: /var/log/kern.log
   Description: Everything related Kernel, since last boot. (dmesg)
6. Crash Logs
   Location: /var/crash
   Description: If Host crashes and its due to the kernel or it affects the flow of kernel, it will get logged here.
   These are basic logs that can help to analyze the conditions and re-creation of the scenario of the machine state in which the event has occurred.

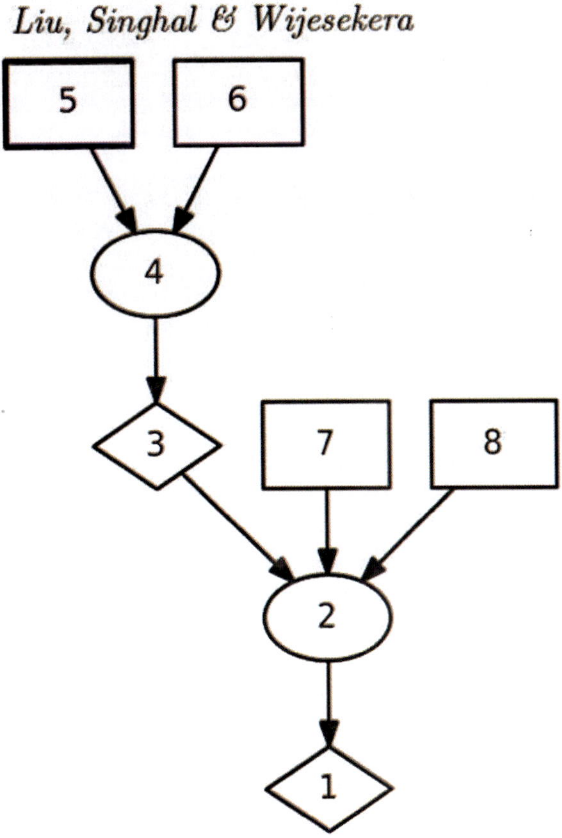

Fig. 9 Example logical evidence graph [11]

You can also find other details on the troubleshoot documentation of XCP-ng official docs.

Evaluation of Cloud Forensics Model(CFM) Associated to XCP-ng Liu et al. [12, 13] explained the application of the MulVAL logic-based network security analyzer to reconstruct the used steps by the attacker using different evidence collected by investigator fro the respective network.

The rules, which are based on expert knowledge, are used as hypotheses by an investigator to link chains of evidence that are written in the form of Prolog predicates to create attack steps.

Attack scenarios are reconstructed in the form of acyclic graphs as defined below:

Figure 9 shows an example logical evidence graph. In Fig. 9, fact, rule and consequence fact nodes are represented as boxes, ellipses, and diamonds, respectively (Fig. 10).

Consequence fact nodes (Nodes 1 and 3) codify the attack status obtained from event logs and other forensic tools that record the postconditions of attack steps. Fact nodes (Nodes 5, 6, 7, and 8) include network topology (Nodes 5 and 6), computer

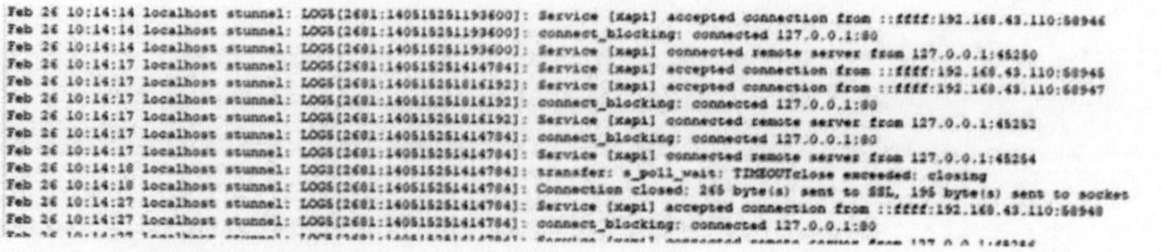

Fig. 10 Secure.log file—snapshot

configuration (Node 7), and software vulnerabilities obtained by analyzing evidence captured by forensic tools (Node 8). Rule nodes (Nodes 2 and 4) represent rules that change the attack status using attack steps. These rules, which are based on expert knowledge, are used to link chains of evidence as consequences of attack steps. Linking a chain of evidence using a rule creates an investigator's hypothesis of an attack step given the evidence.

Now, as per the concept of logical evidence graph,

1. From where the user establishes the connection i.e end-node.
2. Centralized environment, from where vulnerability can be identified.
3. Network Access—End-Point
4. Network Provider.
5. combination of Network and end-node.
6. Location of Attacker.
7. Information of network to launch.
8. If weakness Exists, Launch and see the output.

Every evidence on each step has been provided in the above sections.

6.5 *Identifying Evidence for Reconstruction*

In context to get evidence for a forensics investigation, XCP-ng hypervisor with control center, end-user, and attacker system were configured to log access.

Also, Wireshark was deployed in the attacker as well as Virtual machine i.e One of the User-agent to monitor network traffic.

Figure 8 shows a snapshot of the packets captured by Wireshark. Client system Parrot Linux at IP Address 192.168.43.121 sent numerous SYN Packets to Host VM at IP address 192.168.43.112 and the host sent back the same back to Parrot Linux. Figures 2 and 8 Attacker System Shows the same.

A Prolog-base forensics tool HPing-3 was used to automate the process correlating items of the evidence to reconstruct the DOS Attack. Figure 8 Shows a snapshot the same while initiating DOS and flooding. At run time, the input file instantiated the code to create the attack path as we can see in the same Figure.

There is more evidence we can find inside logs in hypervisor. Possibility of the existence of Secure.log file within /var/log/ in XCP-ng hypervisor that you can access with the help of the Control center's console.

7 Conclusion

The existing circumstance of statistics, placing the proposition. The unique cloud traditions that industry follows are much behind then the today's expectations and administration arrangements are moreover investigated, which induced that there is no one universal size that can-fit-all' cloud infrastructure. Distinct models of cloud infrastructure that common users be able to buy in after relegating over their needs, every Cloud Service Provider gives capacities that provide industry, a much prominent or electricity on their cloud infrastructure. In Ordinary classical Cloud Forensics Investigation where proof is appropriated over quite a number devices, for instance, *Complex Drives, Prepared Service hosts and Communication Setups*. Investigation for such setups enables the examiner for information recovery concerning previous experiences [14]. In any case, cloud virtualization can perhaps render the CF investigations. Distributed computing additionally for the most phase comprises of multi-tenure. For example, the capability to make use of a comparable programming and interface to organize assets and confine consumer explicit data. The multi-tenancy, the place various purchasers share a similar shared storage and programming application with isolation. Occupancy of storage among different users offers crucial factors of past SaaS. The extensive literature, the plethora of researches that were explored and led to the need for further exhaustive studies Security, concerns of privacy along with ramification as well as rectification in the process of investigation need to have particularly become subjects of DF research [7, 15, 16]. Modern phase of information additionally recognizes vital aspects which helps in building layout perspective as well as tendencies for possible lookup method of research. Therefore the proposed statement has been determined which focuses on enhancing the techniques methodology and fundamentals that are central for Digital Forensics in the cloud, i.e. Cloud Forensics. Specifically, to gather evidence from the cloud and cloud Service Provider via developing a simulated IaaS.

The improvement dependent on the audit of comparable research [17]. Acquiring evidence in a cloud, potential accompanies that will scrutinize the trustworthiness and unwavering quality of the search and seizure and whether the evidence is tested and maintained in the courtroom. For instance, authority's chain of custody, processing the evidence and its time undertaking, wellsprings for proof with examiner capabilities curtail to the regions bound to prove social occasion development that is quite testing while leading CF. Furthermore, several questions evolved while going through the whole process and somehow related to the main problem statement and concurrent phases [18].

This research has explained that evidence collected from various sources can be used to re-construct a cloud attack scenarios. What sources can be IDPS, Logs,

End-point security devices, SIEM, inhouse scripts, system calls, hypervisor calls, application logs, and cloud service providers API calls?

The Logical Evidence graph shown in this research gives more clarity towards the effectiveness and utility of correlation of evidence from multiple sources to reconstruct attack during investigations [19, 20].

Future Research will implement the automated functioning of this procedure on the stand-alone system and critical interconnected infrastructure.

References

1. Shaunak Mody Security & Forensics Research Group, Auckland University of Technology. (2017). December) A centralised platform for digital forensic investigations in cloud-based environments. *15th Australian Digital Forensics Conference* (pp. 5–6). Perth, Australia: Edith Cowan University.
2. Barrett, D., & Kipper, G. (2010). Virtualisation and forensics–A digital forensic investigator's guide to virtual environments. In S. Liles (Ed.), *Burlington*. USA: Elsevier.
3. Almulla, S., Iraqi, Y., & Jones, A. (2014). A state-of-the-art review of cloud forensics. *Journal of Digital Forensics, Security and Law*, *9*, 7–28.
4. Chantry, D. (2009, January). Mapping applications to the cloud. Retrieved August 28, 2014, from Microsoft Developer Network.
5. Mok, E., Samsudin, A., & Fun, T. (2017). *Implementing the honey encryption for securing public cloud data storage.*. https://doi.org/10.4108/eai.27-2-2017.152270.
6. Moussa, A. N., Ithnin, N., & Zainal, A. (2018). CFaaS: Bilaterally agreed evidence collection. *Journal of Cloud Computing*, *7*, 1. https://doi.org/10.1186/s13677-017-0102-3.
7. Wenhai, S. (2018). *Towards secure outsourced data services in the public cloud.* Virginia Tech: Diss.
8. Alali, F. A., & Yeh, C. L. (2012). Cloud computing: Overview and risk analysis. *Journal of Information Systems, 26*(2), 13–33 (Amazon).
9. Buffington, J. (2014, June). Why store once federated duplication matters to HP—And should to you, too. In Enterprise Strategy Group.
10. Cuthbertson, S. (2013). Mutual assistance in criminal matters: Cyberworld realities. *Cross-border law enforcement: Regional law enforcement cooperation-European, Australian and Asia Pacific perspectives*, *2012*, 127–142.
11. Liu, C., Singhal, A., & Wijesekera, D. (2015). A logic-based network forensic model for evidence analysis. In G. Peterson & S. Shenoi (Eds.), *Advances in digital forensics XI* (pp. 129–145). Heidelberg, Germany: Springer.
12. Liu, C., Singhal, A., & Wijesekara, D. (2016). A probabilistic network forensic model for evidence analysis. In G. Peterson & S. Shenoi (Eds.), *Advances in digital forensics XII* (pp. 189–210). Heidelberg, Germany: Springer.
13. Chunga, H., Parka, J., Leea, S., Kang, C., Chung, H., et al. (2012). Digital forensic investigation of cloud storage services. *Digital Investigation*, *9*(2), 81–95.
14. Abdel Razek, S., El-Fiqi, H., & Mahmoud, I. (2017). Cloud storage forensics: Survey. *International Journal of Engineering Trends and Technology (IJETT)*, *V52*(1), 22–35. ISSN:2231-5381.
15. Kesalkar, B., Bagade, D., Barsagade, M., Jakulwar, N., & Zade, S. Implementation of data deduplication using cloud computing. IJARIIT (ISSN: 2454-132X).
16. Gite, Y., Pawar, A., & Ghumbre, S. (2018). Efficient data backup technique for cloud storage. *International Journal of Engineering Research in Computer Science and Engineering (IJERCSE)*, *5*(3)

17. Birk, D. (2011). Technical challenges of forensic investigations in cloud computing environments.
18. Ademu, I. O., Imafidon, C. O., & Preston, D. S. (2011). A new approach of digital forensic model for digital forensic investigation. *International Journal of Advanced Computer Science and Applications*.
19. Kumar, V., & Garg, K. K. (2012). Migration of services to the cloud environment: Challenges and best practices. *International Journal of Computer Applications*, *55*(1).
20. Bahaweres, R. B., Santoso, N. B., & Ningsih, A. S. (2017, January). Cloud based drive forensic and DDoS analysis on seafile as case study. *In International Conference on Computing and Applied Informatics, 755*(1).
21. Narayana Samy, G., Shanmugam, B., Maarop, N., Magalingam, P., Perumal, S., & Al-Bakri, S. (2018) Digital forensic challenges in the cloud computing environment, pp. 669–676. https://doi.org/10.1007/978-3-319-59427-9_69.
22. Mahmoud Nasreldin Ain Shams UniversityCairo, Egypt. (2017). Evidence acquisition in cloud forensics. In *IEEE 4th International Conference on New Paradigms in Electronics & Information Technology*, 2017-11-05 - 2017-11-08.
23. Jaware, A. P., & Borkar, N. R. (2016). Computer science and engineering. Amravati University INDIA. Implementation of a secure and dynamic multi-keyword ranked search scheme over encrypted cloud data. *IEEE Transactions on Parallel and Distributed Systems, 27*(2).
24. Ning, J., Cao, Z., Dong, X., Liang, K., Wei, L., & Choo, K. -K. R. (2018). CryptCloud+: Secure and expressive data access control for cloud storage. *IEEE Transactions on Services Computing*.
25. BeeGFS The Parallel Cluster File System. BeeGFS (2015).
26. Clarke, N., Tryfonas, T., & Dodge, R. (2012). *Proceedings of the Seventh International Workshop on Digital Forensics and Incident Analysis (WDFIA 2012)*.
27. Cloud Security Alliance. (2013, February). Alliance, cloud sacurity. The notorious nine: Cloud computing top threats in 2013. Top Threats Working Group.
28. Contrail Consortium. (2014). Morin, Christine. Open computing infrastructures for elastic services: contrail approach. In *Proceedings of the 5th International Workshop on Virtualization Technologies in Distributed Computing*. ACM, 2011.

Melanoma Detection Among Various Skin Lesions

Aurobindo Gupta and Sanjeev Thakur

Abstract Melanoma is a type of skin cancer that starts and evolves from the "pigment-producing" cells known as "melanocytes." There has been quite some research done in the area of melanoma classification through image detection and classification using machine learning—specifically deep learning and neural networks. Researchers have used convolutional neural networks (CNN), deep neural networks (DNN); some have even used recurrent neural network (RNN) and transfer learning. The research work has not been up to the mark as of yet, we know this because there has been no news of models being put to clinical testing. Another setback to the process of creating a perfect algorithm and mode is the lack of data regarding melanoma; the largest dataset publicly available is the one provided by ISIC for its 2020 competitions; it has 25,333 images in the dataset, but the issue with these is that they do not contain just the images and data for melanoma, they are a dataset for eight different skin lesions to be detected and classified. In this paper, we have proposed a model for melanoma detection using CNN architecture. We have also discussed the issues with our model and the accuracy achieved by it.

Keywords Melanoma · CNN · Deep learning · Neural network

1 Introduction

"Melanoma" is one of the most dangerous and serious types of cancer (skin), they develop in the cells called melanocytes that produce the pigment melanin. This is the pigment that gives your skin its color. This cancerous disease is also known to form in eyes and even sometimes inside one's body such as our nose or throat. The exact reason for this irregularity is not clear as of yet, but it is speculated and, in some cases, proven to be due to exposure to "ultraviolet radiations" from the sun or

A. Gupta (✉) · S. Thakur
Amity School of Engineering and Technology, Amity University Uttar Pradesh, Noida, India
e-mail: aurogpt10@gmail.com

S. Thakur
e-mail: sthakur3@amity.edu

S. Tiwari et al. (eds.), *Proceedings of International Conference on Big Data, Machine Learning and their Applications*, Lecture Notes in Networks and Systems 150,
https://doi.org/10.1007/978-981-15-8377-3_21

other man-made sources that increase the risk of developing melanoma. The number of people affected by this seems to be increasing day by day at an increasing rate. It is seen to be increasingly affect the population under the age of 40 years especially women. Knowledge and detecting it in its early stages can help reduce the rate of fatality globally and the number of cases of newly effected personal. In 20% cases, even surgery fails to cure it. In total skin cancer-related deaths, 75% happen due to melanoma [1].

This disease is the only one responsible for more than 50,000 cancer-related fatalities around the world. In the year 2015, an estimate figure of 73,000 new cases was seen in the USA and approximate death rate of 2.7. It is a kind of skin lesion that can only be identified by expert dermatologists due to its close similarity to benign lesions. Diagnosis of melanoma has mostly self-processed. When the person finally registers that there is something wrong, it is due to lack of awareness and the lack of medical experts around the world. Even then the accuracy rate of melanoma detection by the experts is not that great [2].

To reduce the pressure and to help the medical experts during the time of digital age where AI and machine learning are making leaps in the medical field, the researchers thought the need to provide the experts with a melanoma detection and classification algorithm or model. In the current research, use of deep learning for detection of malignant melanoma is carried out with a proposed architecture with benchmark datasets [1]. To help with early detection of the disease, the researchers have really hit their stride in the past few years when the publicly provided slightly large datasets were provided like the HAM 10000 and the ISCI2019 set, but the issue with those is that they are not just datasets for melanoma. The HAM 10000 contains the data of seven different skin lesions, more so it is an uneven set of data which does not help with the training of neural network. The ISIC 2019 dataset contains 25,332 images and their concerning data. In the literature review, we see how the researcher has processed their data and how have managed to teach and build a "neural network" for melanoma detection, and the types of models and processes used by them to train the model and then test it on test data and validate its efficiency. We propose a CNN architecture model to classify and identify melanoma. The focus of this paper is melanoma classification.

2 Related Work

We have premediated and studied some of the work previously done from the year 2015–2019. Our study and findings from the research papers have been put in a tabular form for easy access below. The table below encases the name of the authors, the classification techniques, datasets and other works done. The third column has the outcome, accuracies of the research work done by the authors. Some of the points found in our study of the related works of the past few years are that there is not enough data publicly available or easily accessible for research related work which causes issues. Before the ISIC provided us with the HAM 10000 dataset for its 2018 challenge, most of the datasets contained at maximum 1000 images, and on an average most datasets contain 200–500 images. Most of the research work done is on various variations of the CNN architecture by changing the number of neurons or the number of hidden layers or by adding external feature extractors to help the model created with its classification.

To provide variations for the model, the researchers have tried to mix and match datasets to help the model better identify melanoma, e.g., author Soumen Mukherjee et al. [1]. They have augmented image data to increase the size of the dataset. Some authors have still worked with small datasets, e.g., Zabir Al Nazi et al. [3], Savy Gulathi et al. [4], Attia et al. [5].

1. **Mukherjee** et al. [1] used a combined dataset of Dermofit – 1300 images and MEDNODE-170 images to increase the size of the dataset they used data augmentation on the dataset. The model created by them was done using CNN architecture which they implemented using MATLAB. The accuracy achieved by their model was 90.58 and 90.14 for each of the datasets separately and 83.07, when combined.
2. **Zabir Al Nazi** et al. [3] used datasets ISIC2018 and PH2. The author used transfer learning (DCNN used for feature extraction with SVM as its classifier). Images were augmented, and dataset size increased. Max accuracy of 92% was achieved on the PH2 dataset by the created model. The beast feature extractor out of the group was DenseNet201.
3. **Savy Gulathi** et al. [4] used created a CAD system and used pretrained models AlexNet and VGG16 for transfer learning and as feature extractors, respectively. They used the PH2 dataset and achieved an accuracy of 97%.
4. **Felipe Moure** et al. [6] This paper focuses on classification on non-malignant melanomas. Authors used the model ResNet (a CNN model by Microsoft South Korea). The dataset was created by integrating multiple datasets{27,963}, and the final set contains 24 different classes of images. The accuracy achieved was 60% in the 2400 test examples.
5. **Seeja** et al. [7] Authors used CNN-based U-NET algorithm for feature extraction. They used: “local binary pattern (LBP),” “edge histogram (EH),” “histogram of oriented gradients (HOG),”, “Gabor method.” The resultant features from the above models were fed into the “support vector machine (SVM),” “random forest (RF),” “K-Nearest Neighbor (KNN),” “Naïve Bayes (NB)” classifiers for classification. The dataset used was from ISBI 2016 (ISIC data) The accuracy achieved was 85.19%.
6. **Attia** et al. [5] used CNN and RNN and the dataset used for training used “900 images” and tested on “375 images.” From ISBI 2016, the accuracy achieved was 0.98.
7. **Karabulut** [8] et al. used CNN and SVM and used dataset: DermIS and DermQuest{206}LBP(local binary patterns)BDIP(block difference of inverse probabilities) for feature extraction. The accuracy achieved was 70%.
8. **Brinker** et al. [9] The authors used CNN for comparison against dermatologists{157 from different German Uni Hospitals}Dataset used were ISIC

archives2018{2169 melanoma, 18,566 atypical nevi images}. Test set of 100 images for both the model created and the experts. The sensitivity and specificity achieved by the dermatologists were 67.2 and 62.2, respectively, whereas the CNN model created achieved sensitivity of 82.3% and a higher specificity of 77.9%.

9. **Codella** et al. [10]: The authors used "deep learning architecture," "sparse coding," and "support vector machine (SVM)" and used the dataset ISIC, which contained 2624 clinical cases: twofold cross-validation: (I) melanoma against all non-melanoma lesions and (II) melanoma against atypical lesions. They achieved accuracy of 93.1%, first task (sensitivity –94.9%, specificity –92.8%), second task accuracy –73.9% (sensitivity –73.8%, specificity –74.3%). In comparison, prior premade models yield 91.2% accuracy.
10. **Esteva** et al. [11] used CNN and as Dataset they used a set of – 129,450 images consisting of "2032 different diseases" the effectiveness of the authors model was tested against "21 board-certified" dermatologists. Three-way classification accuracy 70%, nine-way classification accuracy 49% was achieved by them.
11. **Fujisawa** et al. [12] used DCNN architecture, which they trained on 4867 images taken from 1842 patients to classify them into 14 diagnosis of different skin lesions, the result of which is then compared against that of dermatologists. The accuracy achieved was 76.5%. The model achieved 96.3% sensitivity and 89.5% specificity. The accuracy of classification by dermatologists was 85.3 and 74.4%, the model achieved greater accuracy, as high as 92.4% ± 2.1%.
12. **Han** et al. [13] used "CNN(Microsoft ResNet-152" model by "Microsoft Research Asia" classifying the clinical images of 12 skin diseases. The dataset they used, "Asan dataset," "MED-NODE," and "atlas site images" ("19,398 images in total"). The outcome of their proposed model had the accuracy of 91%.
13. **Hekler** et al. [14] used CNN and used a dataset which contained 695 lesions that were classified by an expert histopathologist, 595 of the resulting images were used to train the model. The rest used to test it. The mean sensitivity = 76% specificity = 60% accuracy = 68% was achieved in 11 test epochs. The 11 pathologists achieved a mean sensitivity = 51.8 specificity = 66.5 and accuracy = 59.2.
14. **Salido** et al. [15] used CNN(AlexNet), transfer learning. And for data they used PH2. The model created was able to achieve accuracy 93% for classifying melanoma and non-melanoma.
15. **Jafari** et al. [16] DCNN for making the model and dataset: Dermquest, the used dataset has 126 digital images. The accuracy achieved was 98%.
16. **Arkadiusz Kwasigroch** et al. [17] used deep convolutional neural networks(CNN), The authors propose the CNN architecture and used Dataset ISIC(10000). The efficiency of the proposed architecture was found to be 84%, authors used pretrained neural network which caused momentous increase in accuracy, from 70.5% (CNN1) to 84% (CNN2).
17. **Li** ID and **Shen** [18] used a "fully convolutional residual networks" (FCRN). A CNN architecture is proposed for the dermoscopic feature extraction and used

dataset: ISIC 2017. The results of the framework show promising accuracy, i.e., task 1-0.753, task 2-0.848 and task 3-0.912.

18. **Mahbod** et al. [19] used CNNs “AlexNet,” “VGG16,” and “ResNet-18,” as deep feature generators and Dataset ISIC 2017. They achieved an accuracy of 83.83% for melanoma classification and 97.55% for seborrheic keratosis classification.
19. **Majtner and Yildirim** [20] The authors proposed an automatic “melanoma recognition system,” which they based on “deep learning” method combined with “RSurf” features and “local binary patterns” (LBP). They used CNN along with SVM. They used data from ISIC dataset. Achieved classification accuracy for the proposed model is 0.826 sensitivity 0.533 and specificity 0.898.
20. **Afonso Menegolayz** [21] used CNN-based model and dataset: ISBI Challenge 2016. The results favor deeper models, pretrained it on ImageNet, fine-tuning it, achieved AUC of 80.7 and 84.5%.

3 Methodology

The most commonly used methodology for most classification problems involving deep learning are as follows:

1. Image-Acquisition
2. Image-Pre Processing
3. Image-Augmentation
4. Feature Extraction.
5. Image-Classification.

- Image acquisition is process of creation of a digitally encoded depiction of the visual features and characteristics of an object. In layman’s terms, it is the process of acquiring a “digital image” of the desired object.
- Image preprocessing is the work done on an image to improve the image data as per the requirements so as to make it easier to work on the image.
- Image augmentation is the process of manipulating the image, mainly done to cover for the lack of data available. In this process, we basically rotate the image by varying degrees or change the color scheme to make the system think that it is a new image.
- Feature extraction is processed to reduce the count of “features” in a dataset by making “new features” from already present features and then discarding the “original features.” These new features are able to summarize most of the information contained in the original features.

- Image classification—"deep learning" now is one of the best in the business for recognizing objects in a image as it is supposed to be implemented using three or more layers of ANN where each layer is responsible for extracting one or more features of the image being used for classification.

The methodology observed in most of the research work done on melanoma beginning with acquiring of datasets which is one of the biggest hurdles presented in this. A proper large dataset of skin lesion images was only provided publicly in the last 2 years in the ISIC challenge 2018–2019 where a huge set of 10,015 and 25,333 (train) images was provided, but the issue with these datasets is that they are not only the sets containing the images of melanoma, but contain seven skin images of seven different skin lesions. This can be a bit of a hindrance while training a model for melanoma detection. It reduces the accuracy of the model due to uneven balance of the number of images provided for each class.

For example, we have a dataset containing of images from three classes A, B, and C respectively. The distribution of data is of the ratio 6:2:2 respectively. Now, when we train the neural network, it learns that if it gives the classification as A most of the time it would achieve the most accuracy easily which is not the aim of the project. To counter this, we could use weights to inform the NN to evaluate some classes according to the weights or we could use the process of random oversampling or random under-sampling. Oversampling is a process in which we use data augmentation to increase the number of images in classes with low quantity. Under-sampling is the process of reducing the image quantity of large quantity classes to match that of low ones. Some of the other datasets used are: PH2, ISIC2018, ISBI2016, DermQuest, Dermofit, MEDNODE, DermIS, ISIC 2017. This section will describe the different types of sarcasm. Sindhu et al. [13] have identified about six primary classes of sarcasm described as below.

After acquiring the data in the data preprocessing phase, the data is reshaped to the required template, resized, etc. Then, if required the data is sent for augmentation, it is mainly done to increase the number of images (Fig. 1).

Most of the work done is seen to be using CNN for feature extraction and image classification which it does automatically as it was mainly developed for image classification (Figs. 2 and 3).

The CNN gained fame through its work done by researchers with image data. The general CNN architecture is as given below:

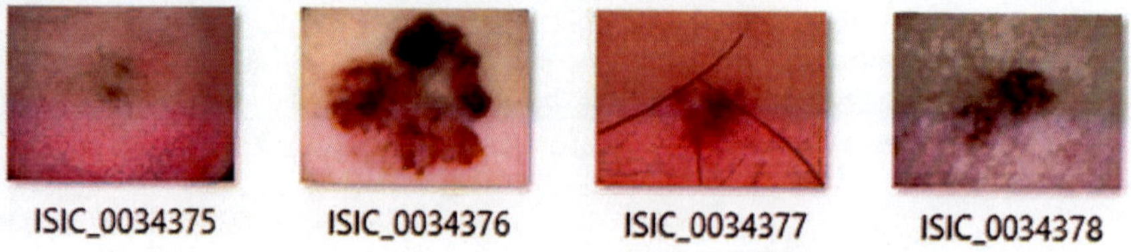

Fig. 1 Images in dataset

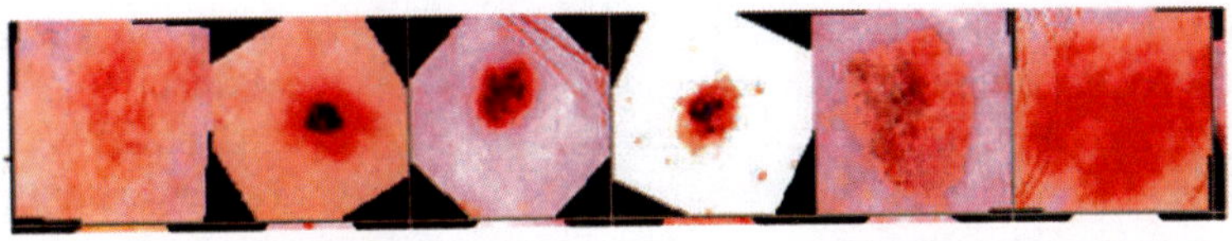

Fig. 2 Images after data augmentation

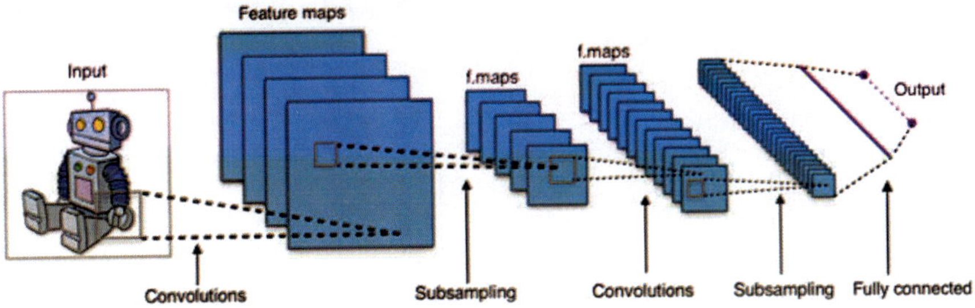

Fig. 3 CNN [22]

"Convolution-> Pooling-> Convolution-> Pooling-> Full connected Layer->Output".

"Convolution" is the process of taking the original data and creating feature maps from it. "Pooling" is the "down-sampling" which is mostly done in the form of "MAX POOLING" where a region is selected and then the max value of that region is taken and then that value becomes the new value for the entire region. Fully connected layers are your typical "neural networks," where all the nodes are completely connected to each other.

Datasets

- **DermQuest**: This dataset consists of 73 images of melanoma and 61 images of non-melanoma lesions. These are really small datasets. The Web site to download this dataset from has been discontinued(http://www.dermquest.com).
- **DermIS**: This dataset consists of 43 images with melanoma and 26 of non-melanoma lesions.
- **Dermofit**: The images are in RGB color. There are 1300 total images of which actinic keratosis 45, basal cell carcinoma 239, melanocytic nevus 331, squamous cell carcinoma 88, seborrheic keratosis 257, intraepithelial carcinoma 78, pyogenic granuloma 24, emangioma 96, dermatofibroma 65, melanoma 76. The access to this dataset is held by University of Edinburgh, and access is restricted to paid members.
- **ISIC2017**: Total of 2000 images are provided which include melanoma 374, seborrheic keratosis 254, and the rest are benign nevi 1372. The images are provided in jpeg format and a csv file with some clinical metadata is also provided.

- **ISCI2018**: This dataset was provided during the ISIC 2018 challenge. The dataset provided can also be found by the name HAM 10000. It contains 10,015 images in total which contains actinic keratoses 327, basal cell carcinoma 514, benign keratosis lesions 1099, dermatofibroma115, melanoma 1113, melanocytic nevi 6705, vascular lesions 142.
- **ISIC2019**: This dataset contained a total of 25,332 images consisting of eight different skin lesions. The dataset contained images of actinic keratoses 867, basal cell carcinoma 3323, benign keratosis lesions 2624, dermatofibroma 239, melanoma 4522, melanocytic nevi 12,875, vascular lesions 253, squamous cell carcinoma 628.

4 Proposed Model

The dataset used for the model proposed in this paper was obtained from the dataset provided for the ISIC2019 challenge. This set contained 25,332 images and csv files containing their detail. The issue with this dataset is that it contained images of eight different skin lesions, not just of melanoma. So, we had to create a code and a model to incorporate that, and we created a model to identify all eight skin lesions. The lesions were "'melanocytic nevi', 'melanoma', 'benign keratosis-like lesions', 'basal cell carcinoma', 'actinic keratoses', 'vascular lesions', 'dermatofibroma', 'squamous cell carcinoma', 'unknown'". These are different variations of skin lesions present in the dataset. Most of the datasets used by the researchers as mentioned above in the literature survey consist of multiple skin lesions, there are not many datasets which are publicly available which mainly focus on melanoma and are large datasets. We are mainly focusing on melanoma detection which is a part of the dataset used.

The dataset was preprocessed to reduce the image size to reduce the load on the model and the to reduce the total running time. The images were read into the system and reshaped to 100,75 pixels to make them more manageable. The images were then converted into set of pixel arrays used NumPy and changing the dimensions of the image again takes place. This helps speeding up the processing speed. Using Sklearn, we split the data into a "training set" and a "testing set."

The model is the created using Keras API to implement the CNN architecture. In it, the first two layers of the model have 32 filters, and the last two contain 64 filters. A 3 * 3 window is used with the padding as "Same." The images have been down sampled twice using Maxpooling. To prevent the model from overfitting to the data, we use DropOut thrice of 0.25, 0.4, 0.5 After the convolutional layer, we use Flatten, on the output to convert it to 1D format. The acquired output is the input into a dense neural network. The intermediate layers in the model use "ReLU" for activation, and the final layer uses "Softmax".

The activation function elaborates and defines the output of a given node when that node is supplied with an input in the neural network; there are different activation functions based on the work they do when they are supplied with a node. In our model, we have used only two of them, namely "ReLU" and Softmax.

ReLU is a recently developed activation function which is an abbreviation for rectified linear units. The formulae for it are quite simple: "max(0, z) max(0, z)" despite how it looks is not "linear" and provides us with the same properties as "sigmoid" with advanced functional capabilities.

Softmax turns the number taken as inputs into probabilities whose sum is equal to 1. It gives a vector as an output that represents the probability distribution of the list of outputs. It is used when we have multiple classes and is ideally used in the output layer where we are trying to attain the probabilities to define the class of each input.

We use the optimizer "adam" and learning rate annealing. The optimizer adam is an extension of the "Stochastic gradient descent" which has in recent times seen a increased range of use in deep learning. The creators of this optimizer explained the characteristics as: straightforward to implement, efficient, less memory required. Invariant to diagonal rescale of gradients, well suited for large dataset problems, appropriate for non-stationary objects, for problems with sparse gradients. It can be seen as a combination of RMSprop and stochastic gradient descent with a momentum.

The "learning rate" is one of the key hyperparameters to set in the neural network. We used the process of learning rate annealing for our model. It starts with a relatively high learning rate and then gradually starts lowering the learning rate during training. The logic behind this approach is that we would like to go over the initial parameters and reach a range of good parameters, and at that moment we would like a learning rate small enough so that we can explore the "deeper, but narrower parts of the loss function." Picking the correct learning rate is very essential as a learning rate with small value will require more processing time, and a large value would not give satisfactory results

We activated the model and ran it for 100 epochs with a batch of the size 128 (Figs. 4 and 5).

5 Result

We activated the model with "model.fit()" and ran it for 100 epochs having a batch size of 128. The model was run on the training data on a on a Intel i5 G7 processor with a 16 GB RAM. It took about 13 h to be executed. At the end of the learning period, the accuracy was around 0.93. As the learning was finished, we tested the model on the test set created when we split the dataset, it gave a low accuracy of 0.681 (Fig. 6).

6 Conclusion

We can see from the above compiled data of related research done in past few years that most of the work is done using neural networks, deep learning, specifically CNNs. It is so because the advancement in the field of AI models has been created

```
Model: "sequential_4"
_________________________________________________________________
Layer (type)                 Output Shape              Param #
=================================================================
conv2d_14 (Conv2D)           (None, 75, 100, 32)       896
_________________________________________________________________
conv2d_15 (Conv2D)           (None, 75, 100, 32)       9248
_________________________________________________________________
max_pooling2d_7 (MaxPooling2 (None, 37, 50, 32)        0
_________________________________________________________________
dropout_10 (Dropout)         (None, 37, 50, 32)        0
_________________________________________________________________
conv2d_16 (Conv2D)           (None, 37, 50, 64)        18496
_________________________________________________________________
conv2d_17 (Conv2D)           (None, 37, 50, 64)        36928
_________________________________________________________________
max_pooling2d_8 (MaxPooling2 (None, 18, 25, 64)        0
_________________________________________________________________
dropout_11 (Dropout)         (None, 18, 25, 64)        0
_________________________________________________________________
flatten_4 (Flatten)          (None, 28800)             0
_________________________________________________________________
dense_7 (Dense)              (None, 128)               3686528
_________________________________________________________________
dropout_12 (Dropout)         (None, 128)               0
_________________________________________________________________
dense_8 (Dense)              (None, 9)                 1161
=================================================================
Total params: 3,753,257
Trainable params: 3,753,257
Non-trainable params: 0
_________________________________________________________________
```

Fig. 4 Model summary

which can automatically extract features and learn from them. Most of the work has been done on some variation of the CNN, e.g., using different no. of hidden layer, using external feature extractors to help with the classification, etc. Some have even used the pretrained models of CNN like the AlexNet, etc. Lack of publicly available data can be seen from the study, some of the data sets have just 100–200 images which is really low, considering we need to train a neural network. The data was augmented by processing and changing the degree of rotation to make the system feel as if it was a new image, there are even some libraries create specifically for data preprocessing and augmentation. There were a number of external models used for feature extraction by the authors, e.g., "local binary pattern (LBP)," "edge histogram (EH)," "histogram of oriented gradients (HOG)," "Gabor method." These authors used these architectures to support the classification model in identifying the image.

```
executed in 13h 43m 4s
Epoch 91/100
20264/20264 [==============================] - 434s 21ms/step - loss: 0.1647 - mae: 0.0195 - acc: 0.9396
Epoch 92/100
20264/20264 [==============================] - 432s 21ms/step - loss: 0.1728 - mae: 0.0201 - acc: 0.9354
Epoch 93/100
20264/20264 [==============================] - 432s 21ms/step - loss: 0.1704 - mae: 0.0200 - acc: 0.9359
Epoch 94/100
20264/20264 [==============================] - 433s 21ms/step - loss: 0.1696 - mae: 0.0198 - acc: 0.9379
Epoch 95/100
20264/20264 [==============================] - 450s 22ms/step - loss: 0.1657 - mae: 0.0197 - acc: 0.9377
Epoch 96/100
20264/20264 [==============================] - 438s 22ms/step - loss: 0.1712 - mae: 0.0199 - acc: 0.9369
Epoch 97/100
20264/20264 [==============================] - 459s 23ms/step - loss: 0.1599 - mae: 0.0190 - acc: 0.9389
Epoch 98/100
20264/20264 [==============================] - 455s 22ms/step - loss: 0.1681 - mae: 0.0191 - acc: 0.9397
Epoch 99/100
20264/20264 [==============================] - 467s 23ms/step - loss: 0.1568 - mae: 0.0186 - acc: 0.9424
Epoch 100/100
20264/20264 [==============================] - 579s 29ms/step - loss: 0.1704 - mae: 0.0193 - acc: 0.9383
```

Fig. 5 Screenshot of model running

```
5067/5067 [==============================] - 29s 6ms/step
The accuracy of the model is 0.681
The Loss in the model is 1.950
```

Fig. 6 Screenshot of result on test set

The classification is done by the CNN architecture mostly, and Keras has provided us with inbuilt functions that make it easy to personalize and create a neural network with CNN architecture. The low accuracy of our designed model did not sit well with us, on further study it was determined that the culprit was the unbalanced dataset. As the number of images for different skin lesion was different, this caused the model to learn from the training set in a skewed way which reduced the accuracy of the created model. We can see the uneven form of the dataset in the figure above. We had our doubts in the beginning, as we thought that the large size of the dataset would cover for the uneven numbers of lesion data, but we were proven wrong.

References

1. Mukherjee, S., Adhikari, A., & Roy, M. (2019). *Using cross-platform dataset with deep learning CNN architecture*. Springer Singapore.
2. Lakhtakia, R., Mehta, A., & Nema, S. K. (2009). Melanoma: A frequently missed diagnosis. *Medical Journal Armed Forces India (MJAFI), 65*(3), 292–294. https://doi.org/10.1016/S0377-1237(09)80036-1.
3. Uddin, M. S., & Bansal, J. C. (2020). *Proceedings of International Joint Conference on Computational Intelligence*, (vol. 669). Springer Singapore.
4. S. Gulati and R. K. Bhogal, Detection of malignant melanoma using deep learning, vol. 1045. Springer Singapore, 2019.

5. Attia, M., Hossny, M., Nahavandi, S., & Yazdabadi, A. (2017). Skin melanoma segmentation using recurrent and convolutional neural networks. In *Proceedings—International Symposium on Biomedical Imaging* (pp. 292–296). https://doi.org/10.1109/isbi.2017.7950522.
6. C´ıcero, F. M., Oliveira, A. H. M., & Botelho, G. M. (2016). Deep Learning and Convolutional Neural Networks in the Aid of the Classification of Melanoma. Sibgrapi (pp. 1–9).
7. Seeja, R. D., & Suresh, A. (2019). Deep learning based skin lesion segmentation and classification of melanoma using support vector machine (SVM). *Asian Pacific Journal of Cancer Prevention, 20*(5), 1555–1561. https://doi.org/10.31557/APJCP.2019.20.5.1555.
8. Karabulut, E. M., & Ibrikci, T. (2016). Texture analysis of Melanoma images for computer-aided diagnosis. (1), 26–29. https://doi.org/10.15242/iae.iae0416011.
9. Brinker, T. J., et al. (2019). Deep learning outperformed 136 of 157 dermatologists in a head-to-head dermoscopic melanoma image classification task. *European Journal of Cancer, 113,* 47–54. https://doi.org/10.1016/j.ejca.2019.04.001.
10. Codella, N. C. F., et al. (2017). Deep learning ensembles for melanoma recognition in dermoscopy images. *IBM Journal of Research and Development, 61*(4–5), 1–15.
11. Valle, E., et al. (2017, November). Data, depth, and design: Learning reliable models for skin lesion analysis. https://doi.org/10.1016/j.neucom.2019.12.003.
12. Esteva, A., et al. (2017). Dermatologist-level classification of skin cancer with deep neural networks. *Nature, 542*(7639), 115–118. https://doi.org/10.1038/nature21056.
13. Fujisawa, Y., et al. (2019). Deep-learning-based, computer-aided classifier developed with a small dataset of clinical images surpasses board-certified dermatologists in skin tumour diagnosis. *British Journal of Dermatology, 180*(2), 373–381. https://doi.org/10.1111/bjd.16924.
14. Han, S. S., Kim, M. S., Lim, W., Park, G. H., Park, I., & Chang, S. E. (2018). Classification of the clinical images for benign and malignant cutaneous tumors using a deep learning algorithm. *Journal of Investigative Dermatology, 138*(7), 1529–1538. https://doi.org/10.1016/j.jid.2018.01.028.
15. Hekler, A., et al. (2019). Deep learning outperformed 11 pathologists in the classification of histopathological melanoma images. *European Journal of Cancer, 118,* 91–96. https://doi.org/10.1016/j.ejca.2019.06.012.
16. Salido, J. A. A., & Ruiz, C. (2018). Using deep learning for melanoma detection in dermoscopy images. *International Journal of Machine Learning and Computing, 8*(1), 61–68. https://doi.org/10.18178/ijmlc.2018.8.1.664.
17. Jafari, M. H., Nasr-Esfahani, E., Karimi, N., Soroushmehr, S. M. R., Samavi, S., & Najarian, K. (2017). Extraction of skin lesions from non-dermoscopic images for surgical excision of melanoma. *International Journal of Computer Assisted Radiology and Surgery, 12*(6), 1021–1030. https://doi.org/10.1007/s11548-017-1567-8.
18. Mitkowski, W., Kacprzyk, J., Oprzędkiewicz, K., & Skruch, P. (2017). *Trends in Advanced Intelligent Control, Optimization and Automation: Proceedings of KKA 2017-The 19th Polish Control Conference*, Kraków, Poland, 18–21 June, 2017. *Advanced Intelligent Systems Computing, 577*, 848–856. https://doi.org/10.1007/978-3-319-60699-6.
19. Li, Y., & Shen, L. (2018). Skin lesion analysis towards melanoma detection using deep learning network. *Sensors (Switzerland), 18*(2), 1–16. https://doi.org/10.3390/s18020556.
20. Ferreira, C. A., Cunha, A., Mendonça, A. M., & Campilho, A. (2019). Convolutional neural network architectures for texture classification (Vol. 1, pp. 783–791). https://doi.org/10.1007/978-3-030-13469-3.
21. Masood, A., Al-Jumaily, A., & Anam, K. (2015) Self-supervised learning model for skin cancer diagnosis. In *International IEEE/EMBS Conference on Neural Engineering NER*, vol. 2015-July, pp. 1012–1015. https://doi.org/10.1109/ner.2015.7146798.
22. https://adventuresinmachinelearning.com/convolutional-neural-networks-tutorial-tensorflow/

Emotion Detection from Audio Using SVM

Rajneesh Kumar and Akash Punhani

Abstract Emotion detection from speech audio stream has been one of the most attentive areas of research for many researchers. Many features are extracted of the audio file are extracted in spectral and perceptual domain. Features like pitch, energy, linear predictive cepstral coefficient (LPCC), formant, mel-frequency cepstral coefficient (MFCC) and their respective statistical values are calculated. In this experiment, we took the mean of MFCC and delta MFCC. And trained SVM on a total eight emotions. The SVM attained an accuracy of 85% on male and female audio files and 83% on whole (male plus female) audio files.

Keywords Speech · Emotion recognition · SVM · MFCC · d-MFCC

1 Introduction

Emotions are one of the most crucial parts of communication. Emotions bring life to the communication and help make the exchange of information crystal clear. Emotion detection using the means of computer machine has been an attentive area of research for many researchers and scholars. Each one of them has done their part in bringing the notable results to get the emotions from a simple audio stream.

Emotion detection from speech has been done by converting the speech first into text and then the text thus obtained is analyzed for emotional content. Extracting the emotional content from audio using speech-based features has been done by many researcher and scholars. Like the work done by Nicholson et al. [1] used neural network and attained 50% accuracy to detect eight emotions. In [2], Kang et al. used pitch, energy and speaking rate and trained MLB, NN and HMM models, out which HMM attained higher accuracy of 89% to detect six emotional states followed by NN having 69% and MLB attained 56% accuracy. In [3], Nogueiras et al. used pitch, energy, articulation and spectral shape as features to train a HMM model which attained 80% accuracy over seven emotional state classification. In [4], Danishman

R. Kumar (✉) · A. Punhani
Department of Computer Science and Engineering, Amity University, Noida, Uttar Pradesh, India
e-mail: rajneeshbaisoya@gmail.com

S. Tiwari et al. (eds.), *Proceedings of International Conference on Big Data, Machine Learning and their Applications*, Lecture Notes in Networks and Systems 150,
https://doi.org/10.1007/978-981-15-8377-3_22

et al. used F0 formant, total energy and MFCC values as input speech features and trained a SVM model on EmoDB that obtained 63.5% accuracy to classify seven emotional states. Xiao et al. [5] extracted a total of 68 features and used hierarchical classification to attained accuracy as high as 70% on Berlin dataset to classify six emotional states. In [6], Shen et al. classified five emotions from Berlin emotional database. They used LPCC, MFCC, pitch, energy and linear prediction coefficient and mel-cepstrum coefficient (LPCMCC). Their SVM model obtained 82% accuracy. Ingale and Chaudhari [7] extracted energy, pitch, LPCC and MFCC features. Their SVM accuracy for speaker independent was 75% and for speaker dependent was 80%.

Since the growth of deep learning algorithms, many researchers used them for their research and these algorithms gives better results with right type and right amount of input features. In [8], Venkataramanan and Rengaraj extracted features such as log-mel spectrogram, MFCCs, pitch and energy and trained deep learning models like convolutional neural network (CNN) which attained 68% overall accuracy. In [9], Luo et al. used CNN and long short-term memory (LSTM) networks in parallel. And trained the final model over mel-spectrogram features which obtained overall accuracy of 65%.

These were some of the notable research work done by now and still counting. In this paper, eight emotional states classification has been done using minimum features vector trained using SVM classifier.

2 System Implementation

2.1 Features Used

2.1.1 Mel-Frequency Cepstral Coefficients (MFCC)

MFCC is remarkable common feature when it comes to the audio analysis. Because of its unique values of each audio sample, it is broadly used in the speech recognition, speaker recognition, music and in any matter which includes the analysis of audio.

MFCC is said to depict the behavior of human ear. The sensitivity of human ears to a speech signal having low frequency and higher frequency is represented in numerical values as MFCC coefficient. And as the emotional content of speech is basically based on frequency of speech audio, so MFCC features are best bet to detect emotional content from speech signal.

Extracting the MFCC is pretty straightforward. It had been explained pretty well in [10]. It includes some basic mathematical calculations which are shown in Fig. 1.

First start with pre-emphasis, in which the speech is passed through high pass filter. The output audio is similar to input but sharper with little lower volume. This new audio is then divided into frames of 20–30 ms with an overlap of 1/3–1/2 of size of frame. This step is crucial to obtain FFT. The frames thus obtained are multiplied with

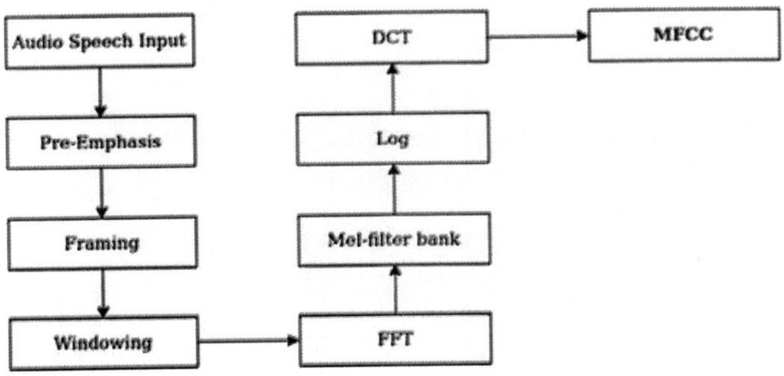

Fig. 1 Extracting MFCC

a hamming window. Then, FFT equations are applied on these hamming windows. The values thus obtained are passed over mel-filter bank. The log of these values obtained from mel-filter bank is taken. And in the end, discrete cosine transform (DCT) of these log values is calculated. And thus we get the MFCC features.

2.1.2 Delta Coefficient

Delta coefficient are used as additionally to extraction of MFCC features. The accuracy of model is increased using the delta coefficients along with any feature like MFCC here. And here is the equation to calculate delta coefficient:

$$d_t = \frac{\left(\sum_{n=1}^{N} n\left(c_{(t+n)} - c_{(t-n)}\right)\right)}{\left(2\sum_{n=1}^{N} n^2\right)} \tag{1}$$

where d_t is a delta coefficient, from frame t computed in terms of the static coefficients c_{t+n} to c_{t-n}. A common value for N is 2.

2.2 *Dataset*

For this paper, the Ryerson Audio-Visual Database of Emotional Speech and Song (RAVDESS) dataset [11, 12] was adopted. This dataset contains both male and female files of speech and song, their audio-only (16-bit, 48 kHz, in .wav format), video-only (no sound) and both audio-video (720p, H.264, AAC, 48 kHz, in .mp4 format) of 24 actors. In which, 12 are males and 12 are females, speaking two statements in neutral North American accent. From this, only speech audio files of male actors were used. Total only speech male audio files were 720. The length of each file was different. So first the unvoiced part or silence part of the audio was removed. This dataset

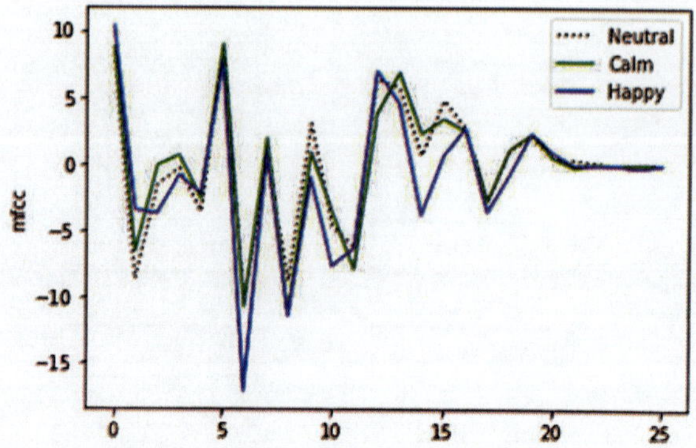

Fig. 2 MFCC plot of neutral versus calm versus happy

contains total eight emotions, viz., sad, fearful, happy, angry, calm, disgust, neutral and surprised. These emotions are also spoken in two different levels of intensities, viz., normal and strong. And then the feature extraction process was applied.

RAVDESS consists of total 7356 files. Out of which, audio-only files contain speech files consists of 1440 files, 60 trials per actor. Song files are 1012, 44 trials per actor (only 23, Actor_18 has no song files).

Audio-visual and video-only contains speech files total 2880 in count, 60 trials per actor in two modalities (audio-visual, video-only). And song files are 2024, 44 trials per actor in two modalities (audio-visual, video-only).

2.3 *Feature Extraction*

In this paper, we took only MFCC features and their deltas employing standard values. Total 25 MFCC features were derived from the dataset file. Above the 25 features, the raise in performance of classifier is nominal.

After taking the MFCC and their deltas, the mean of all frames for each MFCC coefficients was taken and same for their delta values (Figs. 2, 3, 4, 5, 6 and 7).

2.4 *Support Vector Machine (SVM) Classifiers*

SVM is supervised-learning technique that associates training algorithm used to analyze the data used for classification and regression. SVM can classify data linearly and nonlinearly as well. It is easy to train and gives the results fast.

SVM generally perform best for binary classification. But recently SVM are also being used for the multi-class classification given distinct features set for each class to get better results. The SVM performance measures depend on the *C* and the kernel used. At present, SVM supports linear, polynomial and Gaussian radial basis function (RBF).

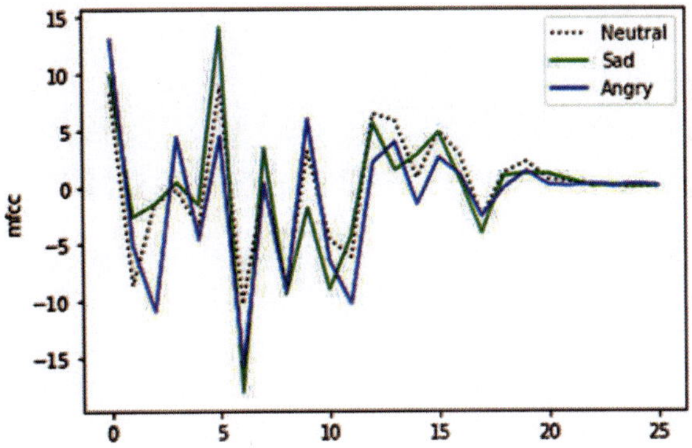

Fig. 3 MFCC plot of neutral versus sad versus angry

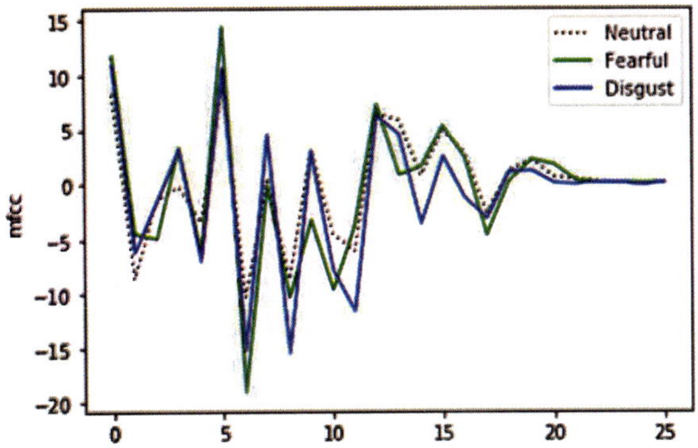

Fig. 4 MFCC plot of neutral versus fearful versus disgust

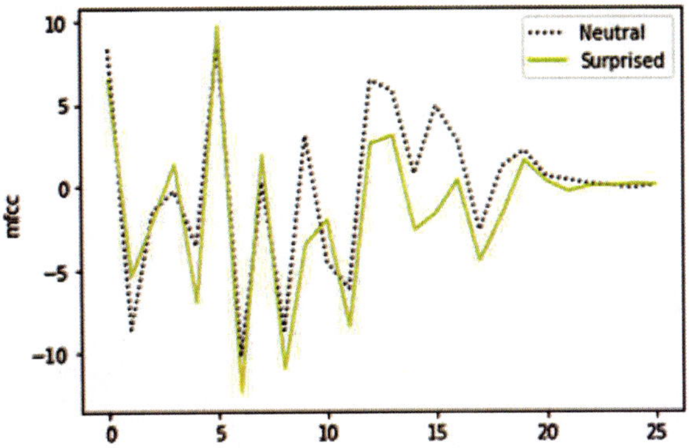

Fig. 5 MFCC plot of neutral versus surprised

3 Experimentation and Results

For this paper, the Gaussian radial basis function (RBF) was used with value of $C = 1$. From the RAVDESS dataset, only audio speech of males was taken that sums

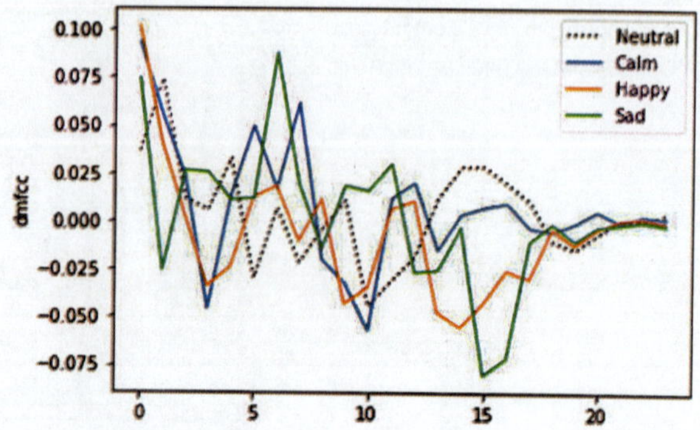

Fig. 6 d-MFCC plot of neutral versus calm versus happy versus sad

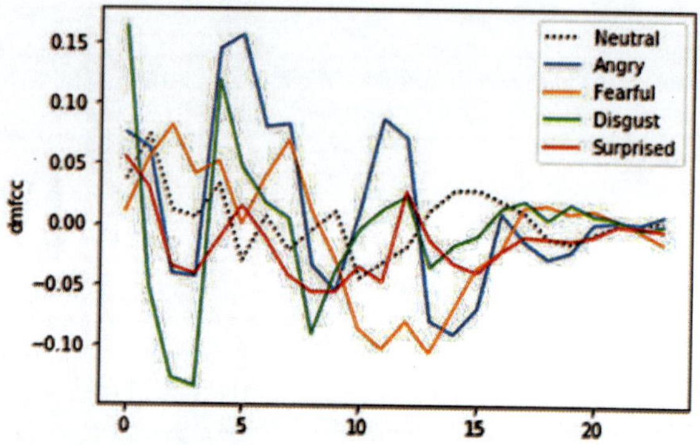

Fig. 7 d-MFCC plot of neutral versus angry versus fearful versus disgust versus surprised

up to 720 file of males only from all eight emotions. The dataset was divided into training and testing in 8:2 ratios (Tables 1, 2 and 3).

The SVM was trained with above configuration on the features described in Sect. 2. The correct result of overall classification of emotional state of male and female audio exceeded 85 and 83% on whole (male and female) audio files.

Table 1 SVM normalized confusion matrix of all eight emotions (male)

Emotion	Sad	Fearful	Happy	Angry	Calm	Disgust	Neutral	Surprised
Sad	**89**	0	6	0	0	0	11	0
Fearful	0	**90**	0	0	0	0	0	0
Happy	12	0	**75**	12	0	0	0	0
Angry	0	12	18	**100**	12	24	0	0
Calm	0	0	0	0	**91**	0	4	4
Disgust	0	0	0	0	0	**83**	0	0
Neutral	0	0	0	0	0	0	**40**	0
Surprised	0	0	0	0	0	0	0	**93**

Table 2 SVM normalized confusion matrix of all eight emotions (female)

Emotion	Sad	Fearful	Happy	Angry	Calm	Disgust	Neutral	Surprised
Sad	**78**	0	0	0	6	11	0	0
Fearful	7	**93**	13	0	0	0	0	20
Happy	0	0	**85**	0	0	0	0	4
Angry	12	0	0	**94**	0	6	0	0
Calm	0	0	0	0	**91**	0	5	0
Disgust	5	0	0	0	5	**84**	5	20
Neutral	0	0	0	0	0	0	**40**	0
Surprised	0	0	0	0	0	0	0	**82**

Table 3 SVM normalized confusion matrix of all eight emotions (male and female)

Emotion	Sad	Fearful	Happy	Angry	Calm	Disgust	Neutral	Surprised
Sad	**82**	12	12	0	0	0	0	0
Fearful	6	**88**	0	6	0	0	0	6
Happy	0	0	**80**	4	0	0	0	0
Angry	0	0	4	**88**	0	4	0	4
Calm	0	0	0	0	**90**	0	10	0
Disgust	11	0	0	6	11	**94**	6	17
Neutral	0	0	0	0	0	0	**50**	0
Surprised	0	0	12	0	0	0	0	**71**

As it can be seen from the confusion matrix that the SVM model has been able to classify angry better and its performance to classify calm, sad, fearful and surprised has been amazing for individual gender. But in case of whole, the recognition accuracy is above 90% for calm and disgust. For emotional states, happy, sad, angry and fearful are also encouraging.

This work only used one feature set from the various features comprised in a speech audio signal. There is a very fine article [13] that reviewed all features a speech or music audio file can have physically and perceptually.

4 Conclusion and Future Work

MFCC is a strong feature in matter of classifying emotional state as the values are unique from each emotional state. But along with the deltas of MFCC the performance of machine learning model can increase. And here the MFCC along with the delta coefficients has been very helpful to classify eight emotional states with very encouraging accuracy.

Table 4 Model comparison table

Author	ML model	Features	Accuracy	Emotional states
J. Nicholson	Neural network	Speech power, pitch, LPC, delta-LPC	50	8
B. Kang et al.	MLB, NN and HMM	Pitch, energy and speaking rate	HMM = 89, NN = 69, MLB = 56	6
A. Nogueiras et al.	HMM	Pitch, energy, articulation and spectral shape	80	7
T. Danishman et al.	SVM	MFCC, total energy and F0 formant values	63.5	7
Z. Xiao et al.	Hierarchical classification	Pitch, energy, etc. (total 68 features)	70	6
P. Shen et al.	SVM	Energy, pitch, LPCC, MFCC and linear prediction coefficient and mel-cepstrum coefficient (LPCMCC)	82	6
A. Ingale et al.	SVM	Energy, pitch, LPCC and MFCC	75	6
K. Venkataramanan et al.	CNN	Log-mel spectrogram, MFCCs, pitch and energy and their first and second derivatives	68	7
Z. Luo et al.	CNN and LSTM	Mel-spectrogram features	65	7
Proposed	**SVM**	**Statistic mean of MFCC and its delta**	**85**	**8**

As the neural networks are now being used to solve almost every problem because of their ability to give better results. In the future work, the deep learning model will be implemented using same features and new features that can effectively classify emotions will be extracted and will be used in deep learning neural networks (Table 4).

References

1. Nicholson, J., Takahashi, K., & Nakatsu, R. (2000). Emotion recognition in speech using neural networks. *Neural Computing and Applications, 9*(4), 290–296. https://doi.org/10.1007/s005210070006.
2. Kang, B. S., Han, C. H., Lee, S. T, Youn, D. H., & Lee, C. (2000). Speaker dependent emotion recognition using speech signals. In *6th International Conference on Spoken Language*

Processing: ICSLP 2000, ICSLP, pp. 6–9.

3. Nogueiras, A., Moreno, A., Bonafonte, A., & Mariño, J. B. (2003). Speech emotion recognition using hidden Markov models. *Speech Communication, proof*, 2–5.
4. Danisman, T., & Alpkocak, A. (2008). Emotion classification of audio signals using ensemble of support vector machines. In *Lecture Notes in Computer Science (including subseries Lecture Notes in Artificial Intelligence (LNAI) and Lecture Notes in Bioinformatics (LNBI))* (Vol. 5078, LNCS, pp. 205–216). https://doi.org/10.1007/978-3-540-69369-7_23.
5. Xiao, Z., Dellandrea, E., Dou, W., & Chen, L. (2008). Automatic hierarchical classification of emotional speech (pp. 291–296), January, 2008. https://doi.org/10.1109/ism.workshops.2007.56.
6. Shen, P., Changjun, Z., & Chen, X. (2011). Automatic speech emotion recognition using support vector machine. In *Proceedings of 2011 International Conference on Electronic and Mechanical Engineering and Information Technology, EMEIT 2011* (Vol. 2, pp. 621–625). https://doi.org/10.1109/EMEIT.2011.6023178.
7. Ingale, B., & Chaudhari, D. S. (2016). Speech emotion recognition. *International Journal of Engineering and Technology, 8*(1), 311–323.
8. Venkataramanan, K., & Rengaraj, R. (2017). Emotion recognition from speech. *Cognition Technology,* no. 9783319436647, 409–428. https://doi.org/10.1007/978-3-319-43665-4_20.
9. Luo, Z., Xu, H., & Chen, F. (2019). Utterance-based audio sentiment analysis learned by a parallel combination of CNN and LSTM, November, 2018. Accessed: April 29, 2019. [Online]. Available: https://arxiv.org/abs/1811.08065.
10. Pal Singh, P. (2014). An approach to extract feature using MFCC. *IOSR Journal of Engineering, 4*(8), 21–25. https://doi.org/10.9790/3021-04812125.
11. Livingstone, S. R., & Russo, F. A. (2018). *The Ryerson audio-visual database of emotional speech and song (RAVDESS)*.
12. RAVDESS. https://zenodo.org/record/1188976#.Xqp3mfnhUw8.
13. Alías, F., Socoró, J. C., & Sevillano, X. (2016). A review of physical and perceptual feature extraction techniques for speech, music and environmental sounds. https://doi.org/10.3390/app6050143.

Plant Disease Detection Using Image Classification

Ankit Chhillar and Sanjeev Thakur

Abstract Indian agriculture yields have been facing major problems that lead to the fall in overall production. The major factor can be listed as proper irrigation, availability of water, fertile land, and plant disease. Farmers invest a huge sum of money on disease management, regularly without effective technical support, causing poor disease control, contamination, and harmful outcomes. Early detection of disease not only helps in imposing quick protective major but as well as increase overall farm yield. In earlier times, the process of disease detection was done manually with the help of experts, costing a fortune for the local farmers, and sometimes, unavailability of any expertise help leads to a huge loss in farmlands. Digital image processing works as a time and cost-efficient alternative to the manual process. This process is proven to be much faster and gives better results than the manual method. Deep learning and CNN are widely used for the purpose of image classification (Eldem et al. in A model of deep neural network for iris classification with different activation functions, 2018; Abu et al. in Int J Eng Res Technol 12(4):563–569, 2019) [1, 2]. This paper focuses on developing an artificial intelligence (M. Approach in A modern approach) [3]-based plant disease detection technique based upon computer vision (Sladojevic et al. in Comput Intell Neurosci 2016, 2016) [4]. The dataset for the study was downloaded from the kaggle which contains 14 different categories of plants. For this paper, we mainly focused on "4" category of plants (corn, pepper, potato, and tomato). "50" images from each category were taken, and the maximum accuracy which was achieved through the algorithm is "96.54%". This algorithm is suitable getting maximum accuracy from a small dataset.

Keywords Plant disease detection · CNN · Deep learning · Neural network

A. Chhillar · S. Thakur (✉)
Amity School of Engineering and Technology, Amity University, Noida, Uttar Pradesh, India
e-mail: sthakur3@amity.edu

A. Chhillar
e-mail: ankitchhillar001@gmail.com

S. Tiwari et al. (eds.), *Proceedings of International Conference on Big Data, Machine Learning and their Applications*, Lecture Notes in Networks and Systems 150,
https://doi.org/10.1007/978-981-15-8377-3_23

1 Introduction

The historical backdrop of agriculture in India goes back to Indus Valley Civilization and even before that in certain spots of Southern India. India positions second worldwide in agriculture yields. India positions first on the planet with the highest net cropped region followed by the USA and China. The monetary commitment of farming to India's GDP is consistently declining with the nation's broad-based economic growth. All things considered, farming is demographically the broadest economic sector and assumes a huge job in the general financial structure of India. Farmers produce various crops throughout the year, keeping the demands of the citizen in check. Farmers also produce vegetables for the source of income and to meet the daily needs of the consumers. "Potato" and "tomato" are the most consumed vegetables in India. Meeting the day-to-day need of these vegetable places a huge burden upon farmers.

Farmers and experts are continuously working on strategies for producing more yield from the farmlands to meet the needs of the consumers. From a long time, farmers have been facing various problems regarding the loss in the quantity of the yield to sometimes total loss of the plantation. Disease being major causes affecting the efficient production of the farm goods. Regular monitoring and use of different chemicals for crop prevention should be done. Taking helps from experts regularly or using the manual technique for disease detection and prevention is very time-consuming and cost a lot of fortune. Sometimes it may happen that there might be unavailability of experts in the particular area, and not getting proper observation and treatment can lead to crop loss. In these situations, image processing has an important role in disease identification.

Digital image processing makes use of different image processing algorithms with accurate and efficient disease detection. Convolutional neural network (CNN) [5] and deep learning [6] play an important role as it can be used for the feature extraction or it can be termed as extraction of identification features from an image which then further be used for the purpose of classification. This study focuses on the detection of disease from the set of "4" plants named as "corn," "pepper," "potato," and "tomato" with different infections affecting the plants. Classification of the different disease from the listed "4" category is done.

Some diseases affecting the crops are "corn: northern leaf blight, pepper: bacterial spots, potato: early blight, tomato: sectorial leaf spot."

2 Related Work

We have premediated and studied some of the work previously done from the year 2015–2019.

1. Eldem et al. in [1] present a deep neural network model for iris flower classification. Utilizing deep neural network which utilizes the length and width of

the petal and sepal feature as input they try to classify "3" different types of iris flower. Experiments were done by altering the epoch number or the activation function used for the classification. They were able to achieve "96%" accuracy using the proposed methodology.

2. Abu et al. in [2] present a study on image classification based on deep learning and tensor flow. The model was trained using python on the tensor flow framework. Tensor flow being a symbolic library is used for machine learning application. The study was performed on trying to classify the "5" different types of flower. Using deep neural network, they were able to get an accuracy of "90.585%" for rose flower prediction and up to 90% for the other flowers in the dataset.
3. In [5], Dhillon and Verma presented a review of CNN methodologies and its applications. Deep learning shows impressive performance for image classification and includes a detailed description of various CNN models like LeNet, AlexNet, ZFNet, GoogleNet, VGGNet, ResNet, ResNeXt, SENet, DenseNet, Xception, and PNAS/ENAS.
4. Wahab et al. in [7] proposed an efficient chili disease detection technique. An effective way for early detection chili disease through leaf features inspection is discussed. Through their study, they determine that in farms the chemical is applied periodically without considering the requirements of the plants. Using this technique, they will be able to ensure that the chemical will only be applied when the disease is detected in the plants. The proposed model first performs feature extraction on the captured chili leaf images, and using the image classification techniques to detect disease in the plant.
5. Khirade in [8] discuss methodology for plant disease identification. They discuss the use of image segmentation and feature extraction algorithms used for the purpose of plant disease detection. Throughout their research, they use various image processing techniques like image acquisition, image preprocessing, image segmentation, and feature extraction from images. They also use k-means clustering for classification. They conclude that the use of ANN methods such as self-organizing feature map, backpropagation algorithm, SVMs has proven effective.
6. Cassava is a large source of for human food and more vulnerable to disease. Ramcharan et al. in [9] proposed a deep learning algorithm for cassava disease detection. Training a deep learning CNN is to predict cassava disease. Through their algorithm, they achieve an overall accuracy of 93%, concluding that the learning approach offers a fast, effective, and affordable that can easily be deployed.
7. Ferentinos [10] used convolutional neural network for the purpose of plant disease detection. They used different deep learning methodologies for diagnosis of simple leaves images for healthy and disease plants. Training was performed with a total 87,848 plant datasets. They were able to get a very high success and a very useful early warning tool.

8. Sheikh et al. in [11] propose image processing technique for maize and peach leaf disease detection. Using convolutional neural network, they achieved a satisfactory result through which the disease can be identified at an early stage and protective measures can be taken to prevent agriculture loss.
9. Valdoria et al. [12] developed a mobile application for terrestrial plant disease detection. They made use of image processing using deep learning. Results were trained using classification models to identify the disease. Accuracy of the algorithm was tested using F1 score testing method. They were able to achieve satisfactory results, thus supporting the functionality, usability, accuracy, functionality, and portability performance of the application.

3 Type of Disease

Plant diseases are caused by microorganisms or infections and can seriously harm mild leaf, fruits and can result in death of the plant. Listed below are some of the major plant disease:

Bacterial: Any group of tiny single-celled life frames that live in monstrous numbers in practically every condition on Earth, from remote ocean vents unnecessarily far underneath Earth's surface to the stomach related tracts of people.
Fungus: They are among the most comprehensively appropriated living creatures on Earth and have a fantastic characteristic and clinical significance. Various parasites are free-living in soil or water; others structure organism or worthwhile relationship with plants or animals.
Virus: Irresistible specialist of little size and fundamental association that can copy just in living cells of animals, plants, or organisms. The name is from a Latin word meaning "disgusting fluid" or "harmful substance (poison)."

4 Proposed Methodology

This paper presents an image-based plant disease detection technique. This study mainly focuses on vegetable plant for the purpose of disease detection. The plants are listed as corn, pepper, potato, and tomato. Both the healthy and infected plant leaf images were used for training the classification model:

4.1 Flowchart of the Methodology

See Fig. 1.

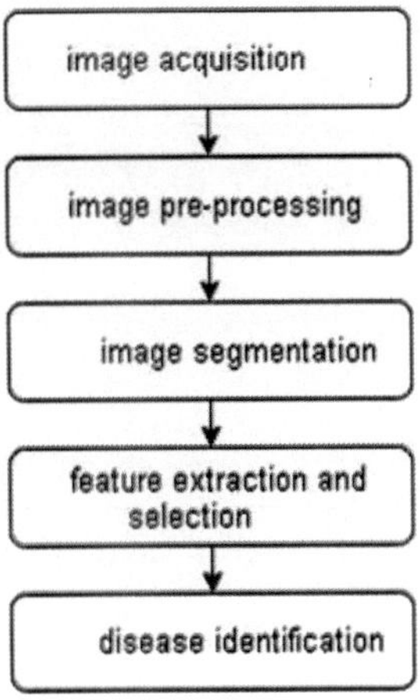

Fig. 1 Images in dataset

4.2 Dataset

Dataset for the study was downloaded from profile of a kaggle user named Samir Bhattarai. This dataset contains 87k RGB images of healthy and infected crop leaves from 38 different classes. For this paper, we selected 4 different classes (corn, pepper, tomato, potato) out of the 38 classes. From the 87k picture, we used 50 images from 4 class and different subclasses. The training and classification were done using 950 total images.

4.3 Image Acquisition

Image acquisition is being the starting step towards the process of image classification. Collecting and creating the dataset are the foremost task which is performed in this step. Image dataset was collected from kaggle containing images in RGB format. The dataset contained various classes from which we choose to study on 4 classes (corn, pepper, tomato, potato). Image was in jpg format. We took 950 total number of images for the classification task.

4.4 Image Preprocessing

From total 87k pictures, we took total of 950 images and split them into two parts. These two parts can be termed as test and train containing 80 and 20% of the total dataset. The split was done using python function split at a test size of 0.2 and a random state of 42. This will split the dataset into train which contains the images for the training purpose and test which include the images which will be used for

Fig. 2 Plant disease dataset

the testing purpose of the algorithm. The four different classes of the dataset contain various subclasses like corn-maize (cercospora-leaf spot-Gray-leaf spot and healthy), pepper-bell (bacterial-spot and healthy), potato (early blight and healthy), tomato (bacterial-spot and healthy) leaf images. 50 images from each subcategory resulting in 950 total number of mages. For accurate classification, all the images in the dataset should have same dimension, and for the study, we have taken the images with a dimension of 256 * 256. Some pictures from the dataset are shown in Fig. 2 [20].

4.5 *Image Segmentation*

This process includes altering the representation of the image into something meaning which can be easily analyzed. This process divides the image into multiple segments (set of pixels or arrays) and typically used to locate the edge and the boundaries of the object in the image. In this paper, the imported images from the dataset are sending to an image_to_array function for creating the image data to a numpy array. Changing the dimension of the images is again performed in this step before the conversion. After that the images are converted into a numpy array using inbuilt numpy function. Numpy array helps in speeding up the processing speed and requires a less amount of time. Exceptions are also checked, and if found an error message will be displayed.

4.6 *Binary Labeling and Augmentation*

Multi-label image classification is done in this process with the help of multilabel binarizer python library. It is a one-versus-all fashion means there are multiple object categories in an image. This step is done to create the list of different classes of images from the dataset (Fig. 3).

Augmentation is done for the purpose of increasing the diversity of data. This process does not involve the collection of new images from the dataset but itself transforms the previous images.

```
[[0.5376   0.569    0.6445  ]
 [0.6445   0.676    0.751   ]
 [0.511    0.542    0.6177  ]
 ...
 [0.3157   0.4443   0.3733  ]
 [0.2311   0.3555   0.271   ]
 [0.4443   0.5645   0.4756  ]]

[[0.4443   0.4756   0.5513  ]
 [0.547    0.5776   0.6533  ]
 [0.5957   0.6265   0.6934  ]
 ...
 [0.3333   0.4622   0.391   ]
 [0.4756   0.6      0.511   ]
 [0.2844   0.409    0.3066  ]]
```

```
['Corn_(maize)___Cercospora_leaf_spot Gray_leaf_spot'
 'Corn_(maize)___Common_rust_' 'Corn_(maize)___Northern_Leaf_Blight'
 'Corn_(maize)___healthy' 'Pepper,_bell___Bacterial_spot'
 'Pepper,_bell___healthy' 'Potato___Early_blight' 'Potato___Late_blight'
 'Potato___healthy' 'Tomato___Bacterial_spot' 'Tomato___Early_blight'
 'Tomato___Late_blight' 'Tomato___Leaf_Mold' 'Tomato___Septoria_leaf_spot'
 'Tomato___Spider_mites Two-spotted_spider_mite' 'Tomato___Target_Spot'
 'Tomato___Tomato_Yellow_Leaf_Curl_Virus' 'Tomato___Tomato_mosaic_virus'
 'Tomato___healthy']
```

Fig. 3 Image numpy array and dataset label

Fig. 4 Original, rotated, width-shift, height-shift, zoom

In this algorithm, we perform various augmentations for increasing the diversity of the data: rotation-range = 25, width-shift range = 0.1, height-shift range = 0.1, shear range = 0.2, zoom range = 0.2, horizontal-flip = True, fill-mode = nearest (Fig. 4) [20].

4.7 CNN Architecture

The architecture of the CNN model starts with taking a basic sequential model and adding multiple convolutional layers to it. First we set the dimensions of the channel to −1. Checking of the image data format is performed, and if found equal to the channel first, then we will change the input shape of the image by supplying the depth, height, and width initialized in the beginning.

For convolutional layer, 1 of the model contains 32 feature with a 3 * 3 window Conv2D (32, (3 * 3)), padding "same" and the input shape is equal to the default input shape. ReLu is used as the activation function, batch normalization (axis = chanDim). Max pooling is performed for the pool size of 3 * 3, and a dropout of 0.25 is performed at the end of the first convolution layer. For convolution layer, 2nd Conv2D (64, (3 * 3)) filter size 64, input size 3 * 3, padding "same," batch normalization (axis = chanDim). The 3rd convolution layer Conv2D (64 (3 * 3)) have a filter size of 64. Max pooling is performed for the pool size of 3 * 3, and a dropout of 0.25 is performed at the end. In 4th (last) convolution layer, we increase the filter size to 128 and rest of the parameters are same as 2nd layer. The 5th layer is same as 3rd layer with a filter size of 64. After the last convolution layer, we have added flatten and dense layer (128) followed by activation, normalization, and a dropout of 05. Flatten is layer which is used to flatten the input shape $n * c * h * w$ to a simple vector output of shape $n * (c * h * w)$. Dense layer defines the output shape produces by the layer.

At last, a softmax activation function is applied. The total number of trained parameters is 872,019

$$\sigma(z)_j = \frac{e^{z_j}}{\sum_{k=1}^{k} e^{z_k}} \quad \text{for } j = 1, \ldots, k$$

where

z Input Vector,
k Real Number.

4.8 Activation Function

In artificial intelligence machine learning, activation function [13] defines the output of that given node when supplied with an input. There are different activation function based upon their functioning when supplied with an input. For the study, we studied three activation functions (ReLu, sigmoid, tanh). Comparison and choosing are the best from the above-listed activation:

ReLu: An ongoing creation which represents rectified linear units. The recipe is misleadingly straightforward: max (0, z) max (0, z). In spite of its name and appearance, it is not direct and gives indistinguishable advantages from sigmoid however with better execution.

$$f(x) = \max(0, x)$$

where

x Input to Neuron.

Sigmoid: Sigmoid accepts a genuine incentive as info and yields another incentive somewhere in the range of 0 and 1. It is anything but difficult to work with and has all the decent properties of initiation works: it is non-straight, persistently differentiable, monotonic and has a fixed yield go.

$$f(x) = \frac{1}{1 + e^{-x}}$$

where

x input to neuron.

Tanh: Tanh squashes a genuine esteemed number to the range [−1, 1]. It is non-straight. Be that as it may, in contrast to sigmoid, its yield is zero-jogged. Along these lines, by and by the tanh nonlinearity is constantly wanted to the sigmoid nonlinearity.

$$\tanh(x) = 2\text{sigmoid}\ (2x) - 1$$

where

x input to neuron (Table 1 and Fig. 5).

Table 1 Activation functions

Name	Plot	Equation	Derivation
ReLu	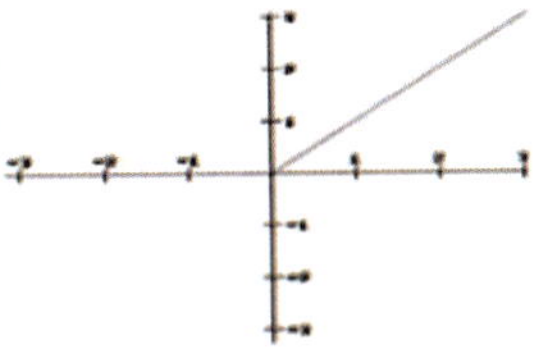	$f(x) = \max(0, x)$	$f'(x) = \begin{cases} 0 \text{ for } x < 0 \\ 1 \text{ for } x \geq 0 \end{cases}$
Sigmoid	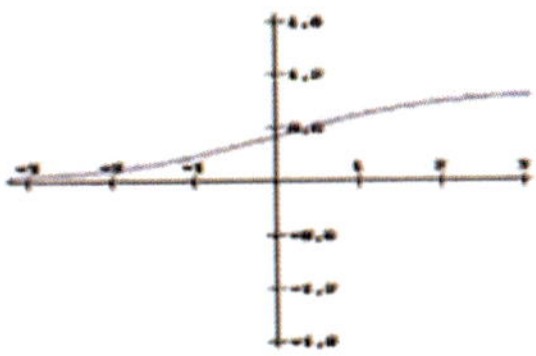	$f(x) = \frac{1}{1+e^{-x}}$	$f'(x) = f(x)(1 - f(x))$
Tanh	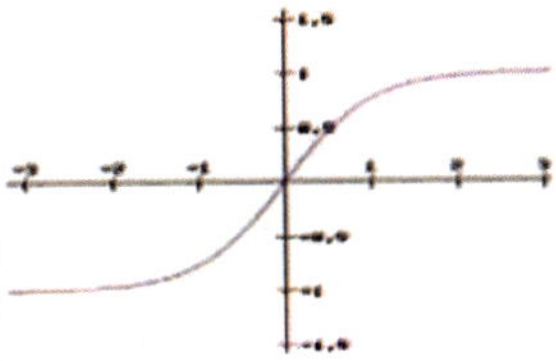	$\tanh(x) = 2\text{sigmoid}(2x) - 1$	$f'(x) = 1 - f(x)^2$

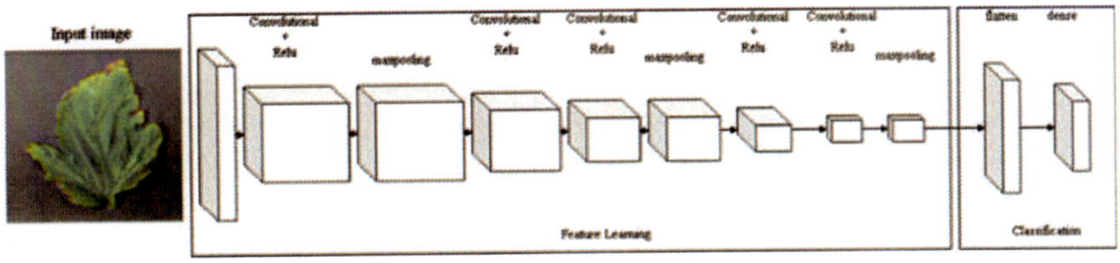

Fig. 5 Proposed CNN architecture

4.9 Learning Rate

The learning rate [14] is a hyperparameter that controls the changes in the model because of the estimated error each time, the model weights are refreshed. Picking the suitable learning rate is quite a challenge as too small value of learning rate will require more processing time and a value to large will not give satisfactory results. For the training process, the learning rate is set to 1e−4.

4.10 Optimizer

Optimizer [15] is used for changing attributes or the weight parameters in order to minimize the loss function. Optimizers reduces the losses and provides the most accurate results possible.

The exhibition of stochastic slope descent (SGD) relies fundamentally upon how learning rates are tuned and diminished over time [16]. Through investigation we discovered that Adam (5) enhancer gives most precise yield for our concern.

$$m_t = (1 - \beta_2) \sum_{i=1}^{t} \beta_2^{t-1} \cdot g_i^2$$

where

m Moving average at t iteration,
g Gradient on current mini-batch.

4.11 Model Training

This process of image classification focuses on extracting the feature from the input dataset for the purpose of classification and prediction. Taking batch size of 32, we trained the model multiple times.

The learning rate will affect the validation accuracy. For model training process, augmentation is done for the “x” and “y” training dataset and supplied with the batch size. Validation data which contains “x” and “y” dataset is also used for generation of validation accuracy. Steps per epoch define number of batches samples used in one epoch.

It also represents the ending of one epoch and starting of another. It is calculated by dividing the length of “x train” dataset by the batch size. Steps per epoch will increase if we increase the total number of images in the dataset.

5 Result

The proposed methodology has been tested for different categories of crops and got satisfactory outcomes. The total overall accuracy achieved by the algorithm is "96.54%".

6 Analysis

We trained the model with a total "950" number of images, and we altered the epoch for each run. The number of epochs used was "10", "25", and "50". The accuracy for "950" images at "10" epoch is "95.40%", at "25" epoch is "96.40%", and at "50" epoch is "96.54%".

We also tested the model by taking input image as "greyscale" and found out that the accuracy of the model with "RGB" is better than "greyscale." Separate testing of dataset at 50 epochs (Table 2):

6.1 *Accuracy and Loss Graph*

Over fitting happens when a quantifiable model or AI calculation gets the clamor of the information. Intuitively, over fitting happens when the model or the estimation fits the data unnecessarily well [17]. Under fitting happens when a truthful model or AI calculation cannot get the shrouded example of the data. Intuitively, under fitting happens when the model or the estimation does not fit the data all around ok [18] (Figs. 6, 7 and 8).

6.2 *Confusion Matrix*

It [19] is summary of predictions made by the classifier. The number of predictions can correct or incorrect which are both included in Table 3. The model is evaluated

Table 2 Separate results

Disease	Val accuracy	Training accuracy	Val. loss	Tra. loss
Corn	0.9062	0.9594	0.2437	0.0960
Pepper	0.9438	0.9250	0.2399	0.4726
Potato	0.8778	0.9889	0.2773	0.0379
Tomato	0.9220	0.9588	0.2246	0.1068

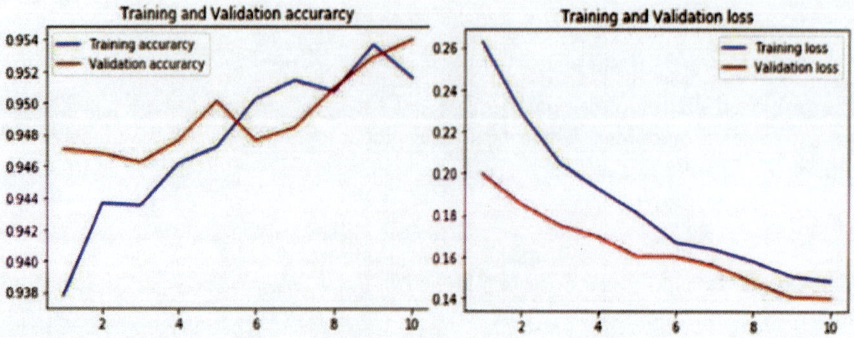

Fig. 6 Accuracy and loss curve at 10 epoch

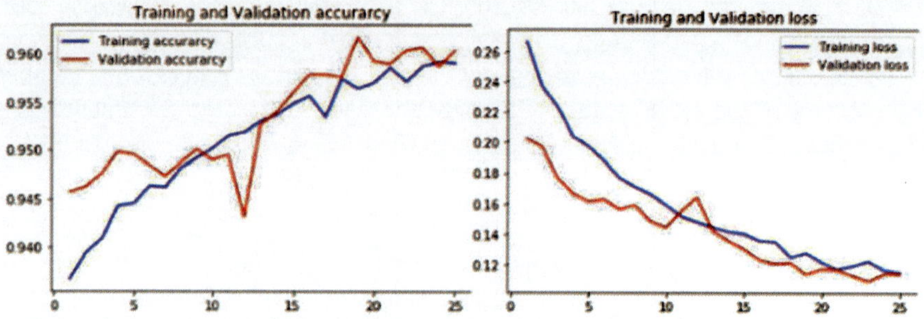

Fig. 7 Accuracy and loss curve at 25 epoch

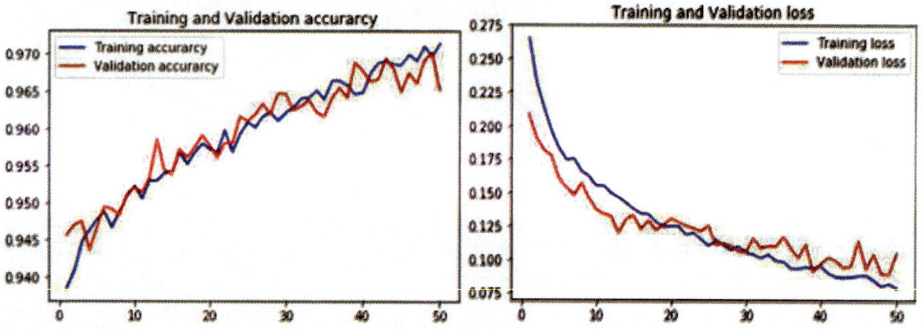

Fig. 8 Accuracy and loss curve at 50 epoch

Table 3 Confusion matrix

	Corn	Pepper	Potato	Tomato	Total
Corn	33	0	3	4	40
Pepper	0	15	3	2	20
Potato	2	0	24	3	29
Tomato	4	1	7	89	100

Table 4 Precision recall matrix

Name	Precision	Recall	F1 score	support
Corn	0.825	0.825	0.825	40
Pepper	0.938	0.714	0.811	20
Potato	0.649	0.828	0.727	29
Tomato	0.907	0.880	0.893	100

by calculation of the performance metrics like precision (5), recall (6), accuracy, and F1-score (7).

- **Precision**: Is the proportion of effectively anticipated positive perceptions to the absolute anticipated positive perceptions:

$$\text{precision} = \frac{\text{TP}}{\text{FP} + \text{TP}}$$

where

TP True Positive,
FP False Positive.

- **Recall**: Is the proportion of effectively anticipated positive perceptions to the all perceptions in genuine class.

$$\text{recall} = \frac{\text{TP}}{\text{FN} + \text{TP}}$$

where

TP True Positive,
FN False Negative.

- **F1 score**: Is the weighted average of Precision and Recall (Table 4).

$$fi\ \text{score} = 2\frac{\text{precision} * \text{recall}}{\text{precision} + \text{recall}}$$

6.3 Comparison with Different Models

We have compared our results with some related research paper and created a comparison (Table 5).

Table 5 Accuracy comparison

Work	Accuracy (%)	Work	Accuracy (%)
Ayşe Eldem	96	K. P. Ferentinos	99.53
Mohd Azlan Abu	90.59	H. Sheikh	99.28
Z. Bin Husin	90.9	J. C. Valdoria	80
A. Ramcharan	93	Proposed	96.54

Table 6 Individual accuracy

Disease	Total images	Accuracy (%)
Corn	200	90.63
Pepper	100	94.375
Potato	150	87.80
Tomato	100	92.20
Complete	950	96.54

6.4 Accuracy of Individual

We have also calculated the accuracy for the separate class as well as the accuracy of the mixed dataset shown in Table 6. The test run was done for the individual category in the dataset at 50 epochs.

7 Conclusion

This paper focuses on implementing an image classification model for the detection of plant disease (corn, pepper, potato, tomato) from the dataset of leaf images. This prediction helps in early disease detection, and early protective measures can be taken. This will increase the overall yield of the farm and will reduce the loss of the crop. For disease detection, image classification is at the peak as various researchers are implementing this technique to detect disease easily. Convolutional neural network (CNN) helps in fast and effective segmentation and analysis of the images from the dataset. This system is efficient as it gives accurate prediction through using a limited number of images "950". This system helps in reducing the manual prediction and the cost required for the manual prediction done by the experts. As for the future, we are trying to improve the accuracy of the algorithm, as well we are aiming to develop a mobile application for automatic prediction and provide protective in-hand solutions.

References

1. Eldem, A., Eldem, H., & Üstün, D. (2018). A model of deep neural network for iris classification with different activation functions.
2. Abu, M. A., Indra, N. H., Halim, A., Rahman, A., Sapiee, N. A., & Ahmad, I. (2019). A study on image classification based on deep learning and tensorflow. *International Journal of Engineering Research and Technology, 12*(4), 563–569.
3. Stuart, R. (2003) *Artificial Intelligence A Modern Approach.* vol. *3*, Prentice Hall.
4. Sladojevic, S., Arsenovic, M., Anderla, A., Culibrk, D., & Stefanovic, D. (2016). Deep neural networks based recognition of plant diseases by leaf image classification. *Computational Intelligence and Neuroscience, 2016.*
5. Dhillon, A., & Verma, G. K. (2019). Convolutional neural network: A review of models, methodologies and applications to object detection. *Progress in Artificial Intelligence*, no. 0123456789.
6. Mohanty, S. P., Hughes, D. P., Salathé, M. (2016). Using deep learning for image-based plant disease detection. *Frontiers in Plant Science, 7*, 1–10.
7. Wahab, A. H. B. A., Zahari, R., & Lim, T. H. (2019). Detecting diseases in chilli plants using K-means segmented support vector machine. In *2019 3rd International Conference on Imaging, Signal Processing and Communications* (pp. 57–61).
8. Khirade, S. D. (2015). Plant disease detection using image processing (pp. 1–4).
9. Ramcharan, A., Baranowski, K., Mccloskey, P., Ahmed, B., Legg, J., & Hughes, D. P. (2017). Deep learning for image-based cassava disease detection. *Frontiers in Plant Science, 8*, 1–7.
10. Ferentinos, K. P. (2018). Deep learning models for plant disease detection and diagnosis. *Computers and Electronics in Agriculture, 145*, 311–318.
11. Sheikh, H., Mim, T. T., Reza, S., Shahariar, A. K. M., Rabby, A., & Hossain, S. A. (2019). Detection of maize and peach leaf diseases using image processing. In *2019 10th International Conference on Communication and Network Technology* (pp. 1–7).
12. Valdoria, J. C., Caballeo, A. R., Condino, J. M. M., & Fernandez, B. I. D. (2019). iDahon: An android based terrestrial plant disease detection mobile application through digital image processing using deep learning neural network algorithm. In *2019 4th Interantional Conference on Information Technology* (pp. 94–98).
13. Agostinelli, F., Hoffman, M., Sadowski, P., & Baldi, P. (2015). Learning activation functions, (2013), 1–9.
14. Luo, H., & Hanagud, S. (1997). Dynamic learning rate neural network training and composite structural damage detection. *AIAA Journal, 35*(9).
15. Recognition, I., & Postalcioglu, S. (2019). Accepted manuscript to appear in *IJPRAI International Journal of Pattern Recognition and Artificial Intelligence.*
16. Schaul, T. (2011). No more Pesky learning rates.
17. Hawkins, D. M. (2004). The problem of overfitting. *Journal of Chemical Information and Modeling, 44*, 1–12.
18. Allamy, H. K. (2016). Methods to avoid over-fitting and under-fitting in supervised machine learning (comparative study).
19. Parker, J. R. (2001). Rank and response combination from confusion matrix data. *Information Fusion, 2.*
20. https://www.kaggle.com/vipoooool/new-plant-diseases-dataset

Sarcasm Detection Technique on Twitter Data with Natural Language Processing

Shubham Chaudhary and Misha Kakkar

Abstract Sarcasm is defined a form of irony that a user uses to express its feeling via social networking platforms such as Twitter or Facebook. Sarcasm can be used in different situations such as mockery or constructive criticism. Verbal sarcasm is rather easy to determine, but detecting sarcasm in the written text is equally difficult. Determining sarcasm in the text improves the sentiment analysis process and also the decision-making process. Sentiment analysis refers to the processing of determining the negative and positive emotions of the Internet user over a specific topic. The dataset used in this research is collected from Twitter via Web scrapping. Four machine learning algorithms such as linear SVC (accuracy = 83%, f1-score = 0.81), Naïve Bayes (accuracy = 74%, f1-score = 0.73), logistic regression (accuracy = 83%, f1-score = 0.81) and random forest classifier (accuracy = 80%, f1-score = 0.77) were implemented.

Keywords Natural language processing · Sarcasm detection · Sentiment analysis · Web scrapping · Term frequency · Inverse document frequency machine learning · Linear SVC · Naïve Bayes · Logistic regression · Random forest classifier

1 Introduction

With the advancement of the social networking Web sites and blogs [1], it has become a common routine to write one's own thoughts, so that others can also read these thoughts and know what a person is trying to convey. According to a report by Statista there are about 2.8 billion active Facebook users, 333 million active Twitter users, who read and write content around the Internet on a regular basis [2]. These thoughts may sometime have some kind of sarcasm attached to it. Sarcasm is defined as irony or mockery on a particular topic. It basically is a sentence with a negative view on

S. Chaudhary (✉)
Data Science, Amity University, Noida, Uttar Pradesh, India
e-mail: shubhamsehrawat517@gmail.com

M. Kakkar
Computer Science and Engineering, Amity University, Noida, Uttar Pradesh, India

S. Tiwari et al. (eds.), *Proceedings of International Conference on Big Data, Machine Learning and their Applications*, Lecture Notes in Networks and Systems 150,
https://doi.org/10.1007/978-981-15-8377-3_24

any topic with a positive sentiment like "It's so refreshing to wait here for 2 hours" or a positive view with the help of negative sentiment like "That was such a bad presentation  ". Sarcasm can also be defined as a way of mocking a person or making fun of that person.

The principle reason behind the concept of finding the sarcasm in the written text is to help the sentiment analysis process more accurate and advance. Sentiment analysis is the process of finding the polarity of a particular text, whether the text is conveying a positive message or the negative message, and as it is known that text with a hint of sarcasm in it, indicates a completely opposite message to the one that the text is trying to convey. Hence, the wrong interpretation of sarcasm in the text possess a very difficult challenge in analyzing the sentiments.

The process of analysis of sarcasm and sentiment is done on the textual data; hence, it is important to understand the basic of the NLP, that is natural language processing. Most of the ML algorithm work on the numerical data and not on the string data, hence, it is important to convert the string or text data to the numerical data. This conversion can be done with the help of the natural language processing techniques, which makes it easier to implement the machine learning algorithms on the text data [3]. NLP algorithm includes cleaning of the data and converting the text to numerical matrix and making the data compatible for analysis.

2 Related Work

A tweet, post, review, feedback or a post from any person on any social media platform depend on their individual characteristics such as age, gender, location and above all the psychology of the person [4]. Many researchers have previously performed many other investigating analysis techniques to detect sarcasm or irony in the text, which is what it will be discussed in this section.

Forslid [5] have used techniques such as support vector machine, decision tree, etc. algorithms to design a machine learning model that will classify the text into sarcastic and non-sarcastic labels. The accuracy of the machine learning model that used Twitter dataset was 71%, but the issue with this is the dataset. The dataset has about 120,000 tweets, out of which there are about 100,000 non-sarcastic tweets and 20,000 were sarcastic tweets. This problem is known as biased data which ruins the accuracy and the predicting ability of the ML model.

Peng et al. [6] also used different number algorithms to detect the sarcasm in the data. Naïve Bayes algorithms had 62% of accuracy in detecting the sarcasm. Another technique used was one class SVM that was only 50% accurate in detecting the sarcasm in the text. But the issue lies in the accuracy, the Naïve Bayes model has the accuracy of 99% while detecting the sarcasm tweets but only 24% while detecting the non-sarcastic tweets.

Clews [7] used the technique of rudimentary lexicons to detect the sarcasm tweets. The author collected the data from the Twitter with the help of the Twitter API. The author created two corpora, one was with using hashtag "Sarcasm" with 5000 tweets and other using “Trump” with 1000 tweets. The accuracy of the model was only 35% which is very low as compared to the other models created by different researchers.

Kumar et al. [8] used the different approaches such as noun phrase exact match technique, noun phrase cosine technique or deep learning technique. The researcher also used the Twitter data to detect the sarcasm in the text data. Three different datasets were created and a test dataset to determine the accuracy of sarcasm detection model. The ML approach has f1-score of 0.83 and the deep learning approach has the f1-score of 0.93.

3 Types of Sarcasm

This section will describe the different types of sarcasm. Vadivu and Sindhu [9] has identified about six primary classes of sarcasm describe as below:

A. Class 1: This is the type of sarcastic text, where the positive sentiment and negative sentiments co-exists. For example: “The mango is not as sweet as it is supposed to be.”
B. Class 2: This is the type of sarcasm where the negative sentiment is followed by the positive sentiment and vice versa. For Example: “Well, the mangos are meant to be sweet, not rotten.”
C. Class 3: In these types of sarcastic comments, the user often indicates a state of confusion. For example: “This mobile has excellent features, for a strangely low price.”
D. Class 4: These types of sarcasm consist of a positive phrase immediately followed by the negative phrase and vice versa. For example: “The delivery of the product was good but the product was not.”
E. Class 5: These types of sarcasm text consist of the comparison between the bad and worst meaning in a situation. For example: “The product was very bad, but at least I won’t be losing any more money on this.”
F. Class 6: The text that shows the sarcasm between two products, falls in this category of the sarcasm. For example: “There is place which sells more delicious mangoes than these.”

Ratawal and Tayal [10] have distinguished the types of sarcasm into different categories, such as:

A. Based on Situation and Sentiment:
In this type of sarcasm, the situation and sentiment used to describe that situation are exactly opposite. If the situation is negative, the sentiment used to describe that situation is positive or vice versa. For example: “I love to wait around here as the flight is delayed.”
B. Based on Mindset of the Person:
In this type of sarcasm, a person describes its emotion based on the mood. Example of these type of sarcasm is **Wit**—“What an AWESOME weather it is,”

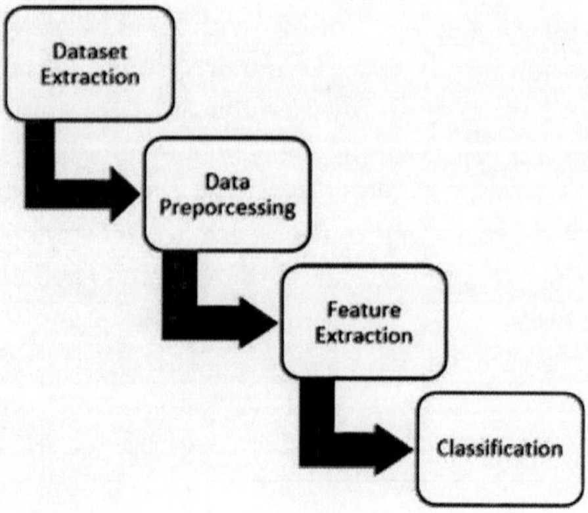

Fig. 1 Sarcasm detection process

here the emphasis is on word Weather, which describe the wit. Other examples are **Snivel** (describes the angry or furious mood of the person), **Rampant** (it occasionally describes the violent nature of the person with the help of the hyperbolic text or sentence), **Prevarication** (in this type of sarcasm text, the person usually ignores or avoid a situation to give an answer).

4 Proposed Methodology

This section will describe the detailed description of the proposed work done for the detection of sarcasm in tweets.

4.1 Dataset

The dataset corpora was formed with the help of tweets. The tweepy Python library and Twitter API was used to fetch tweets based on the hashtag such as #Sarcasm or #Sarcastic and #NonSarcasm or #Non-#Sarcasm. These tweets are then stored in the CSV file with the corresponding sarcasm label; 0—non-sarcastic and 1—sarcastic. The total number of sarcastic tweets collected were 15,000 and non-sarcastic tweets were 15,000 (Fig. 1).

4.2 Data Preprocessing

It is one of the important steps before analyzing any dataset. Data preprocessing includes cleaning of the data such as tokenization, stemming and lemmatization and removal of stop words.

Step 1: Tokenization, the dataset which is in the form of sentences and phrases, is now subjected to tokenization process, where the sentence and the phrases are broken into individual words which is called as tokens. Creating token helps in POS tagging, stop word removal, lemmatization and stemming, etc.
Step 2: Stop word removal, stop words are defined as the recurrent words, which appear in a document for 80% of the time and only constitute to 20% of document meaning. Articles like a, an, the are the best example of the stop words.
Step 3: Lemmatization and stemming, it is the process of converting a particular word to its root form so that the multiple words can be analyzed as single word. For example, words like connected, connecting, connection will all converted to a single root word connect.

These are the three basic data preprocessing techniques in the textual data. Apart from this, there are additional data preprocessing techniques; lowercasing all the text in the text data; it is helpful in case when a same word is written in the different case at different places. Noise removal and null value removal are also a step used of data preprocessing.

4.3 *Feature Extraction*

This step includes analyzing different feature of the text, which will help in classification of the tweets, whether the tweets are sarcastic or not. The different feature extracted from the text are:

Feature 1: The initial feature that was determined from the text data was the sentiment value of the text. The process of sentiment analysis was done on the text and stored as separate feature. Sentiment value of the text retrieved are categorized on the basis of the value.

$$|\text{Sentiment}| = \begin{cases} 0, \ 0 < S < 0.5 \\ 1, \ S = 0.5 \\ 2, \ 0.5 < S < 1 \end{cases}, \tag{1}$$

where

0 Negative sentiment,
1 Neutral sentiment,
2 Positive sentiment,
S Sentiment value.

Feature 2: The next feature that is extracted is positive and negative phrase/words percentage in a text. Suppose that the number of words in the sentence are 10. On performing the data preprocessing steps, the words remaining in the sentence are 7, out of these there are 3 positive words and 3 negative words. So negative words % = 3/7 and positive words % = 3/7.

Feature 3: Once the sentiment value has been identified, the next feature that is extracted in whether there are emoticons present in the text or not. People always have a tendency to use emoticons while expressing any sarcastic comments. With help of emoticon detection, three features are extracted, they are:

$$|\text{Emoticon}| = \begin{cases} 0, & \text{Emoticon Absent} \\ 1, & \text{Emoticon Present} \end{cases} \tag{2}$$

$$|\text{Negative Em}| = \begin{cases} 0, & \text{Negative Em Absent} \\ 1, & \text{Negative Em Present} \end{cases} \tag{3}$$

$$|\text{Positive Em}| = \begin{cases} 0, & \text{Positive Em Absent} \\ 1, & \text{Positive Em Present} \end{cases} \tag{4}$$

Determination of the positive and negative emoticon helps determining whether a particular text is a sarcasm or not. It is often seen that a negative comment of something is often attached with the positive emoji specially a laughing emoji and vice versa.

Feature 4: The next extracted feature is POS tagging. POS tagging help in determining the use of various part of speech in the text such as verbs, nouns and adjective. With the help of POS tagging, four features are extracted which are nouns, verbs, adjectives and miscellaneous. These features will have the total number of words that have been characterized as verbs, nouns, adjective or miscellaneous.

Feature 5: This is the last step in feature extraction. In this step, the textual data is converted into the numerical data, so that it is easier to analyze with machine learning algorithms. The best way to convert the text data to the numerical features is with the help of the TF-IDF [11]. A total to 500 feature vectors is generated from the text data. These feature vectors are then combined with the other features to form the corpus which is then subjected to machine learning algorithms.

4.4 *Classification*

Once all the features are extracted, the next step is to put the data in the machine learning algorithm to detect the tweets, whether they are sarcastic tweets or not. The corpus is first subjected to the principle component analysis. With the help of the PCA, the number of features is reduced by 95%. These features itself carry the gist of the entire corpus.

The dataset is initially split into three different sets, first is for the training of the machine learning model, the second set is for the testing the machine learning model and the third set is for the predicting the tweets with the help of the trained and tested model.

Once this is done, different machine learning algorithms are applied on the dataset. Linear support vector classifier, Gaussian Naïve Bayes, logistic regression and random forest classifier are the four algorithms that are used in this research paper for sarcasm detection.

5 Result

This section deals with the evaluation of different algorithms that were used in classification of the tweets into sarcasm and non-sarcasm tweets. Table 1 determines a quick comparison in the accuracy of different ML algorithms used to detect the sarcasm tweets.

5.1 Linear Support Vector Classifier

The implementation framework of the linear SVC is similar to that of the support vector machine with linear kernel. But linear SVC is more optimized and time efficient than the SVM with linear kernel [12]. Result by the linear SVC is shown in Fig. 2.

Table 1 Accuracy comparison table

Machine leaning algorithm	Training accuracy	Testing accuracy
Linear support vector classifier	0.90	0.83
Naïve Bayes	0.79	0.74
Logistic regression	0.88	0.83
Random forest classifier	0.89	0.80

Fig. 2 Confusion matrix and classification report for linear SVC

```
Confusion Matrix
[[641 120]
 [105 470]]
Accuracy Score : 0.8315868263473054
Classification Report
              precision    recall  f1-score   support

           0       0.86      0.84      0.85       761
           1       0.80      0.82      0.81       575

   micro avg       0.83      0.83      0.83      1336
   macro avg       0.83      0.83      0.83      1336
weighted avg       0.83      0.83      0.83      1336
```

Fig. 3 Confusion matrix and classification report for Naïve Bayes

```
Confusion Matrix
[[496 102]
 [250 488]]
Accuracy Score : 0.7365269461077845
Classification Report
              precision    recall  f1-score   support

           0       0.66      0.83      0.74       598
           1       0.83      0.66      0.73       738

   micro avg       0.74      0.74      0.74      1336
   macro avg       0.75      0.75      0.74      1336
weighted avg       0.75      0.74      0.74      1336
```

5.2 *Naïve Bayes*

It is one of the simplest ML algorithms used for binary classification. The central idea of the Naïve Bayes binary classification is based on the Bayes' theorem. Naïve Bayes makes an assumption that the features are independent and have equal contribution to the target variable [13] (Fig. 3).

5.3 *Logistic Regression*

Logistic regression is a type of parametric classification algorithm. The central idea behind the implementation of the logistic regression is that the target variable is dependent on the feature variables [14] (Fig. 4).

Fig. 4 Confusion matrix and classification report for logistic regression

```
Confusion Matrix
[[642 122]
 [104 468]]
Accuracy Score : 0.8308383233532934
Classification Report
              precision    recall  f1-score   support

           0       0.86      0.84      0.85       764
           1       0.79      0.82      0.81       572

   micro avg       0.83      0.83      0.83      1336
   macro avg       0.83      0.83      0.83      1336
weighted avg       0.83      0.83      0.83      1336
```

Fig. 5 Confusion matrix and classification report for random forest classification

```
Confusion Matrix
[[635 148]
 [111 442]]
Accuracy Score : 0.8061377245508982
Classification Report
              precision    recall  f1-score   support

           0       0.85      0.81      0.83       783
           1       0.75      0.80      0.77       553

   micro avg       0.81      0.81      0.81      1336
   macro avg       0.80      0.81      0.80      1336
weighted avg       0.81      0.81      0.81      1336
```

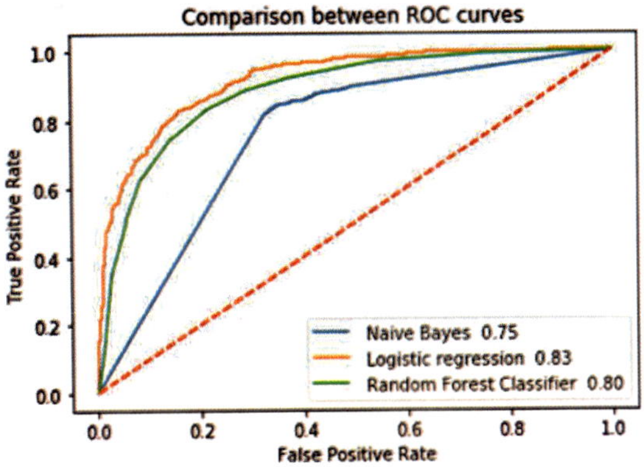

Fig. 6 ROC curve for Naïve Bayes, logistic regression and random forest classifier

5.4 *Random Forest Classifier*

RF classifier is basically a collection on number of the decision trees, which makes random forest a type of ensemble ML algorithm. The fundamental concept behind the implementation of random forest is a relative model of uncorrelated tress, operating together outperforms the individual models [15] (Fig. 5).

The comparison between Naïve Bayes, logistic regression and random forest classifier is done with the help of the ROC curve. The ROC curve depicts that the accuracy of the logistic regression is more than that of the other two machine learning techniques (Figs. 6 and 7).

6 Conclusion

Previous work done on the sarcasm detection models, mainly involves injection words, uni-grams, bi-grams and POS tagging. The approach used in this paper uses the detection of sentiment score, emoticons and negative and positive word

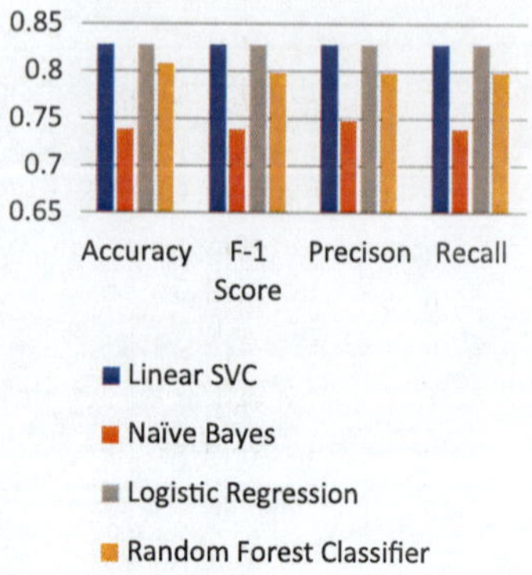

Fig. 7 Machine learning algorithm comparison

percentage in the text which makes the sarcasm detection more efficient compared to other approaches. With the help of better sarcasm detection approaches, the decision-making systems will have a better performance which will benefit the organization to understand the views, thoughts and opinions of their users.

References

1. Aggrawal, N., & Arora, A. (2016). Vulnerabilities issues and melioration plans for online social network over Web 2.0. *CDQM, An International Journal, 19*(1), 66–73.
2. Global social media ranking 2019 | Statistic [Online]. Available: https://www.statista.com/statistics/272014/global-social-networks-ranked-by-number-of-users/. Accessed: March 08, 2020.
3. Gelbukh, A. (2005). Natural language processing. In *Fifth International Conference on Hybrid Intelligent Systems (HIS'05)*, Rio de Janeiro, Brazil, p. 1.
4. Wicana, S. G., & İbisoglu, T. Y., & Yavanoglu, U. (2017). A review on sarcasm detection from machine-learning perspective. In *Proceedings of the International conference on Semantic Computing* (pp. 469–476).
5. Forslid, E. (2015). Automatic irony- and sarcasm detection in social media.
6. Peng, C., Pan, J. W., Edu, C. S., & Edu, J. S. (2015). Detecting sarcasm in text: An obvious solution to a trivial problem.
7. Clews, P. (2017). Rudimentary lexicon based method for sarcasm, *5*(4), 24–33.
8. Kumar, L., Somani, A., & Bhattacharyya, P. (2017). 'Having 2 hours to write a paper is fun!': Detecting sarcasm in numerical portions of text.
9. Vadivu, G., & Sindhu, C. S. (2018). A comprehensive study on sarcasm detection technique in sentiment analysis. *International Journal of Pure and Applied Mathematics, 118*(22), 433–442.
10. Ratawal, Y., & Tayal, D. (2018). A comprehensive study: Sarcasm detection in sentimental analysis. *An International Scientific Journal, 113*(October), 218–226.
11. Qaiser, S., & Ali, R. (2018). Text mining: Use of TF-IDF to examine the relevance of words to documents. *International Journal of Computer Applications, 181*(1), 25–29.
12. Gurvir, K., & Parvinder, K. E. (2017). Novel approach to text classification by SVM-RBF kernel and linear SVC. *International Journal of Advanced Research, 3*, 2015–2018.
13. Kaviani, P., & Dhotre, S. (2017). Short survey on Naive Bayes algorithm. *International Journal of Engineering Science Invention Research & Development, 4*(11), 607–611.

14. Peng, C. Y. J., Lee, K. L., & Ingersoll, G. M. (2002). An introduction to logistic regression analysis and reporting. *Journal of Educational Research, 96*(1), 3–14.
15. Klusowski, J. M. (2018). Analysis of a random forest model. *Journal of Machine Learning Research, 13*(2012), 1063–1095.

CDMD: An Efficient Crop Disease Detection and Pesticide Recommendation System Using Mobile Vision and Deep Learning

Shivam Gupta, Garima Garg, Preeti Mishra, and R.C. Joshi

Abstract The traditional agriculture systems are less suitable for better crop productivity. The use of modern technologies such as mobile vision and deep learning, can help in solving some of the farming related problems for improving the crop growth. In this paper, we primarily focus on developing a user-friendly system for crop disease detection and pesticide recommendation, called CDMD to improve crop productivity. We employ a pre-trained deep learning model Visual Geometry Group (VGG)-13 for learning the features of various diseased and non-diseased images. CDMD also recommends the appropriate pesticide based on the type of disease detected. CDMD provides an average accuracy ~95.12% with loss 0.1607 for 24 different diseases of Plant Village dataset which seems to be promising. A user-friendly android app has also been developed as a prototype of the proposed system. It currently supports both English and Hindi languages as a user interface.

Keywords Alexnet · VGG-13 · Squeezenet · Crop disease detection · Pesticide recommendation

1 Introduction

Agriculture plays a critical role in the economy of every country. Farmers grow a variety of crops every year. One of the main considerations answerable for the harvest devastation and not having appropriate development of the crops is infections or diseases. Crop diseases can reduce both the quality and quantity of crops. Most of the farmers especially in urban areas are still using traditional approaches of farming. The farmers still rely on experts for the usage of pesticides, fungicides, and chemical applications. The excess or inappropriate use of pesticides may reduce the quality of the crops [1]. In fact, some of the crop diseases can't be detected in an early

S. Gupta · G. Garg · P. Mishra (✉) · R.C. Joshi
Department of CSE, Graphic Era Deemed to be University, Dehradun, India
e-mail: scholar.preeti@gmail.com

S. Tiwari et al. (eds.), *Proceedings of International Conference on Big Data, Machine Learning and their Applications*, Lecture Notes in Networks and Systems 150,
https://doi.org/10.1007/978-981-15-8377-3_25

stage by human eyes. Hence, computer vision is one such technique that helps in the identification of the severity of crops.

In this paper, we propose an integrated crop disease detection and pesticide recommendation system, called CDMD which combines the deep learning approaches with the mobile vision for timely detection of the crop diseases.In addition, it provides a crop information system for providing a detailed report about disease and causes of detected disease. It also helps in recommending the appropriate use of pesticides for handling the detected disease. A prototype of CDMD has also being developed for farmers which aims to improve crop productivity and double the farmer's income. The proposed system provides an easy to use interface supporting both Hindi and English dictionary which helps the farmers to interact easily with the system. Currently, CDMD supports the detection of diseases of five different crops such as Apple, Tomato, Grape, Corn, and Potato. The details of various types of diseases considered for each of the crops have been discussed in Section IV. The key contributions of the proposed work are as follows:

- To design and develop an integrated crop disease detection and pesticide recommendation system using deep learning and mobile vision.
- To validate and compare the proposed work with existing techniques using publicly available datasets.
- To develop a prototype (android APP) supporting all the proposed functions.

This paper is organized into five Sections. A detailed discussion on the related work is provided in Sect. 2. Section 3 shows a detailed description of the design methodology and design architecture. In Sect. 4, we analyze and validate our technique using publicly available datasets. It also provides a discussion about the practical application of our architecture and compares the proposed work with relevant existing approaches. Finally, Section V concludes the paper with future directions.

2 Related Work

Kusumo et al. [2] proposed an approach for detecting corn diseases only. They used SVM, Decision Tree (DT) , Random Forest (RF) , and Naive Bayes (NB) for learning and detecting the disease features. Budihal et al. [3] proposed an approach for detecting diseases in tomato leaf only. They used color transformation and applying Otsu's threshold. Using this technique, the authors can detect disease spot accurately. Fine-tuning using GoogleNet enhanced the accuracy from 97.71% to 99.18% and similarly using AlexNet the accuracy is enhanced from 97.35% to 98.66%.

Wang et al. [4] proposed an approach to differentiate between diseases of wheat and grape. The image segmentation was done by using the K-means algorithm, and then 50 features are extracted manually. Hu et al. [5] used the Hyperspectral Imaging technique for the detection of diseases in the Potato crop in which 95% accuracy was gained. Xu et al. [6] proposed a model in which K-Nearest Neighbor is used for the detection of tomato crop diseases and the overall accuracy 82% was achieved.

Tian et al. [7] proposed an approach for the detection of four types of diseases in Apple crop. They have used Cycle-Consistent Adversarial Network (CycleGAN) based on deep learning model, data augmentation, a densely connected neural network (DenseNet) which are used to optimize layers of the pre-trained (YOLO V3) model. They have used low-resolution images and achieved an accuracy of 95.57%.

3 CDMD:Design Architecture

In this Section, we provide a detailed description of CDMD architecture along with the description of detection components. The CDMD architecture is divided into two key detection components as shown in Fig. 1: (i) Crop Health Monitoring System (CHMS) and (ii) Pest and Disease Recommendation System (PDRS). CHMS is responsible for the real-time detection of crop diseases. It employs mobile vision and deep learning technology. The timely detection of crop disease helps the farmers to take preventive measures on time and enhances the productivity of the crops. Each component is divided into sub-components which are discussed below.

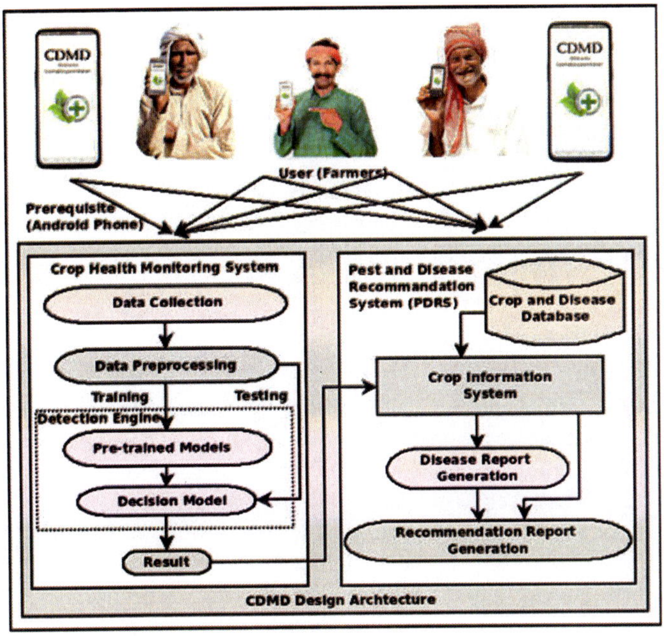

Fig. 1 Proposed CDMD architecture

3.1 *Crop Health Monitoring System*

CHMS component consists of three sub-components i.e., Data Collection (DC), Data Pre-processing (DP) and Detection Engine (DE).

3.1.1 Data Collection

This is the beginning stage for the execution flow of the algorithm. Diseases on a leaf can be distinguished based on the different features like spots, area and on the basis of their color. These features are contrasted with prearranged limits in order to determine the emulate indisposition. DC involves the activity to collect the plant leaf images which should have superiority sharpness. It can be achieved by using a cell phone camera or by using online sources. The color channel of the image is considered to be in the form of Red, Green, and Blue (RGB).

3.1.2 Data Pre-processing

The crop leaf images which are procured from the field may contain dust, spores and water spots as noise. Hence, one of the important tasks of DP is to eliminate the noise in the image. The pixel values are adjusted to enhance the quality of the image. It is one of the prerequisites for making the input images well suitable for training the model. The image enhancement technique upgrades the perception of information in images for human viewers.

Initially, the image is converted from RGB format to Grayscale to reduce the complexity of the model. Once scale conversion is done, image masking is performed on the dataset. Image masking is a process of hiding non-essential portions of an image and it performs many types of image pre-processing like edge detection, background removal, and noise reduction.

Squeezenet Model consists of only few pooling layers. Alexnet model is used which consist of 5 convolutional layers and 3 fully-connected layers. This model helps to extract only the important features using kernels (or filters) and the filter size is not constant which results in loss function.VGG-13 have filter size constant and after applying the max-pooling on each convolution layer the number of filter becomes double which help to decrease the loss function.

3.1.3 Detection Engine

DE is one of the key components which is helpful in detecting the diseases from the images. In the present stage, DE is capable to detect 25 diseases of 5 different crops.CDMD employs a deep learning algorithm for image classification, called Convolution Neural Network (CNN) [8]. The main reason of using CNN for CDMD was that it automatically detects the important features without any human involvement. Because of the 'spatial invariance' property, convolutional networks learn to recognise image features present in the image. DE will provide an output which consist of the name of the disease (if any) though which plant is suffering. This output will send to the next component i.e., Crop Information System.

3.2 Pest and Disease Recommendation System

PDRS component consists of three sub-components i.e., Crop Information System (CIS), Disease Report Generation (DRG) and Recommendation Report Generation (RRG).

3.2.1 Crop Information System

CIS maintains a database which is on cloud of various pesticides with respect to individual diseases that we have considered. There is a need for having such a system specifically in rural areas because it is difficult to find the agronomist and the experienced farmers. This system will be beneficial for non-technical farmers because of having a simple and user-friendly interface. CIS takes the input from the farmer through their android phone and pass that input image to the DE. On the basis of that input, DE will provide the output. Then the output will be searched in the database. The CIS not only provides information about pesticides with respect to each crop but also provides information about the fertilizers. The database includes diseases of five crops which are Apple, Grape, Potato, Tomato and Corn.

3.2.2 Disease Report Generation

DRG generates the report corresponding to the disease detected by the CHMS module. The report contains a detailed description of the detected disease and the causes behind it. For example, if Apple is suffering from *apple scab* disease then the generated report consist of the following contents:-

- Small, olive-green or brown spots on leaves,
- Later form large brown patches,
- Leaf deformation and premature fall off and so on.

Apple scab is a disease caused by the fungus *Venturia inaequalis*. Causes should also be known to the farmers so that prior prevention measures can be taken to avoid the disease in the crop.

3.2.3 Recommendation Report Generation

RRG provides suggestions and prevention methods for detected diseases. It provides recommendations for the farmers to avoid the re-occurrence of the same disease in the future. The recommendation will be in the form of prevention methods which could be biological, chemical and cultural. Those methods can help the farmer to recover the crop from diseases as early as possible without much expenditure. This recommendation system provides mitigation measures in detail to handle the detected disease. It provides a report which includes prevention and treatment measures for that specific type of disease.

4 Experiments and Result

In this Section, we present some discussion on the practical application and advantages of the proposed CDMD architecture. Then, there is comparison of CDMD with some of the previously proposed techniques. The following pre-processing and data augmented techniques are used: Resize (Re), Segmentation (Sg), Crop (Cr), Flipping (Fl), Rotation (Ro), Zooming (Zo), Noise Removal (NR), Background Removal (BR) in the dataset images.

CDMD approach has been implemented in a machine having Ubuntu 18.04.3 LTS, Intel Core i5-7th Gen CPU @ 2.20 GHz 16 GB RAM, 500GB SSD having 4096 MB Nvidia GeForce 1650GTX GPU. The programs are coded using Python 3.7.3 and having libraries as follows: sys, keras-2.2.4, tensorflow-2.0.0, cv2, skimage, matplotlib-3.0.3, numpy-1.16.2, tensorflowhub-0.7.0 to make use of some predefined functions. It is necessary for the system to have GPU.

4.1 Description About Data Set

CDMD has been validated using Plant Village Dataset [7] having 31,048 images of different crop leaves. The whole dataset have been divided into 25 classes and corresponding class labels assigned to them. In CDMD, we applied hold out validation technique to split the dataset. The dataset has splitted into training and testing. The total number of images in the training dataset is 24,838 and in testing is 6210. The details of the dataset are shown in Table 1.

4.2 Discription About Result

In this experiment, three pre-trained models (Squeezenet, Alexnet, VGG-13) have been trained on the dataset described in the previous Section. To train and evaluate the

Table 1 PlantVillage dataset

S.No.	Crop Name	Disease	No. of Images
1	Apple	Apple Scab	631
		Apple Black Rot	622
		Apple Cedar Rust	276
		Apple Healthy	1646
2	Grape	Grape Black Rot	1181
		Grape Black Measles (Esca)	1384
		Grape Leaf Blight	1077
		Grape Healthy	424
3	Potato	Potato Early Blight	1001
		Potato Late Blight	1001
		Potato Healthy	153
4	Corn	Corn Gray Leaf Spot	514
		Corn Common Rust	1193
		Corn Northern Leaf Blight	986
		Corn Healthy	1163
5	Tomato	Tomato Bacterial Spot	2128
		Tomato Early Blight	1001
		Tomato Late Blight	1910
		Tomato Leaf Mold	953
		Tomato Septoria Leaf Spot	1772
		Tomato Two Spotted Spider Mite	1677
		Tomato Target Spot	1405
		Tomato Mosaic Virus	373
		Tomato Yellow Leaf Curl	5358
		Tomato Healthy	1592

performance of these models, we use the concept of transfer learning. These three CNN architectures are trained for the plant disease classification task using two different strategies i.e., Data Augmentation (DA) and without Data Augmentation. DA is a strategy that increases the diversity of data available for training models. Using these two approaches, CNN classifier tries to learn more specific features for plant disease classification. Therefore, a technique has been validated using 5 different CNN classification approaches and those are Squeezenet V3 without DA, Squeezenet V3 with DA, Alexnet V2 without DA, Alexnet and VGG-13 with DA. Their comparison can be seen in Fig. 2. All these approaches uses the same hyper-parameters values i.e., Activation Function = Softmax, Learning Rate = 0.001, Batch

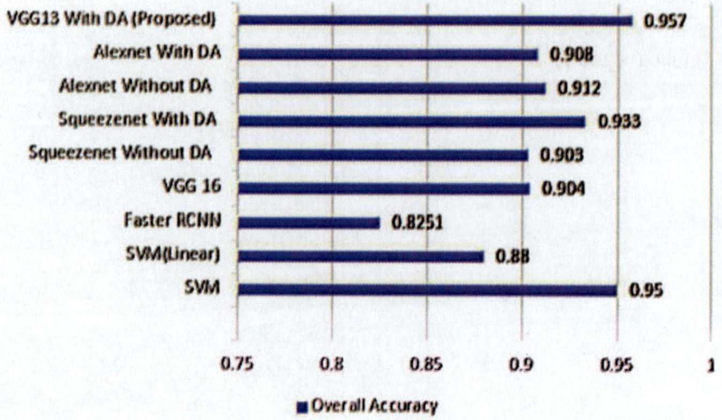

Fig. 2 Comparison graph

Table 2 Experimental results of leaf disease detection

Paper	Architecture	Overall accuracy
Islam et al.[9]	SVM	0.95 (Potato)
Kusumo et al. [10]	SVM (Linear)	0.88 (Corn)
Gutierrez et al. [11]	Faster RCNN	0.8251 (Tomato)
Wang et al. [12]	VGG 16	0.904 (Apple)
CDMD	Squeezenet without DA	0.903 (Overall)
CDMD	Squeezenet with DA	0.933 (Overall)
CDMD	Alexnet without DA	0.912 (Overall)
CDMD	Alexnet with DA	0.908 (Overall)
CDMD	VGG 13 with DA (Proposed)	0.9512 (Overall)
		0.973 (Potato)
		0.9414 (Corn)
		0.9256 (Tomato)
		0.9802 (Apple)

Sizes = 64. We used Adam as an optimizer. To perform transfer learning we just unfreeze some layers of base network and add 2 more dense layers to the model. Note: The accuracy from different CNN Architecture which is shown in Table 2 were obtained on the same above hyperparameters and no. of epochs are reduced to 5 to show the difference in the CNN Architecture and to reduce time complexity.

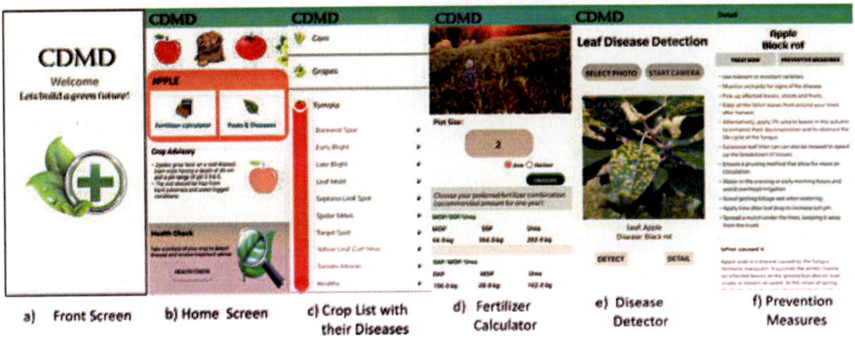

Fig. 3 Step wise execution of CDMD

Table 3 Comparison of proposed architecture with existing Work

Parameters	Islam et al. [9]	Kusumo et al. [10]	Gutierrez et al. [11]	Wang et al. [12]	Proposed Architecture
Dataset	PlantVillage	PlantVillage	PlantVillage	PlantVillage	PlantVillage
Species Name	Potato	Corn	Tomato	Apple	Tomato, Apple, Corn, Potato, Grapes
No. of species	1	1	1	1	5
No. of classes	3	4	7	4	24
No. of images	2055	3852	4331	2086	31048
Pre-processing	BR, RGB Imaging	N/A	Fl, Ro, Cr	Re, Fl, Ro, Zo	Re, Fl, NR, Sg, BR
Validation technique	K-Fold Cross Validation	K-Fold Cross Validation	K-Fold Cross Validation	N/A	Hold out Validation
Machine learning	SVM	SVM(Linear), RF, DT	Faster RCNN	VGG16	VGG 13
Transfer learning	No	No	No	Yes	Yes
Learning rate	N/A	N/A	N/A	0.001	0.001
Activation function	N/A	N/A	N/A	SGD	Softmax
Batch size	N/A	N/A	N/A	N/A	64
Train-test ratio	60%–40%	N/A	90%–10%	80%–20%	67%–33%
No. of iterations	N/A	N/A	N/A	N/A	30
Recom. system	No	No	No	No	Yes
Overall accuracy	95%	88%–90%	82.51%	90.40%	95.21%

4.3 Practical Usage and Deployment of CDMD in Android

Initially, the user is supposed to download CDMD application from online available sources on their smartphones. The step-wise execution of CDMD is shown in Fig. 3. PDRS module is having four features that make CDMD different from other existing work. At the top, CDMD have a collection of icons having 5 different crops and each icon is providing access to the recommendation module. The Health Check module user can provide the image through live capturing of leaf or by importing the image from gallery, which will be passed to the proposed model i.e., Ensemble model. Secondly, the Fertilizer Calculator provides a feature to calculate the fertilizer required for the better productivity of the crops. Quantity of the fertilizer required for every crops is different. For example some crops like apple, tomato, etc. requires fertilizer on the basis of the number of trees while other crops such as grapes depends on the area (hectares or acres). The Pest and Disease module provides a list of crops and the description of their diseases i.e. their prevention measures and the treatment methods. Lastly, the Recommendation module tells about the specific requirements for the crop such as temperature required, pH value, etc. And lastly, it provides a report about the crop.

There exists some comparable work [9, 10, 12] for crop disease detection which have been validated using same dataset i.e., Plant Village Datase as show in Table 3. Plant Village Dataset has 14 different species and each species have their multiple classes. Islam et al. [9] have proposed an approach using SVM architecture for the detection of diseases in potato only among 14 species.The author has made the use of BR, RGB imaging techniques for the pre-processing the potato leaf images and the data has been splitted into 3:2 ratio. K-Fold Cross-validation technique is used for validation and achieved 95% of accuracy. CDMD also uses the same dataset in which 67% of data is used for training and 33% for testing. CNN architecture integrated with a pre-trained model i.e.,VGG 13. The number of species on which CDMD trained are Apple, Grape, Tomato, Corn, and Potato. Each of them has classes which are mentioned in Table 1 with number of images. Pre-processing techniques like Re, F, NR, and BR have been used. For better performance, the transfer learning rate is taken very less i.e., 0.001 and make the use of softmax activation function results in better accuracy. The features which are present in CDMD and lacking in other existing works are fertilizer calculator, a pesticide recommendation system and prevention measures through an android application.

Wang et al. [12] proposed an approach for the detection of diseases in apple using VGG-16 model.The author has performed few pre-processing techniques such as Re, Fl, Ro, Zo, and splitted the dataset into 4:1 ratio. SGD is used as the activation function. The author achieved 90.04% accuracy in detection of apple diseases.It can be concluded that CDMD performance is much better than the existing approaches. The other discussed approaches lack a particular crop information data and recommendation system.

5 Conclusion

CDMD architecture can detect in total 25 diseases of 5 different crops. We employ deep learning, using pre-trained models for the timely detection of crop diseases. As, different pre-trained models are used to compare their accuracy and to reduce the loss function. This architecture not only supports the timely detection of disease but also provide the recommendation system for fertilizer and pesticides for the detected disease. In the future, we would like to enhance our system with IoT to make it more realistic so that it can be used for large fields. Some more modules with advanced features will be incorporated into this system.

References

1. UKEssays. (2016). *The importance of agriculture in developing countries economics essay*. https://www.ukessays.com/essays/economics/the-importance-of-agriculture-in-developing-countries-economics-essay.php?vref=1.
2. Shirahatti, J., Patil, R., & Akulwar, P. (2018). A survey paper on plant disease identification using machine learning approach. In *3rd International Conference on Communication and Electronics Systems (IC- CES)* (pp. 1171–1174), IEEE.
3. Hajawagol, S. D., Budihal, S., Sandhya, R., & Navi, S. R. (2015). Detection of disease in Tomato Leaf. *International Journal of Advanced Computer Engineering and Communication Technology (IJACECT), 4)*.
4. Wang, H., et al. (2012). Application of neural networks to image recognition of plant diseases. In *International Conference on Systems and Informatics (ICSAI2012)* (pp. 2159–2164), IEEE.
5. Hu, Y. H., et al. (2016). Detection of late blight disease on potato leaves using hyperspectral imaging technique. *Guang pu xue yu guang pu fen xi= Guang pu, 36*(2), 515–519.
6. Xu, G., et al. (2011). Use of leaf color images to identify nitrogen and potassium deficient tomatoes. *Pattern Recognition Letters, 32*(11), 1584–1590.
7. Tian, Y., et al. (2019). Detection of Apple Lesions in orchards based on deep learning methods of CycleGAN and YOLOV3-Dense. *Journal of Sensors 2019*.
8. Reddy, S. V. G., Thammi Reddy, K., & Valli Kumari, V. (2018). Optimization of deep learning using various optimizers, loss functions and dropout. *International Journal of Innovative Technology and Exploring Engineering (IJITEE), 8*, 272–279.
9. Islam, M., et al. (2017). Detection of potato diseases using image segmentation and multiclass support vector machine. In *30th Canadian Conference on Electrical and Computer Engineering (CCECE)* (pp. 1–4), IEEE.
10. Kusumo, B. S., et al. (2018). Machine learning-based for automatic detection of corn-plant diseases using image processing. In *International Conference on Computer, Control, Informatics and its Applications (IC3INA)* (pp. 93–97), IEEE.
11. Gutierrez, A., et al. (2019). A benchmarking of learning strategies for pest detection and identification on Tomato plants for autonomous scouting robots using internal databases. *Journal of Sensors 2019*.
12. Wang, G., Sun, Y., & Wang, J. (2017). Automatic image-based plant disease severity estimation using deep learning. In *Computational intelligence and neuroscience 2017*.

Generative Adversarial Neural Networks for Text-to-Image Synthesis of Dog Images

S. Sidharth and B. S. Charulatha

Abstract Computer vision has been making advancements in the recent years with generative algorithms. A set of artificial intelligence algorithms creates their own data based on the training set they are trained on. Out of all generative algorithms that exist, generative adversarial neural networks are making rapid strides due to their ability to mimic the data distribution of the training data that they are trained upon which can be constrained using a text caption for the images. Using this concept of conditioning, the generative adversarial neural network, a text-to-image synthesis model can be created that gives out realistic images. This paper will explore the concept of creating a text-to-image synthesis for dog images that are conditioned based on certain caption words.

Keywords Activation function · CNN · Discriminator · Generator · RNN · Data density

1 Introduction

The ability of the computer to extract information from a single image, set of images, or videos is defined as computer vision. Computer vision has been at the forefront of ground breaking research in the past decade, as its application keeps increasing due to its ability to interpret accurate data and enables a system to take actions on its own, etc. One of the most challenging problems in the world of computer vision is synthesizing high-quality images from text descriptions. No doubt, this is interesting and useful, but current artificial intelligence systems are far from this goal. In recent years, the family of powerful neural network architectures like generative adversarial networks (GAN) have been found to generate good results. Samples generated by existing text-to-image approaches, such as autoencoders, whose input is the same

S. Sidharth (✉) · B. S. Charulatha
Department of CSE, Rajalakshmi Engineering College, Chennai, Tamil Nadu, India
e-mail: sidharths758@gmail.com

B. S. Charulatha
e-mail: charu2303@yahoo.co.in

S. Tiwari et al. (eds.), *Proceedings of International Conference on Big Data, Machine Learning and their Applications*, Lecture Notes in Networks and Systems 150,
https://doi.org/10.1007/978-981-15-8377-3_26

as the output can roughly reflect the meaning of the given descriptions, but they fail to contain necessary details and vivid object parts. Through this work, one can explore architecture that could help to achieve the task of generating images from given text descriptions. Generating photo-realistic images from text has tremendous applications, including photo-editing, computer-aided design, etc.

Deep learning has been used in many applications after its inception due to its ability to mimic distribution of data of various probabilities from various kinds of data, them being images, audio signals, natural language processing to name a few. Out of all the deep learning algorithms, generative adversarial neural network (GAN) is gaining traction in research communities due to its ability to map one data distribution as accurately as possible to another data distribution.

This work will comprise of a two-stage generative adversarial neural network system with each having its own generator and discriminator network. Primarily, a 64×64 image which is a noise initialized with random normal distribution is fed into stage-1 generator network. This noise propagates through the stage-1 network through all the computations and mathematical transformations of the respective network and is fed into the discriminator of the stage-1 network which along with the output of the generator network also has the original image as the input. The discriminator network, which is a discriminative model, has the task of classifying which of the images that is fed to it is original or fake, i.e., it needs to ensure that the images that are fed to it from the generator network are classified as fake while the original images that are fed to it are classified as original. The generator network here also has the text data associated with each image as input. This text is converted into numerical form through recurrent neural network and other mathematical conversions and fed along with the image to generator. For one epoch, with all hyperparameters, there is forward propagation which ensures the mathematical conversion done on the said values that are moved forward through the generator to the discriminator. Once the discriminator discriminates between the original and fake images, there will be backpropagation that takes place where the derivatives computed for each layer of the network, from the output layer to the hidden layers, are used to adjust the weights of the neurons in each layer. Once that is completed, the forward propagation once again takes place followed by the backpropagation, and this process keeps going until a certain number of times specified by the user. This hyperparameter is called epoch.

2 Literature Survey

Conditional generative adversarial neural networks, a class of generative adversarial neural network algorithms, that use a means of constraining the GAN model based on certain parameter is showing good results for generating images from text [1]. When StackGAN is used along with conditional generative adversarial neural networks which is fed with image and text associated with that image as its input, the GAN model produces considerable results as output.

The GAN model here consists of two sets of generator and discriminator network with each trying to outsmart the other. There are two levels to this GAN model; in the first GAN model, the input to the generator network is a noise data which travels through the generator network whose output is fed to the discriminator network that is also has the original image as input to it [2]. This network captures the basic underlying structure of the image. This basic image that is the output of the primary stage is fed as input to the second level which also takes in the text word as input to create a realistic image as output. The model that is used here uses convolutional neural networks, a set of discriminative algorithms that is used for learning the image features of the input image and recurrent neural network which have long short-term memory that is used for remembering the words from previous layers.

The input to the primary generator in the first level is a noise data which when trained after a certain number of epochs, a hyperparameter, mimicking the data density of the original data using a game theory approach, i.e., trying to outperform the discriminator which tries to classify the images from the generator that are fake while the generator's role is to ensure that it tricks the discriminator into classifying an image generated by it as original [3].

The second level which has input from the primary level takes in the image and transforms it using mathematical functions such that a realistic image is created. Both the levels have unsampling and downsampling to ease computation complexities of the network [1].

Martin et al. defined a form of GAN called Wasserstein-GAN that minimizes the approximation of the EM distance [4].

GAN is more poplar under supervised learning. This paper provides the architecture to state that it can be used for unsupervised learning also. This class of CNN is called as deep convolutional GAN [5].

Karra et al. proposed an architecture for generative adversarial networks using style transfer literature. The new architecture leads to an automatically learned, unsupervised separation of high-level attributes and applied for human face detection [6].

3 Experimental Setup

Primarily, the dataset for the entire work is obtained from the Internet.

Following which the network design for the entire neural system has to be envisioned. Once the number of layers, initial hyperparameters, and other factors are decided upon, the network has to be built. Once the network is built, the initial iterations and the primary results are obtained based on which one can assess whether the network will have exploding or vanishing gradients, the network is overfitting or underfitting to resolve which one can increase the dataset, tune the hyperparameters, increase the dropout rate until one can obtain the required result which is checked based on the image that is produced.

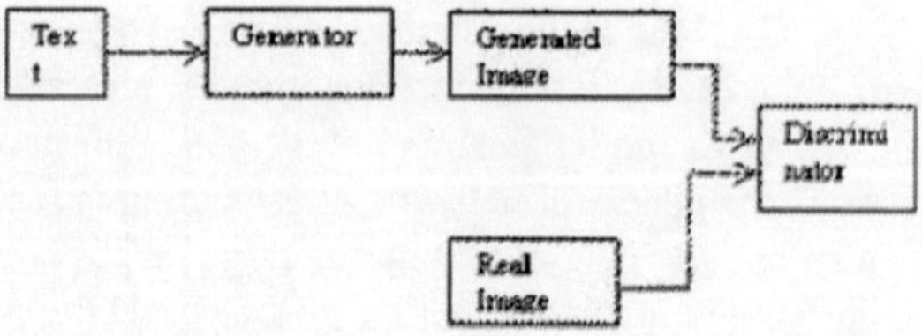

Fig. 1 Stage 0

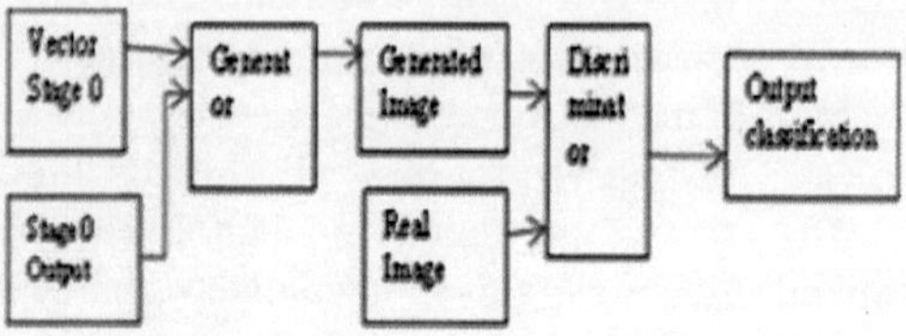

Fig. 2 Stage 1

The input to the stage-0 network as shown in Fig. 1 is the text data and noise with random normal distribution (μ, σ) with mean μ and variance σ. The values are propagated from the generator network to the discriminator network by serious of mathematical matrix calculations and transformations which end with passing the values through an activation function, tanh in the generator and ReLU in discriminator networks. As far as the initialization is concerned with respect to the weights of the neurons and the initial noise data, a random normal distribution is used with Xavier initialization. While the random normal distribution breaks the symmetry, ensuring that the weights in each neuron in the layers are not similar, the Xavier initialization makes sure that the weights in each layer neither explode.

Both the levels of the network have Adam optimization as the optimization function which is a combination of RMSprop and AdaGrad optimization functions. This function has a β_1, i.e., the exponential decay rate of first moment as 0.9 and β_2, i.e., the exponential decay rate of second moment as 0.999. The entire aim of the network is to optimize the cost function of a GAN model which is minmax function (Fig. 2 and Table 1).

4 Results and Discussions

The entire model is trained for 100 epochs with α value of 0.01. Each of the layers is virtual batch normalized for improving the training process [7]. Table 2 beneath will summarize the generator and discriminator loss of the stage-1 network.

The images that were used to create the model were the dog images that were available from the Udacity Github link. Few images from the dataset are in Fig. 3.

Each of these images was converted to $64 \times 64 \times 3$ images and then normalized by being divided by 255 for the stage-0 network and is later converted as $256 \times 256 \times 3$ for the stage-1 network. The data preprocessing stage also had image rotation,

Table 1 Parameters of the network

Generator	Discriminator
Dense layer (128 * 8 * 4 * 4)	Conv2D (128)
Batch norm	Batch norm
Reshape (1, 4, 4, 128 * 8)	Conv2D (128 * 4)
Conv2D (128 * 2)	Batch norm
Batch norm	Conv2D (128 * 8)
Conv2D (128 * 2)	Batch norm
Batch norm	Conv2D (64 * 2)
Conv2D (128 * 4)	Batch norm
Unsampling (16, 16)	Conv2D (64 * 2)
Conv2D (128 * 4)	Batch norm
Batch norm	Conv2D (64 * 8)
Conv2D (128)	Batch norm
Batch norm	Conv2D (64 * 4)
Conv2D (128)	Batch norm
Batch Norm	Conv2D (64 * 8)
Elementwise Layer	Batch norm
Unsampling (32, 32)	Input layer
Conv2D (128 * 2)	ExpandDims (1)
Batch norm	ExpandDims (1)
Unsampling (64, 64)	ExpandDims (1)
Conv2D (128)	Conv2D (4, 4)
Conv2D (128)	Batch norm
Unsampling (64)	Logits
Logits	tanh (Logits)

scaling, and other affine transformations so that the neural network is open to different variations of the same image. The text caption associated with each image (1) is shown in Fig. 4, similarly each image class had about 40 text word documents with each having 10 captions for training the recurrent neural network of the model [8].

The network had the graph as in Fig. 5 as the generator and discriminator loss.

The graph in Fig. 5 is a classic example to show as to how the generator and discriminator are competing to outperform each other. The final result for the text of "Dog is Watching" yielded a result as shown below in Figs. 6, 7 and 8.

The entire model can be optimized more with further enhancement on the StackGAN part, with better architectures. There is still a need for a quality metric with which one can ascertain the quality of the GAN model quantitatively which is still an area of active research. Finally, there is a need for creating ways by which the computations become faster and simultaneously ensuring mode collapse does not occur.

Table 2 Loss for generator and discriminator of stage 1

Epochs	Discriminator loss	Generator loss
1	0.7645	1.4673
10	1.5323	1.2316
20	1.5578	1.0212
30	1.4145	1.7789
40	2.2345	1.5349
50	1.3679	0.7474
60	1.3596	1.3500
70	1.3375	0.6640
80	1.1813	0.9992
90	1.3787	0.7672
100	1.2409	0.7381

Fig. 3 Training dog images

Fig. 4 Text caption for image (1)

```
Dog is standing
Dog is seeing
Dog is viewing
Dog is waiting
Dog is roaming
Dog is running
Cat is thinking
Cat is cute
Cat is tilting head.
```

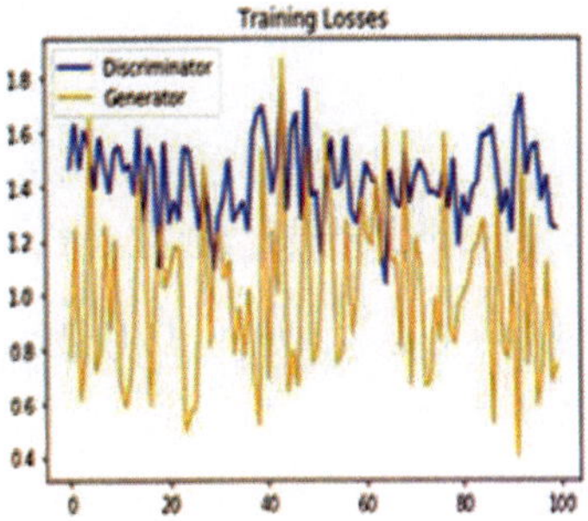

Fig. 5 Generator and discriminator loss

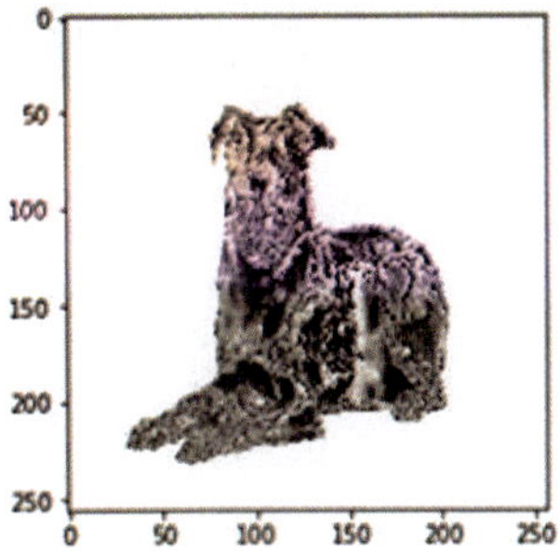

Fig. 6 Image after epoch 60

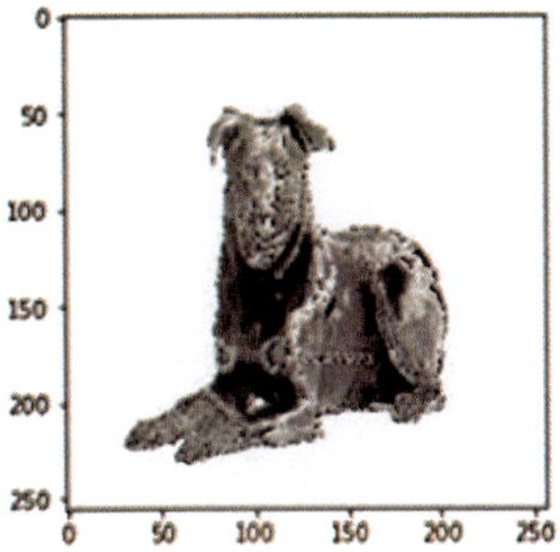

Fig. 7 Image after epoch 80

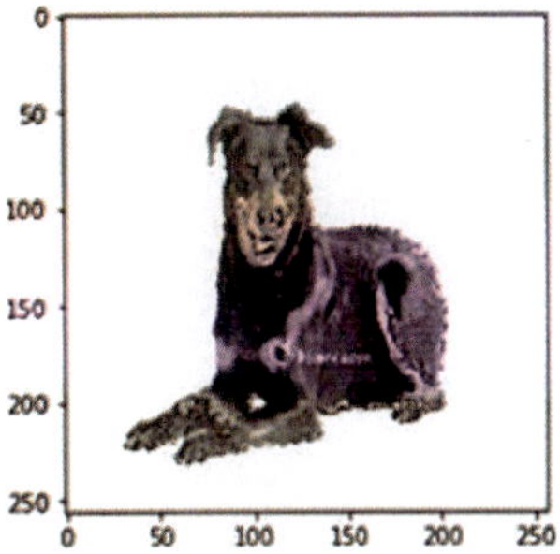

Fig. 8 Image after epoch 100

5 Conclusion and Futurework

In this work, we have explained the working of generative adversarial neural network and have ensured we bound it by using conditions, with the conditions being the text words. We have also used StackGAN along with RNN for the generation of realistic images; the future works for this work include creating a metric based on which one can assess the quality of image that is produced. One can also try creating more stable architectures for creating much better quality images with better results. The entire architecture takes lots of time for training because it is computationally expensive; therefore, there is a need for creation of methods or algorithms that can ensure that the entire GAN process completes much faster.

References

1. Mirza, M., & Osindero, S. (2014). Conditional generative adversarial nets. arXiv:1411.1784v1 [cs.LG], November 6, 2014.
2. Zhang, H., Xu, T., Li, H., Zhang, S., Wang, X., Huang, X., & Metaxas, D. (2017). StackGan: Text to photo-realistic image synthesis with stacked generative adversarial networks. arXiv:1612.03242v2 [cs.CV], August 5, 2017.
3. Goodfellow, I. J., Pouget-Abadie, J., Mirza, M., Xu, B., Warde-Farley, D., Ozairy, S., et al. (2014). Generative adversarial nets. arXiv:1406.2661v1 [stat.ML], June 10, 2014.
4. Arjovksl, M., Chintala, S., & Bottou, L. (2017). Wasserstein GAN. arXIV:1701.07875v3 [stst.ML], December 2017.
5. Radford, A., Metz, L., & Chintala, S. (2016). Unsupervised representation learning with deep convolutional generative adversarial networks. arXiv:1511.06434v2 [cs.LG], ICLR 2016.
6. Karra, T., Laine, S., & Aila, T. (2019). A style-based generator architecture for generative adversarial networks. arXiv:1812.04948v3 [cs.NE], March, 2019.
7. Salimans, T., Godfellow, I., Zaremba, W., Cheung, V., Radford, A., & Chen, X. (2016). Improved techniques for training GANs. arXiv:1606.03498v1 [cs.LG], 10, 2016.
8. Abel, L., Pierson, V., & Meltem Tolunay, E. (2018). Dank learning generating memes using deep neural networks. arxiv:1806.04510v1 [cs.CL], June 8, 2018.

Human Activity Recognition by Utilizing Local Ternary Pattern and Histogram of Oriented Gradients

Arati Kushwaha and Ashish Khare

Abstract Due to increasing demand of automated monitoring systems, human activity recognition (HAR) has become one of the most active research areas. The goal of the proposed work is to develop a view invariant algorithm to recognize human activities from realistic unconstrained videos. In this study, we proposed a robust and discriminative feature vector for human activity recognition based on integration of local ternary pattern (LTP) and histogram of oriented gradients (HOG). The LTP descriptor is used to grab texture features from the sequence of frames followed by computation of HOG to represent the extracted texture feature by local oriented histogram. Since LTP is an efficient feature descriptor to grab structural arrangement of pixels in frame sequences, HOG is useful for measuring shape of an object locally. Therefore, by utilizing the properties of these two descriptors, an efficient feature descriptor has been constructed for multi-view and realistic environment. The classifier used for the classification of activities is multiclass support vector machine (SVM). The proposed work has been tested on three publically available human activity recognition datasets that are—UT interaction, CASIA, and UCF101. Performance of the proposed method is compared with several existing methods. The experimental results have shown the usefulness of the present work.

Keywords Human activity recognition · Histogram of oriented gradients · Complex background · Local ternary pattern · Support vector machine

A. Kushwaha · A. Khare (✉)
Department of Electronics and Communication, University of Allahabad, Allahabad, India
e-mail: ashishkhare@hotmail.com

A. Kushwaha
e-mail: aratikushwaha.jk@gmail.com

S. Tiwari et al. (eds.), *Proceedings of International Conference on Big Data, Machine Learning and their Applications*, Lecture Notes in Networks and Systems 150,
https://doi.org/10.1007/978-981-15-8377-3_27

1 Introduction

Due to increasing demands of automatic surveillance-based security systems, human activity recognition is a growing research area. It has applications in automated monitoring systems, human computer interaction (HCI), robot vision, automated surveillance systems, video analytics, healthcare assessment, entertainment, education, etc. The purpose of developing human activity recognition systems is to automate the visual existing monitoring system that can help human operators to identify unusual events of interest. Even a number of researches have been done in this area but still it is a challenging area of research due to presence of complex and cluttered background, moving camera, noise, varying illumination, object occlusion, and changes in view point [1–4].

In the present work, we have proposed a framework for human activity recognition based on integration of LTP and HOG features. For solving the activity recognition problem, single feature has been used in many literatures, but the use of single feature is not sufficient to represent information related to variety of objects. By combined use of multiple features, chances of identification of object can be increased, as different features carry information of the object in different aspects. Combination of multiple features has been used in object classification [5], image retrieval [6–8], activity recognition [9], and object tracking [10] and has given better performance than the use of single feature. This fact has motivated us to utilize integration of more than one feature in the proposed work. This paper proposes the integration of LTP and HOG descriptor to compute feature vectors for activity recognition problem. LTP has been used to extract complex structural arrangement of pixels, and HOG features have been used to compute shapes feature of object. In this way, the combination of these two features will extract more fine and unique information from the frame sequences. Then, for classification of activities in respective category, multiclass support vector machine (SVM) has been used.

Remaining part of the paper is as follows: Sect. 2 discusses literature review. Section 3 discusses the basics of features used for activity recognition along with their properties. The detailed discussion of the proposed method is given in Sect. 4. Section 5 consists of experimental result and discussions. Finally, the work has been concluded in Sect. 6.

2 Literature Review

The past decade has witnessed a number of algorithms for human activity recognition. The detailed discussion of several methodologies developed for activity recognition is available in survey papers [1–3]. Most of the existing methods are facing with the challenges like presence of complex background, object occlusion, scaling of object, processing in real time, etc. The basic processing task of activity recognition is to grab the efficient discriminating feature vectors from frame sequences and further utilize these feature vectors for activity recognition. Therefore, it is necessary to represent human object by appropriate and robust feature. A large number of feature descriptors such as LBP [11], LTP [12], HOG [13], scale-invariant feature transform (SIFT) [14], and space-time interest points (STIP) [15] have been utilized in development of algorithms for human activity recognition [16–21]. Schuldt et al. [16] proposed an activity recognition method based on STIP feature descriptor. Nigam et al. [18] have used an integrated feature descriptor for activity recognition based on moment invariants and uniform local binary pattern. The dynamic texture pattern such as LBP and LTP is being utilized for recognition of human activities [21–24], based on assumption that, the human activities are modeled as a moving texture pattern. LBP is a feature discriminator [11], which is invariant to gray-level transform and easy to compute. Rotation invariance in LBP can also be achieved [23, 24]. But the major limitations of LBP are its highly sensitive nature to noise and its failure to discriminate between multiple patterns. Tan et al. [12] provided the concept of LTP as an extension of LBP. LTP grabs more fine details than LBP, and it is less sensitive to noise. Yeffet et al. [21] have proposed LTP-based activity recognition approach, to classify activities in respective category. In the present work, we have proposed combination of local ternary pattern (LTP) and histogram of oriented gradients (HOG) for activity recognition.

3 The Feature Descriptors and Their Properties

3.1 Local Ternary Pattern (LTP)

Tan et al. [12] have given the basic concept of LTP for texture analysis. LTP operator originally works in 3 × 3 pixel cell of an image. Each pixel is thresholded by center pixel and encoded by values (0, 1 and −1). The LTP descriptor divides the image into two binary patterns, i.e., positive and negative components. The LTP code with a threshold t is computed as

$$S_{\text{LTP}}(G_c, G_i) = \begin{cases} 1, & G_c \geq G_i + t \\ 0, & |G_c - G_i| < t \\ -1, & G_c \leq G_i + t \end{cases} \quad (1)$$

The properties of LTP, which are useful in computer vision applications, are

1. LTP grabs complex structural arrangement of pixels in the images.
2. LTP descriptor is suitable for the problem near constant gray value areas.
3. LTP grabs more detailed information than the other local patterns.
4. Since each neighborhood pixels are encoded by three values (1, 0, and −1) using center pixels, it is less sensitive to noise.
5. LTP descriptor has more discriminative capability for texture classification than the other local patterns.

3.2 *Histogram of Oriented Gradients (HOG)*

In 2005, Dalal et al. [13] have given the concept of HOG, which gives shape information of object. HOG descriptor provides local shape information of object by computing intensity gradient distribution in each cell after dividing the image into cells. Gradients of an image are useful because the edges and corners have large magnitude of gradients, and it is known that edges and corners contain more information about object shape than flat regions. To compute HOG descriptor, firstly the whole image is divided into cells and then for each cell 1D histogram of oriented gradients has been computed locally over the pixels of each cell. Important properties of HOG descriptor which makes HOG descriptor advantageous for recognition are

1. HOG is a dense descriptor used to extract features from whole image without detecting any keypoints.
2. It describes images by set of local oriented gradient histogram.
3. It measures the shape of object based on distribution of local intensity gradients without previously knowing the corresponding gradient.
4. HOG descriptor efficiently capture the foreground shape information.
5. HOG descriptor is invariant to photometric and geometric transformations.
6. Edge and gradient structures are captured by HOG descriptor in a localized manner.

3.3 *Advantage of Combining LTP and HOG Feature*

Realistic unconstraint videos consist of complex background and foreground. Arrangement of pixels in a frame should be encoded in the manner to grab the complex details of each frame sequences for the recognition of activities. The texture feature of an image gives the pixels structural arrangement. A number of feature descriptors are used to grab texture feature from sequence of frames. The combination of features makes feature vector more discriminative than the single feature. In this work, LTP has been used to extract texture and the extracted texture feature is further processed by HOG descriptor to compute the shape information of objects. The usefulness of combining LTP and HOG is as follows

1. Since LTP extracts more detailed information than LBP and invariant to noise, LTP feature descriptor has been utilized to extract complex structural intensity values from sequence of frames.
2. HOG descriptor has been utilized as a shape descriptor to measure the shape of the object in localized manner without any segmentation.
3. Combination of LTP and HOG descriptor gives significant discriminative feature to achieve high accuracy for activity recognition.

4 The Proposed Method

The goal of the proposed method is to present an algorithm for recognizing human activities based on combination of LTP and HOG. Following are the steps of the proposed method

1. Convert sequence of frames from color images to grayscale images.
2. Compute texture feature of each grayscale frames by LTP and store them separately.
3. Compute shape feature by HOG descriptor of LTP codes for each frame.
4. Perform classification.

As a first step of the proposed work, we converted sequence of frames into grayscale frames. Then, the texture feature of each frame has been computed by LTP. It grabs the complex structural arrangement of pixels of the frame sequences and is invariant to noise. HOG descriptor has been utilized for extraction of shape feature from texture feature computed by LTP descriptor. HOG descriptor measures the shape feature of frames by utilizing the description of intensity gradient distribution. The resultant of HOG descriptor has been used as feature vector for activity classification. In this work, we have utilized multiclass SVM classifier for activity recognition within high-dimensional feature space. SVM is used as classification approach which is based on supervised machine learning approach. SVM is comparatively easy to use, cost effective and requires less computation time than the other classifiers [25]. In this approach, first labels are assigned to each example frames of trained data and then we train the multiclass SVM using these labeled data. After training SVM, each category of activities is separated by hyperplane in the feature space with different labels. For testing, similar feature vectors are computed for query videos and classified using trained SVM model. The overall framework of the proposed method is shown in Fig. 1.

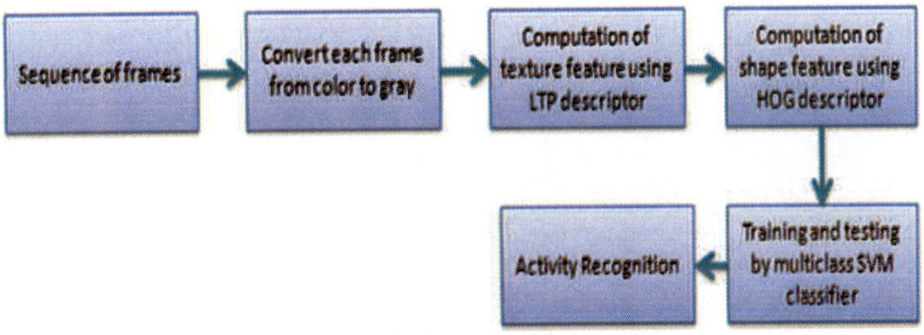

Fig. 1 Overall framework of the proposed method

5 Experimental Result and Discussion

The proposed method has been tested on a number of publically available datasets of human activity recognition. In this study, we have shown and discussed the results on three benchmark activity recognition datasets—UT Interaction [26], CASIA [27], and UCF101 [28]. The objective of UT interaction dataset was to recognize complex activities such as human–human interaction. It was created to provide challenges like aerial view and for wider area of activity videos. This dataset consists of six activity categories: shaking hand, hugging, pointing, kicking, pushing, and punching with static background. CASIA dataset is a realistic multi-view human activity recognition dataset. It was captured by outdoor video cameras with different viewing angles. This dataset consists of total 1446 videos with 25 frames per second frame rate and resolution 320 × 240 pixels. This dataset contains video clips of eight single person activities such as running, walking, jumping, bending, fainting, wandering, crouching, and punching a car and seven categories of video clips of two person activities such as fighting, robbing, overtaking, following, meeting and parting, following and gathering, meeting and gathering. UCF101 human activity dataset is one of the most challenging datasets for recognizing human activities. It has 101 activity classes with 13,000 video clips, 320 × 240 frame size, and 25 FPS frame rate. UCF101 videos are realistic unconstraint videos and typically include camera motion, varying lighting condition, partial occlusion, cluttered background, low-quality frames, etc. In this study, we have performed the testing of the proposed work on 10 activity classes of UCF101: biking, band marching, playing flute, fencing, ice dancing, billiards, tennis, cricket bowling, horse racing, and diving. Figure 2 shows sample frames of activities from UT interaction, CASIA, and UCF101 datasets. The proposed work has been compared and tested with the other existing activity recognition methods proposed by Schuldt et al. [16], Laptev et al. [17], Nigam et al. [18], Moussa et al. [19], Seemanthini et al. [20], and Yeffet et al. [21].

Recognition rate (RR) has been considered as a quantitative performance evaluation criteria for the comparison of the proposed method with other existing methods. Recognition rate is also called as accuracy and is defined as follows:

$$\mathrm{RR} = \frac{C_{\mathrm{A}}}{T_{\mathrm{A}}} \times 100\ \text{(in percent)} \tag{2}$$

where C_{A} is total number of correctly classified activity sequence and T_{A} is total number of activity sequences to be tested.

The performance measure values, in terms of recognition rate, of the proposed method and other existing methods, for all the considered datasets, are shown in Table 1.

The comparison results of UT interaction, CASIA (single person), CASIA (interaction), and UCF101 with the other existing methods are shown in Table 1. Best values in Table 1 has been shown in bold face. From Table 1, it can be observed that the proposed method achieves highest value of recognition rate for UT interaction

Fig. 2 Sample frames from the datasets **a** UT interaction and **b** CASIA **c** UCF101

Table 1 Performance of the proposed method and other state-of-the-art methods

Method	Accuracy (%) for UT interaction	Accuracy (%) for CASIA (single person)	Accuracy (%) for CASIA (interaction)	Accuracy (%) for UCF101
Schuldt et al. [16]	91.35	84.25	64.10	91.83
Laptev et al. [17]	83.33	27.72	17.79	89.97
Nigam et al. [18]	86.19	38.77	27.00	63.35
Moussa et al. [19]	98.19	27.78	19.64	89.93
Seemanthini et al. [20]	80.92	44.92	30.20	88.89
Yeffet et al. [21]	99.05	91.87	**95.66**	91.66
Proposed method	**100**	**95.04**	93.00	**92.94**

dataset (100%), CASIA (single person) (95.04%), UCF101 (92.94%) and the second highest value of recognition rate for CASIA (interaction) (93.00). Only the method proposed by Yeffet et al. [21] has highest value of recognition rate 95.66% for the dataset CASIA (interaction). The capability of extracting finer details by LTP and efficient object recognition capability of HOG is the reason behind good accuracy of the present work. From the recognition results, it can be clearly seen that the proposed method is found suitable for aerial view of wider angle scenes and for different viewing angle scenes. This shows that proposed method is also useful for surveillance systems. The recognition results show the usefulness of the proposed work for complex activities, scenes from different viewing angle and real scenes.

6 Conclusion

In this study, we presented a framework for human activity recognition in real and multi-view environment. We have proposed a feature descriptor based on integration of LTP and HOG. To represent each activity, we have computed feature vector by combining texture and shape features. Texture features are computed by LTP descriptor of sequence of gray-level frames. Then, the shape features are extracted from HOG descriptor of the resulting LTP codes. Since LTP grabs more fine details than other texture descriptors and is less sensitive to noise, HOG descriptor extracts shape information from the entire image as local feature. Therefore, the proposed method exploits the advantages of LTP and HOG descriptor. The resulting feature vectors from HOG descriptor are then utilized as feature vector for further activity recognition. For the recognition of human activities, multiclass SVM classifier has been used. Testing of the proposed method has been done on three publically available datasets—UT interaction, CASIA, and UCF101. Recognition rate has been taken as performance measure. From the quantitative measure values, it can be clearly observed that the proposed method is useful for single person activities as well as for high-level activities. Comparison of the proposed method has been done with the other existing activity recognition techniques [16–21]. The experimental result demonstrates that the proposed method outperforms on the other existing activity recognition techniques in realistic unconstraint environment that typically includes camera motion, varying lighting condition, partial occlusion, cluttered background, low-quality frames, etc.

References

1. Aggarwal, J. K., & Ryoo, M. S. (2011). Human activity analysis: A review. *ACM Computing Surveys (CSUR), 3*(16), 1–43.
2. Ke, S. R., Thuc, H., Lee, Y. J., Hwang, J. N., Yoo, J. H., & Choi, K. H. (2013). A review on video-based human activity recognition. *Computers, 2*(2), 88–131.
3. Zhang, H. B., Zhang, Y. X., Zhong, B., Lei, Q., Yang, L., Du, J. X., & Chen, D. S. (2019). A comprehensive survey of vision-based human action recognition methods. *Sensors, 5,* 1005–1020.
4. Binh, N. T., Nigam, S., & Khare, A. (2013). Towards classification based human activity recognition in video sequences. In *Proceedings of 2nd International Conference on Context Aware Systems and Applications* (pp. 209–218).
5. Khare, M., Binh, N. T., & Srivastava, R. K. (2014). Human object classification using dual tree complex wavelet transform and Zernike moment. In *Transactions on Large-Scale Data-and Knowledge-Centered Systems, XVI* (pp. 87–101).
6. Srivastava, P., Binh, N. T., & Khare, A. (2014). Content-based image retrieval using moments of local ternary pattern. *Mobile Networks and Applications, 19*(5), 618–625.
7. Srivastava, P., & Khare, A. (2017). Integration of wavelet transform, local binary patterns and moments for content-based image retrieval. *Journal of Visual Communication and Image Representation, 42,* 78–103.
8. Srivastava, P., & Khare, A. (2018). Utilizing multiscale local binary pattern for content-based image retrieval. *Multimedia Tools and Applications, 77*(10), 12377–12403.
9. Ahmad, M., & Lee, S. W. (2008). Human action recognition using shape and CLG-motion flow from multi-view image sequences. *Pattern Recognition, 41*(7), 2237–2252.
10. Khare, M., Srivastava, R. K., & Khare, A. (2017). Object tracking using combination of Daubechies complex wavelet transform and Zernike moment. *Multimedia Tools and Applications, 76*(1), 1247–1290.
11. Ojala, T., Pietikäinen, M., & Harwood, D. (1996). A comparative study of texture measures with classification based on featured distributions. *Pattern Recognition, 29*(1), 51–59.
12. Tan, X., & Triggs, W. (2010). Enhanced local texture feature sets for face recognition under difficult lighting conditions. *IEEE Transactions on Image Processing, 19*(6), 1635–1650.
13. Dalal, N., & Triggs, B. (2005). Histograms of oriented gradients for human detection. In *Proceedings of IEEE Computer Society Conference on Computer Vision and Pattern Recognition* (Vol. 1, pp. 886–893).
14. Lowe, D. G. (2004). Distinctive image features from scale-invariant keypoints. *International Journal of Computer Vision, 60*(2), 91–110.

15. Laptev, I. (2005). On space-time interest points. *International Journal of Computer Vision, 64*(2–3), 107–123.
16. Schuldt, C., Laptev, I., & Caputo, B. (2004). Recognizing human actions: A local SVM approach. In *Proceedings of the 17th International Conference on Pattern Recognition (ICPR)* (Vol. 3, pp. 32–36).
17. Laptev, I., Marszalek, M., Schmid, C., & Rozenfeld, B. (2008). Learning realistic human actions from movies. In *2008 IEEE Conference on Computer Vision and Pattern Recognition* (pp. 1–8).
18. Nigam, S., & Khare, A. (2016). Integration of moment invariants and uniform local binary patterns for human activity recognition in video sequences. *Multimedia Tools and Applications, 75*(24), 17303–17332.
19. Moussa, M. M., Hamayed, E., Fayek, M. B., & El Nemr, H. A. (2015). An enhanced method for human action recognition. *Journal of Advanced Research, 6*(2), 163–169.
20. Seemanthini, K., & Manjunath, S. S. (2018). Human detection and tracking using HOG for action recognition. *Procedia Computer Science, 132,* 1317–1326.
21. Yeffet, L., & Wolf, L. (2009). Local trinary patterns for human action recognition. In *2009 IEEE 12th International Conference on Computer Vision* (pp. 492–497).
22. Vili, K., Guoying, Z., & Matti, P. (2008). Texture based description of movements for activity analysis. In *International Conference on Computer Vision Theory and Applications (VISAPP)* (Vol. 1, pp. 206–213).
23. Fernández, A., Ghita, O., González, E., Bianconi, F., & Whelan, P. F. (2011). Evaluation of robustness against rotation of LBP, CCR and ILBP features in granite texture classification. *Machine Vision and Applications, 22*(6), 913–926.
24. Ahonen, T., Matas, J., He, C., & Pietikäinen, M. (2009). Rotation invariant image description with local binary pattern histogram Fourier features. In *Scandinavian conference on image analysis* (pp. 61–70). Berlin, Heidelberg: Springer.
25. Chang, C. C., & Lin, C. J. (2011). LIBSVM: A library for support vector machines. *ACM Transactions on Intelligent Systems and Technology (TIST), 2*(3), 27.
26. Ryoo, M. S., & Aggarwal, J. K. (2009). Spatio-temporal relationship match: Video structure comparison for recognition of complex human activities. In *2009 IEEE 12th International Conference on Computer Vision* (pp. 1593–1600).
27. Wang, Y., Huang, K., & Tan, T. (2007). Human activity recognition based on transform. In *2007 IEEE Conference on Computer Vision and Pattern Recognition* (pp. 1–8).
28. Soomro, K., Zamir, A. R., & Shah, M. (2012). UCF101: A dataset of 101 human actions classes from videos in the wild. arXiv preprint arXiv:1212.0402.

Study and Analysis of Various Load Balancing Techniques for Software-Defined Network (A Systematic Survey)

Aashish Kumar and Darpan Anand

Abstract The interconnections between persons, organizations and various entities are possible only because of information communication technology (ICT). ICT changes the era into digital world. In this digital world, all things are possibly connected to each other through ICT and makes possible the concept of Internet of things (IoT). To implement these concepts, the computer network plays a vital role. But, the traditional IP network is difficult to manage, complex and rigid to configure the network according to the change of the organization's policies. The traditional computer network is also difficult to handle the fault, load and changes because the data plane and control plane are integrated. Therefore, the idea of software-defined networking (SDN) Fei et al. (IEEE Commun Surv Tutor 16(4):2181–2206, 2014, [1]) is a flourishing field to solve the issues of traditional computer network. SDN vows to change the traditional network by segmenting the planes of networking into two separate planes as data plane and control plane. SDN makes it simpler to create and present abstraction and simplifying in networking. There are different functionalities provided by the SDN, for example, traffic engineering, load balancing, routing, intrusion detection, security and so on. Load balancing is one of them. This paper is analyzing the concept and evolution of SDN and further emphasis on the concept of load balancing. Further, analysis and comparison of various load balancing techniques are also given in this text. This work gives the concept, evolution, analysis of the load balancing in SDN which may help to make it better, efficient and cost effective.

Keywords Software-defined network · Load balancing · Communication network · Network performance

A. Kumar
Chandigarh University, Chandigarh, Punjab, India
e-mail: aashishcs024@gmail.com

D. Anand (✉)
Department of CSE, Chandigarh University, Chandigarh, Punjab, India
e-mail: darpan.e8545@cumail.in

S. Tiwari et al. (eds.), *Proceedings of International Conference on Big Data, Machine Learning and their Applications*, Lecture Notes in Networks and Systems 150,
https://doi.org/10.1007/978-981-15-8377-3_28

1 Introduction

Computer network is used for communication which is becoming the bottleneck not only for the business but also for other communities. Because of this communication, the network load is increasing day-by-day. To serve the smooth communication complex techniques, different traffic patterns and new technologies make the existing system more complex and difficult to manage. Management of the existing network is not only the problem for network engineering but the requirement of new services with effective and efficient data transfer are the motivation to change the existing network architecture. There is a requirement of data plane (required to forward data) and control plane (required for control path) for any network architecture. In traditional network architectures, these both planes are coupled to each other at a specific hardware which is generally very costly and vendor specific. These problems are solved by the concept such as network function virtualization (NFV) [2] and software-defined network (SDN). In NFV, the network architecture uses the concept of virtualization for classes network node functions. While, SDN separate the data plane from control plane. For these planes, there are separate hardware software and protocols are developed. To manage these both planes, there are another plane which is management plane which uses for configuration of the network as per user's requirements as load balancer, traffic engineering, firewall, etc. This paper is developed to explain of the major functionality (load balancing) of the management plane. We will discuss NFV, SDN, OpenFlow [3], load balancing in SDN and researcher views and comparison in detail.

1.1 Network Function Virtualization

A traditional network is based on the concept of "one node one service" which means one service is deployed over one specific hardware. Due to this traditional network leads to the problem of scalability, flexibility, lack of speed, delay and other network latency issues. These issues will be harmful for business and organization. So we require a platform to overcome the above mention issues and NFV is suitable to solve the issues with virtualization of network resources. For network virtualization, NFV uses three mechanisms are virtualization [4], softwarization [5] and orchestration [6]. NFV is also useful to replace costly, dedicated and specific purpose hardware with generic hardware. There are following features of NFV as

- Replaces purpose built hardware to generic hardware.
- Easy scale-up and down.
- Virtual machine creates building block.

NFV is a specific type of network architecture where network services are installed on generic hardware through software inspite of specific and dedicated hardware. There are three major components of NFV are shown in Fig. 1

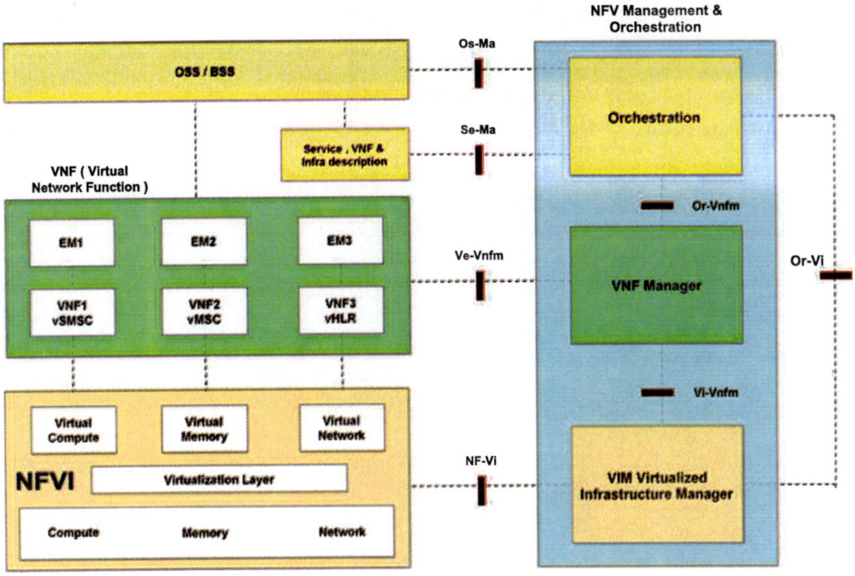

Fig. 1 Network function virtualization architecture

- NFV infrastructure building block (NFVI)
- Virtual network function (VNF)
- Management and orchestration (MANO).

NFVI consists of software as well as hardware components to provide the environment to execute various VNF services. This component can be subdivided into the following:-

- ***Hardware resources***:- Hardware resource consists of computing, storage and networking resources like router switches wireless network, etc. It provides storage capacity, processing and network connectivity to the virtualization layer.
- ***Virtualization layer***:- It provides an intermediary between hardware resources and virtual resources as shown in Fig. 1 The following function performed by the virtualization layer as:-
  - It abstracts and logically partitioned physical resources.
  - It provides the platform to deploy the software and services for implementation of virtual network function.
- Virtualization layer also known as hypervisor.

After decouple of the network hardware and network services, there is a requirement to deploy these network services over various hardware without any concern of its manufacturing, therefore, we require a virtual machine over hardware to deploy these services. Therefore, these services function called as NFV. Virtual network function (VNF) layer can further divided into two parts:-

- Virtual network function (VNF).
- Element management system (EMS).

MANO layer deals with all the assets in the infrastructure layer, NFVI and VNF. It is also responsible for the creation and deletion of resources and manages their allocation for VNFs. MANO has three entities as:-

- ***Virtualized Infrastructure Manager***: It controls and manages the interaction of hardware resources (such as computing, processing) with virtualized resources. VIM performs the following task:-
  - Manages the infrastructure resources and allocation.
  - Allocates the VMs on hypervisor, compute resources, storage and relevant network connectivity.
  - Gathers the information of fault tolerance.
- ***VNF Manager***: It executes VNF life cycle management which includes the following phases:
  - Installation.
  - Updates.
  - Query.
  - Scale-up/down.
  - Termination.
- ***Orchestrator***:- It manages and orchestrated NFV infrastructure. It is also responsible to manage whole NFV architecture.

1.2 Software-Defined Network

While NFV is used to building the virtual platform for virtual services and architecture but there are some problems exist in it as dynamic IP allocation, change in routing, bandwidth management, end-to-end reachability, etc. To deploy these services and change infrastructure, NFV require huge time. To solve these problems, SDN has to be developed. SDN solves the problem network reachability, making IP network more programmable and making routing more flexible and agile.

1.3 Evolution of Software-Defined Network

The advancement of SDN has appeared in Fig. 1

- ***Network Control Point (NCP)***: In 1980, the first step toward the development of SDN was taken as development of network control point [7] was public switched telephone network that separate controller from forwarding devices.

- ***Active Network***: Durning the decade of 1990s, new technology "Active network" [8] has developed. The idea of Active network was proposed by parallel processing community. In this technique, the processing logic was embedded with header of each packet. The processing code was executed at it destination (switches). This idea is failed due to resistance of existing network infrastructure as it was passive and non-interoperable.
- ***Internet Engineering Task Force (IETF)***: It tried to solve the above issues and during this work, they developed the concept to decouple both the planes and both of these planes are coupled with an interface which is called as forward and control separation (ForCES) [9] in 2004.
- In the same year (2004), a new concept of routing control platform (RCP) [10] was developed in which a centralized controller has developed with a global view. It is used to interact with router for routing through IGP [11] and BGP [12] where BGP to distribute routes and IGP to have global view.
- ***Ethane*** is the predecessor of the OpenFlow. The concept of centralized controller, Ethane switches and names and policy languages were introduced. Basically, Ethane[13] works on two principles as

  - Flow setup.
  - Forwarding.

  The Ethane project at Stanford's university prompts the advancement of OpenFlow. An OpenFlow was the first southbound API.

The Open Networking Foundation was established in 2011 to advance the concept of OpenFlow and SDN. Various companies use software-defined network like Google, Verizon, Microsoft, Yahoo, Facebook, etc. OpenFlow protocol is used in SDN technologies (Fig. 2).

Some of the main components of SDN architecture (as shown in Fig. 3) are

- Data plane.
- Control plane.
- Management plane.

Data Plane A data plane is also known as a forwarding plane. It is responsible to process and route the packets as flow entries available in the flow table. They are dumb router and switches. Various switches support OpenFlow are Juniper MX-Series, PC engines, Pronto 3220/3290, HP curve, etc.

Control Plane Control plane contains controllers. It takes a decision where data is to be sent. The main functions of the control plane are configured the system, manage and exchange the routing table information. Some of the controllers are NOX, POX, Floodlight and Beacon, etc. It can deployed at one place and distributed at various places.

Management Plane The management plane is used to manage and access the network devices.

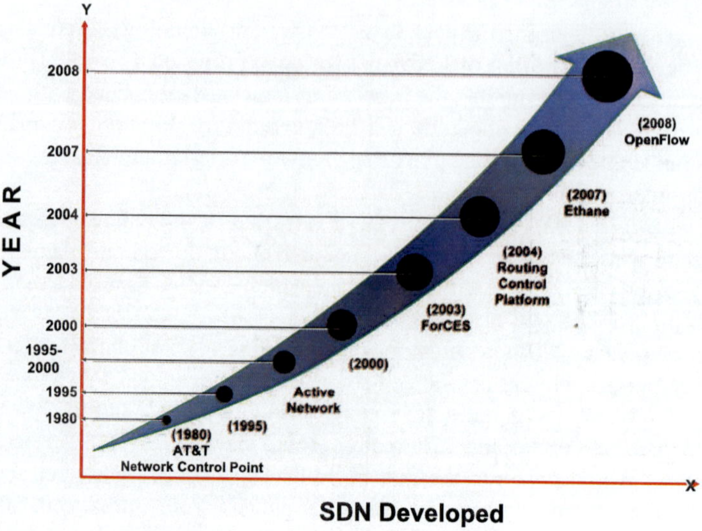

Fig. 2 Evolution of SDN

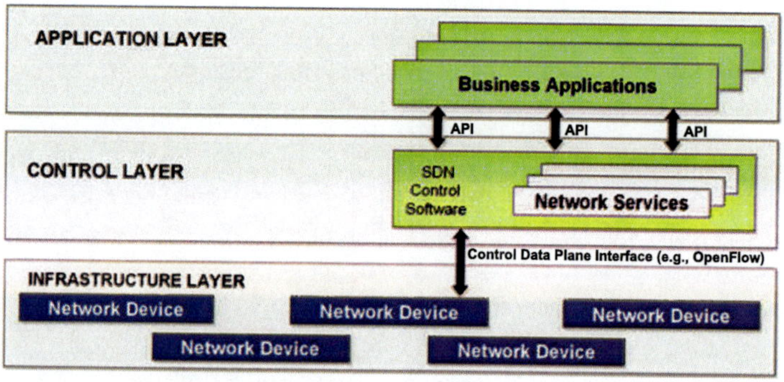

Fig. 3 Software-defined architecture

The communication interface between the controller and the management plane is provided by northbound API. It allows admin to deploy network services like traffic engineering, QoS, intrusion detection, firewall, etc. Northbound API helps the developer to write an application without any detailed knowledge of control functions. Southbound API is mainly responsible for data transfer control messages between routers/switch and controller. OpenFlow considered the primary standard in SDN, which was southbound API. Other southbound API are ForCES, Open vSwitch [14] Database (OVSDB) [15] and OpFlex [16], etc.

Benefits of SDN:-

- Directly Programmable: Through management plane, we can configured the controller (brain of the network).

- Centralized Management: All the functions and forwarding plane are managed by the controller. Controller has overall perspective of the network. It manages where data is to be forwarded or drop the packet. It acts as logically centralized controller.
- Enabling innovation.
- Delivering agility and flexibility.

There was need of protocols that are open standard and not vendor specific. This issue was solved by OpenFlow.

1.3.1 OpenFlow

The most commonly used southbound API in the SDN is OpenFlow and developed, governed by the OpenFlow Networking Foundation (ONF). Initially, it was present only on the Internet. It was a group of people who informally met at Stanford university. On December 31, 2009, first rendition of OpenFlow, i.e., 1.0.0 was developed. Further, the release of next version is given in Table 1

Main building blocks of Openflow are:-

- Flow table.
- Port.
- Messages.

1.3.2 Flow Table

The flow table is the main building block for OpenFlow switches. The packets that enter into switches must pass through one or more flow tables. The sequence of packets is called flow. Flow entries that contain the header fields, counter and flow entry actions are maintained in flow table (as shown in Fig. 4)

Match Fields: When the particular packet reached at switch, it matches the values with flow table header fields.

Table 1 OpenFlow version

OpenFlow specifications		
Version	Year	Features
1.0	December 2009	Single table, fixed match fields
1.1	March 2011	Allow multiple tables, group tables, VLAN full support MPLS
1.2	December 2011	Support for Ipv6 and multiple controllers
1.3	June 2012	Meter table added for Qos
1.5	August 2013	Bundles, flow monitoring, optical ports, eviction
1.6	January 2015	Egress table, scheduled bundles

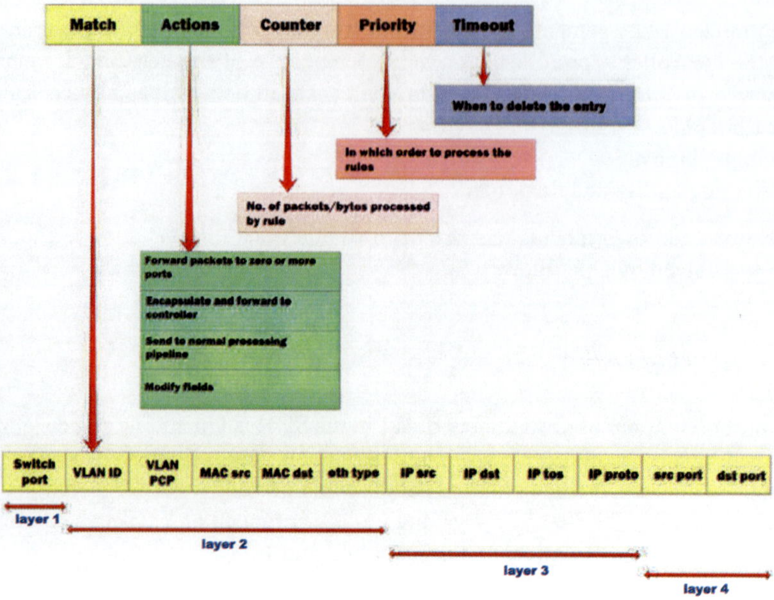

Fig. 4 Flow table

Actions:- The various actions taken by the switch are:-

- It could be sending packets to a particular port, all ports aside from incoming port.
- Dropping the packet.
- Changing values in the header fields and the packet is sent to the controller.

Counters: It is responsible to collect features such as number of bytes, packets received on per port, per-flow entry, per-flow table basis, duration of flows and number of dropped packets, etc.

1.3.3 OpenFlow Port

OpenFlow standard specifies the OpenFlow port. There various kinds of ports are:-

- Physical port.
- Logical port.
- Reserved port.

OpenFlow defined ports are responsible to forward the packets to the controller, flooding or specifying legacy switch methods by using processing/computation. Brief summary of ports are given in Table 2

Table 2 Ports and Description for SDN

Ports	Description	
Physical ports	Specify the hardware interface of a switch A physical port can be either ingress port or outport port	
Logical ports	Do not corresponds to hardware interface of a switch It represents high level abstraction such as link aggregation group and tunnels A logical port can be either ingress port or outport port	
Reserved ports	OpenFlow defined ports for specifying forwarding actions Various reserved ports are:	
	Type	Meaning
	All	All ports that could be used for packet forwarding
	Controller	To OpenFlow controller
	Table	OpenFlow table
	In_port	Send packet out through ingress port
	Any	General port description does not corresponds to a specific port
	Local	Local networking stack
	Normal	Forward to traditional non-OpenFlow pipeline
	Flood	Forward the packet to all the ports

1.3.4 OpenFlow Messages

Messages that transfer between OpenFlow controller and switches are governed by the OpenFlow controller and switch. The various types of messages are

- Controller initiates messages that traded among controller and switch so as to control and give an overall perspective of the network.
- Symmetric messages are send without solicitation in both way (as shown in Table 4).
- Asynchronous messages are the messages that are initiated by controller without asking from the switch (as shown in Table 3).

Table 3 Description of messages for asynchronous controller

Message	Description
Switch to controller	
Packet-in	Send packet to the controller
Flow-removed	Informs controller regarding the expulsion of flow entries
Port-status	Tells about modify at port
Error	Tells about fallacy situation

Table 4 Symmetric controller message details

Message	Description
Symmetric	
Hello	These are traded among the switch and controller during association startup setup
Echo	Messages are either send by switch or controller upon association startup
Vendor	Provides additional functionality

The various available OpenFlow controllers are POX, BEACON, OpenDayLight, FloodLight and Ryu, etc. SDN is a concept that is implemented by OpenFlow.

There are many technologies that come under SDN and one of them is OpenFlow. Basically, OpenFlow is a communication protocol between forwarding devices and controllers. Now, the problem with new architecture is congestion because a lot of data and different patterns of data generated by the users. To solve the issue of scalability and congestion, the concept of load balancing in SDN is developed.

1.4 Load Balancing

It is a procedure used to disperse the load among servers or any other computing devices. Maximum utilization of the resource, maximum throughput, minimize response time, avoid overload and avoid crashing is the aim of load balancing. Apart from them, it also provides from failover, i.e., if one server fails it redirects the request to another server. Traditionally, the load balancer comes software installed on the hardware. Therefore, it was vendor specific and costly. The software load balancer runs on a virtual machine or the hardware.

1.5 Kinds of Load Balancing

- Transport layer load balancer.
- Application layer load balancer.

Transport layer load balancer: In this technique, load balancer uses informations such as source IP address, destination IP and ports defined at packet header of the data and also known as layer-4 load balancing.

Application layer load balancer distributes the load across server depending upon application layer protocols such as HTTP , COOKIES or data.

1.5.1 Load Balancing Algorithm

To perform various tasks, load balancer uses a specific algo which is called load balancing algorithm.

- Round Robin.
- Least Connection.
- Source IP hash load balancing.
- Least time.

Round Robin is a very simple technique. In this, different servers are configured to give the same services and have a unique IP address. Suppose there are N servers, if client requests come to the load balancer, it first sends to the first server, the second request comes goes to the third server and so on till Nth server after that same process is repeated. The problem with this algorithm is that it does not care about link bandwidth and the processing power of the server. There are two variant of Round Robin:- Weighted Round Robin and dynamic Round Robin. In the weighted Round Robin, a weight is assigned to each server by the admin. Larger the weight, higher the request goes to that particular server, whereas, in dynamic Round Robin, a weight is appointed to every server dependent on the server's present burden and ability to process the requests. In the least connection algorithm, both the server's response time and active connection are considered. Neither Round Robin nor weighted Round Robin taken consideration of current server load, but the least connection algorithm takes care of a load of a server. When the request comes it forward to the server with the least number of the active session at the current time.

Source IP hash load balancing: In this technique, it calculates two hash values upon the following:

- Service's IP address and port (X).
- Incoming URL, the domain name, IP address of the source, the IP address of the destination, or the IP address of the source and destination, depending on the configured hash method (Y) (Fig. 5).

This procedure computes another hash value dependent on the hash value of X and Y. Finally, it selects the server which has highest hash values. The steps in IP hashing load balancing are shown in Fig. 10. Load balancer can handle the following kind of traffic:

- HTTP.
- HTTPS.
- TCP.
- UDP.

1.5.2 Load Balancing in SDN

There are two ways to deal with the load in SDN as distributed and centralized approach.

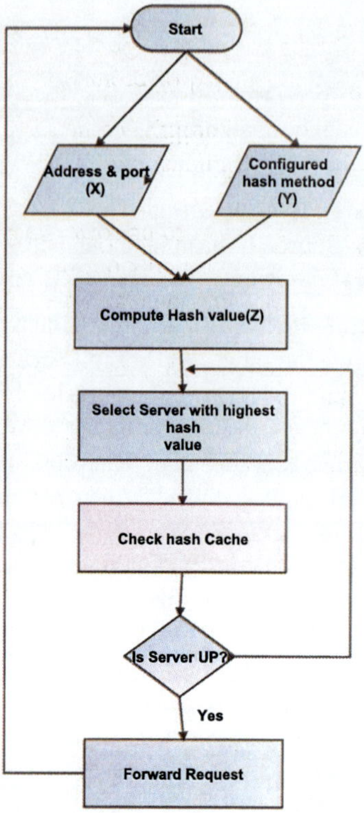

Fig. 5 Flow chart of Source IP hashing load balancing technique

In a centralized approach, there is a super controller which balances the load among other controllers (regular controllers). The problem with this approach is that if super controller fails, the whole network gets down. Other problems are scalability and availability issues. The drawbacks of the centralized approach were solved by the distributed approach.

In the distributed approach, there are many controllers balances the loads among them, but the problem is that communication overhead. We will discuss various algorithms (based on distributed and centralized approach) given by researcher in the next section.

2 Related Work

Aly [17] has proposed an algorithm known as controller adaptive load balancing model (CALB). In this algorithm, there is one super controller and n slave controller associated with it. Each slave controller is associated with the number of switches. It provides both the load balancing and fault tolerance. Author tested CALB algorithm on two scenarios:

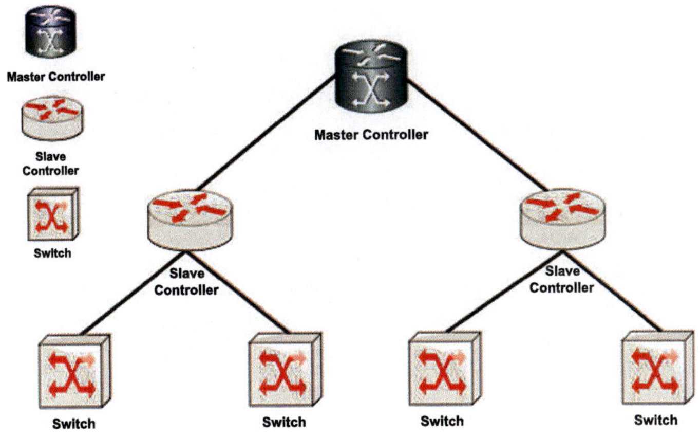

Fig. 6 One master two slave controller

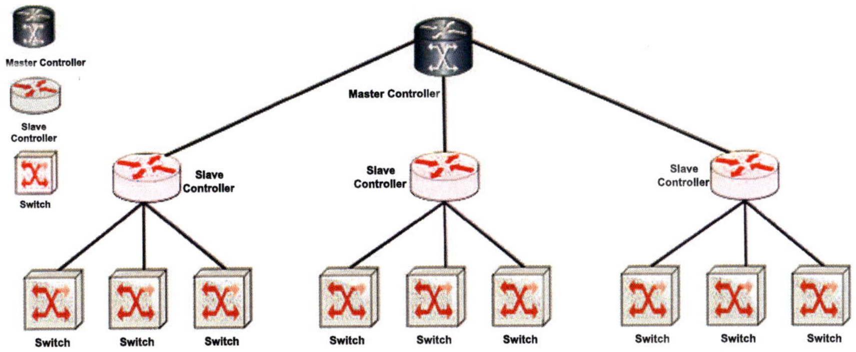

Fig. 7 One master three slave controller

- One master two slave controller is shown in Fig. 6.
- One master three slave controller is shown in Fig. 7.

The master controller has the following task to perform:

- Route the packets.
- Intermittently gathers the load of every controller in a clusters.
- Modify the switch-controller mapping table that is available on the master controller.

Switch migration is performed by the master controller (based on overall load). Overall workload can be calculated as:

$$\text{Workload Percentage} = \text{Current Workload} \div \text{Total work load} \tag{1}$$

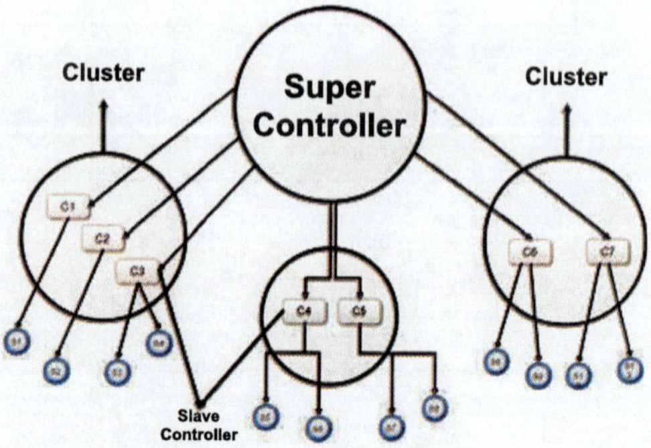

Fig. 8 Dynamic cluster architecture

Messages are exchanged among controller through ZMQ protocol. The master controller is responsible to allocate the faulty controller's load to the reinforcement controller.

Hadar Sufiev et al. [18] given a multi-controller load balancing model known as dynamic cluster as shown in Fig. 8. In the model, two kinds of controller are super controller and regular controller. There is an equal number of regular controllers in a cluster. The author has defined a term called cluster vector (CV) which contains the address of the controller in a cluster. For each cluster, there will be a separate CV. All these clusters are managed by a single super controller. The load balancing occurs at two levels, one at the super controller and the second one at regular controller. The super controller periodically checks the cluster whether they reached a threshold. If a cluster passes threshold then partition algorithm runs. It rearranges the cluster according to the loads. A new CV is sent to the cluster. At the second level, each controller in a cluster checks whether it passes the threshold or not. If passes then it will send some loads to another controller in the same cluster.

Li-Der Chou et al. [19] proposed genetic load balancing algorithm, as shown in Fig. 9. Author has choosen arithmetic average of coefficient to measure the performance of the algorithm. More lower the CV, the algorithm has better load balancing result. The fitness function is given by the formula:

$$\min\left(\frac{\sqrt[2]{\frac{\left(\sum X[j]^2-\left(\frac{\sum X[j]}{K}\right)^2\right)}{K}}}{\frac{\sum X[j]}{K}}\right) \tag{2}$$

K is the number of servers, N is the number of flows, X is the each server work load. For the evolution purposes, he choose Roulette-wheel selection, single point cross-over and single point mutation. He compares the genetic-based algorithm with the load-based, random-based and Round Robin algorithm. He found that the proposed algorithm has low CV, i.e., 15.41% but major disadvantages are scalability, stability and only one parameter is taken for the evaluation of the proposed algorithm.

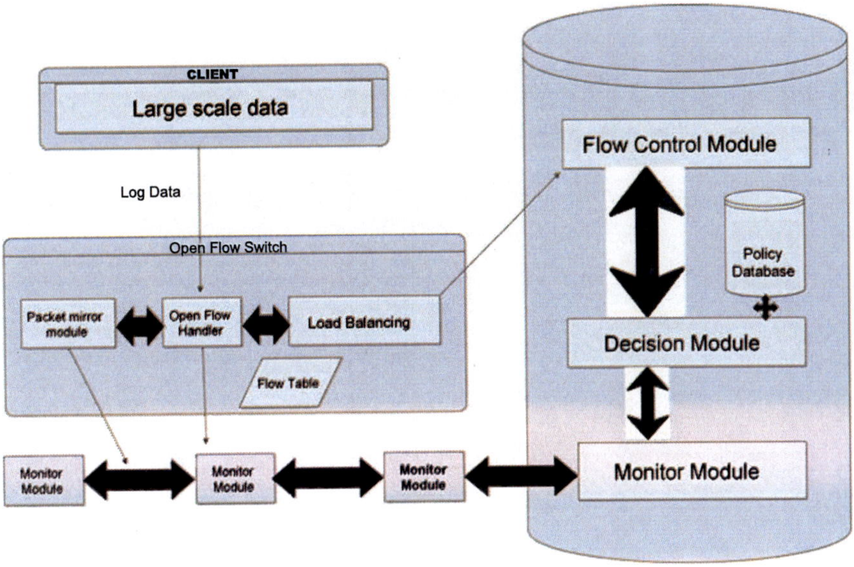

Fig. 9 Genetic load balancing technique

Zhou et al. [20] gave a new algorithm named dynamic and adaptive algorithm (DALB). The architecture has distributed controllers as shown in Fig. 10. DALB algorithm runs on each controller. In his proposed architecture, there are various components that are load measurement, load collection, maker and migration switch migration and switch election and DALB. In the load measurement component, controller collects load parameters such as flow table entries numbers, packets average arrival rate of at every switch and controller to switch round trip time. In load collection, DALB algorithm, the load is collected from other controllers, to check whether it passes threshold or not. This algorithm is an overhead and takes $O(n)$. In the worst case. The decision-maker concludes whether to perform load movement activity or not. The algorithm is tested on parameters such as throughput and average arrival time of packet.

$$L \propto M \tag{3}$$

L Load of controllers.
M Number of PACKET IN Messages.

When controller is overloaded, election of switch to migrate is done according to the following formula

$$L_{\text{controller}} = w_1 * N + w_2 * F + w_3 * R \tag{4}$$

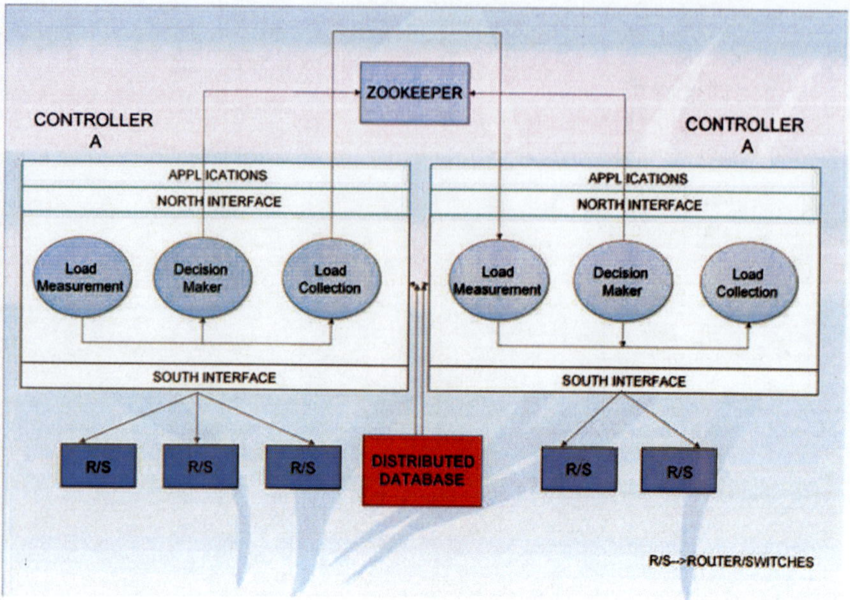

Fig. 10 DALB technique

weight coefficients are w_1, w_2, w_3, the number of flow entries = N, From each switch, average message arrival rate = F, round trip time = R.

Hu et al. [21] proposed an efficiency switch migration algorithm known as SMDM (as shown in Fig. 11). SMDM can be split into three stages. First, it quantifies the load diverseness of each controllers and concludes to perform relocation or not. At that point, it predicts migration cost and migration efficiency. These two things decide migration movement activities to be performed or not. Load of each controller is given by the following equation

$$L_{c_i} = \sum f_{s_k} * d_{s_k c_i} \tag{5}$$

Load diversity between controller c_i and c_j is

$$h_{c_i c_j} = L_{c_i} / L_{c_j} \tag{6}$$

Migration cost is mainly consist of:

- Increment in load cost
- the cost of message exchanging.

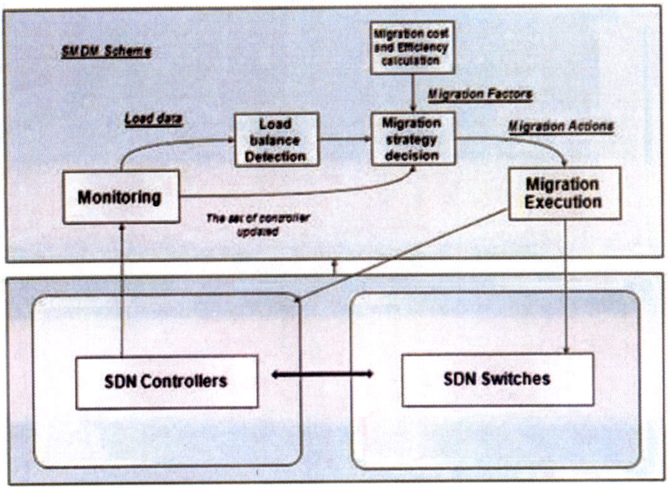

Fig. 11 SMDM

When switch s_k is relocated from controller c_i to controller c_j, migration cost is defined as:

$$r_{s_k c_j} = r_{mc} + r_{\mathrm{lc}} \tag{7}$$

where r_{mc} is the message exchanging cost for switch s_k migration and r_{lc} is the increment in load cost.

PACKET IN message is the load to the controller. There are five modules in the architecture namely monitor component, load balance detection component, migration efficiency calculation component, migration strategy decision component and migration execution component. The algorithm performance is assessed by the response time, migration cost and migration execution time and contrast with MUMA and DNMA.

Wang et al. [22] proposed an algorithm-based distributed architecture (distributed controllers). The proposed model has three principle parts: a load collector, load balancer and switch migrator. Traffic is generated by the Cbench to the controller. Load collector periodically collects the load from the other controllers. Every controller has the accompanying information with it, i.e., its identification, information of the load, related switches and update time. A controllers threshold limit is 70% of the controller bandwidth. When the controller surpasses the edge, the load balancer decides the target controller (lightest burden) to get an over-burden controllers load. Then, the switch migrator passes the messages to its associated switch. The architecture is shown in Fig. 12

Zhang et al. [23] proposed a technique named hybrid routing as shown in Fig. 13. It largely uses destination-based routing features and small features of explicit routing. In this approach, destination-based forwarding entry stored in static random access

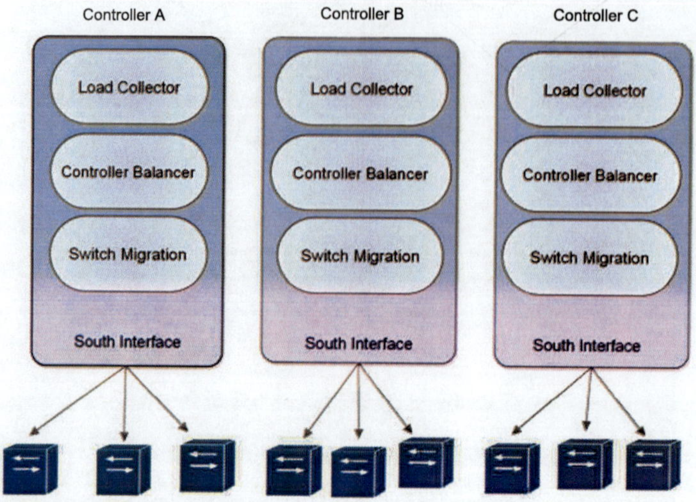

Fig. 12 Architecture of distributed SDN system

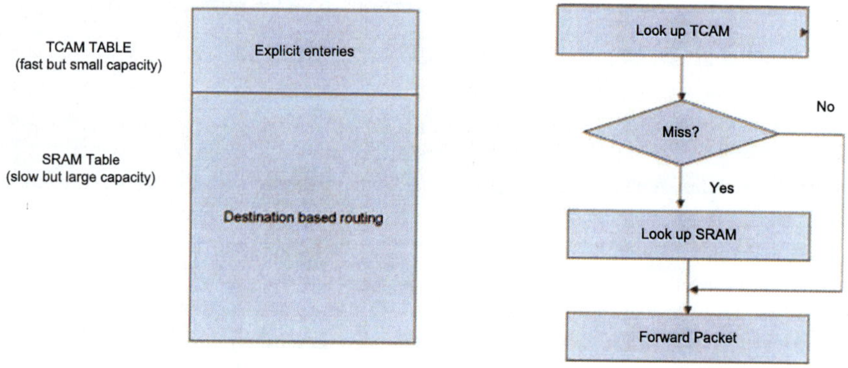

Fig. 13 Wang et al. architecture

memory (SRAM) and spares TCAM assets, whereas explicit routing information stored in TCAM table. When packet matching is done, a router first look into TCAM table, if there is matched associated action is performed. Otherwise, if there is miss (entry not found in TCAM table) then it searches into SRAM table. In this approach, the complexity of the explicit routing decreases, and performance of the architecture is greatly improved.

He et al. [24] given swarn optimization-based load balancing model, as shown in Fig. 14. In this approach, fog computing is integrated into the software-defined network and it applied software-defined fog computing network (SDFCN) into Internet of vehicle (IoV). The efficiency is increased by utilizing the reverse of the mutation

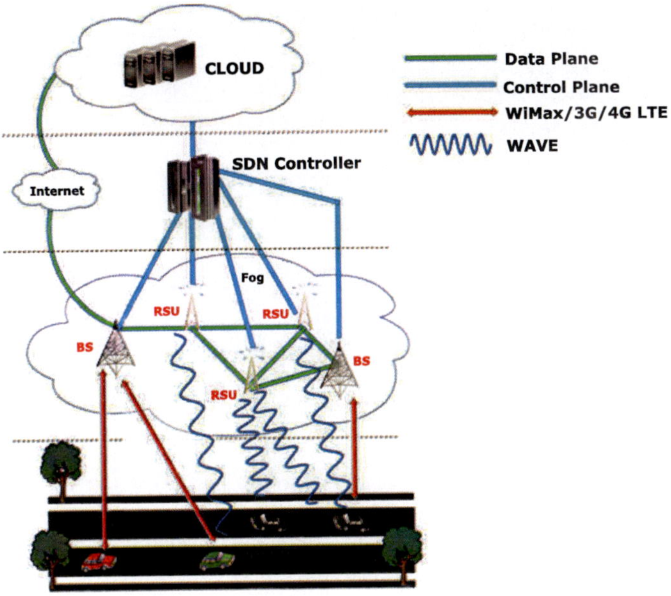

Fig. 14 Software-defined cloud/fog architecture

particle flight and linear inertia weight. The proposed technique improves the quality of service parameters and reduces the latency.

Yong et al. [25] had presented another methodology for load adjusting in the SDN data center, as shown in Fig. 15. They have applied plug-n-server an SDN design. It comprises of three sections are server and clients, OpenFlow switch, the SDN controller and decision platform.

The controller has additionally contained three-section are traffic discovery component, dynamic load scheduling component, load computation component and management of flow. The traffic detection component monitors the traffic of the network at the current time and the load calculation component calculates the load and the advantage of this technique that it has given high throughput as compared to the traditional method (hash-based technique). But due to a single controller, it has a scalability problem, bottleneck problem and low availability. Latency is not measured.

Zhong et al. [26] given server response time (LBBSRT)-based load balancing model. It first collects the response time of each server then chooses the server which has a stable response time. It is easy and easy to implement. It balances the load efficiently and average low response time. Due to the presence of one controller, the proposed algorithm suffers from scalability, availability and bottleneck problem.

Rangisetti and Tamma [27] proposed SDN-based LTE RAN framework and also gives QoS aware load balance (QALB) algorithm. To balance the load, the algorithm calculates the load of the neighboring cell, QoS profile of user equipment and estimated throughput. For more than 80% of the cell, the proposed algorithm provides

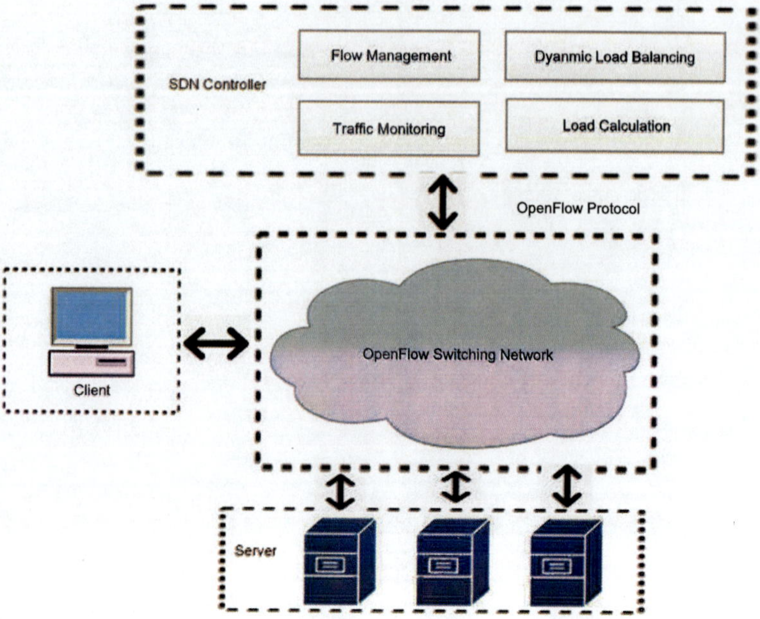

Fig. 15 Yong et al. architecture

better QoS data rates. In this static and mobility, scenarios are also examined. QALB minimizes average OLR. Jitter and delay are not considered, and due to presence of one controller, the proposed algorithm suffers from scalability, availability and bottleneck problem.

Sayali [28] has proposed a technique that finds the least loaded path for data transmission. Through this algorithm, overall network load get optimized. It uses the six parameters such as bandwidth ratio, hop count, latency, packet overhead trust and packet loss. At the point, when clients send the information to the target device, load balancer first checks whether one way or different path accessible. If one path available it will send otherwise if multiple paths available it first select a least loaded path based on six parameter. To choose a path, the author uses ad hoc on-demand distance vector (AODV). To generate traffic, he uses dynamic source routing protocol (DSR). The proposed technique was compared with a randomly based technique and it shows good results over the randomly based technique. It uses an ns2 simulator.

Filali et al. [29] given an optimization approach to deal with the load. The primary aim of this technique is to reduce latency between controllers and switches so as to adjust the load. In this methodology, the first problem is figured as an optimization problem to reduce the response time of the controller. After that, the problem is solved by one too many matching game problems with minimum resource utilization. Later then, author has proposed an algorithm that assigns the switches to the controllers. The results of simulation show that the model minimizes the response time. The other factors such as throughput and jitter are not considered. Controllers load c_j(at any instance of time t) can be computed as:

$$L_{c_j}(t) = \sum_{i=1}^{S_{c_j}} l_i(t) \tag{8}$$

3 Comparison and Analysis

The comparison of various available load balancing is providing in two ways one is the advantage and disadvantage of the techniques. While other is, the decision criteria for load balancing. This criteria is depending on some parameters. Both the comparisons are illustrating in the next sections.

3.1 Advantages and Disadvantages of Load Balancing Techniques

Table 5 illustrates the analytical comparison of various software-defined network load balancing approach. The table is also explaining the advantages and disadvantages of each and every technique.

3.2 Comparative Analysis of Load Balancing Techniques

This analysis is important to compare the available techniques and parameters used for taking load balancing as given in Table 6.

4 Conclusion

This paper gives an systematic review of the load balancing approach can be applied in SDN. We have also discussed load balancing in SDN based on the various parameters such as server response time and bandwidth. We have also discussed a 13 researcher views on load balancing in SDN. In this paper, comparative chart has been given which give brief description of various researcher algorithm. It gives techniques used, parameter, advantages, challenges, etc. Network function virtualization, SDN, OpenFlow, Load balancing in SDN have been discussed. We have divided load balancing in SDN into centralized and distributed approach. We have also integrate

Table 5 Pros and cons of load balancing techniques

Author name	Algorithm used	Advantages	Disadvantages	Basis of decision
Sufiev et al. [18]	Dynamic cluster	Reduced latency super controller does not depend on regular controller	Single point of failure scalability issues	Hybrid flow method
Chou et al. [19]	Genetic-based load balancing	Avoid bottleneck save cost	High complexity and computation time scalability issues	Round Robin random based
Hu et al. [21]	Switch migration-based decision making (SMDM)	Low response time	Not tested in real large-scale wireless network	MUMA and DNMA
Wang et al.[22]	Based on distributed architecture	Efficient adjustment of traffic flow reduced consumption overhead solve reliability problem	Communication overhead	–
Zhang et al. [23]	Hybrid routing	Achieve near-optimal load balancing increase throughput reduced TCAM	Latency is not considered	Explicit routing
He et al. [24]	Swarm optimization	Decrease latency increased QoS	Security issues energy consumption	–
Yong et al. [25]	Load balancing technique based on SDN	High throughput	Scalability issues Availability issues Latency is not considered	Hased based
Zhong et al. [26]	Load balancing-based on server response time	Easy to implement low response time	Low availability low scalability system bottleneck	–
Rangisetti and tamma [27]	QoS aware load balancing algorithm (QALB)	Improved GBR satisfaction better QoS data rates	Jitter and delay are not considered scalability issues	–
Filali et al. [29]	Optimization algorithm	Minimize latency between SDN controller and switches	Jitter and throughput is not considered	–

Table 6 Comparative analysis of load balancing techniques

S. No.	Author	Criteria	Parameter
1	Chou et al. [19]	Fitness function	Coefficient of variation
2	Zhou et al. [26]	When controller's Load is greater than Controller threshold (CT) For adjusting CT, adaptive CT algorithm has been proposed	Request/s Average PACKET_IN Arrival Throughput
3	Hu et al. [21]	Load diversity between Controller ci and controller Cj is greater than threshold	Response time Migration cost Migration execution time
4	Zhang et al. [23]	Optimization problem	Normalized throughput
5	Wang et al. [22]	Load in load collector module is greater than threshold (70% of the bandwidth)	ICMP packet testing
6	Zhong et al. [26]	Minimum value of response time	Response time
7	Filali et al. [29]	Optimization problem	Response time Resource utilization
8	Aly et al. [17]	CALB algorithm	Throughput, response time

SDN into various applications such as fog computing and LTE. So the load balancing in SDN has global view of overall architecture, which provide good performance and QoS services. Various researchers have given their algorithm, their algorithm is based on certain parameter and none of them satisfy all the need of today's network. Various work has to be done to satisfy existing challenges. The existing challenges are energy saving, communication over head, single point of failure, etc.

References

1. Fei, H., Hao, Q., & Bao, K. (2014). A survey on software-defined network and openflow: From concept to implementation. *IEEE Communications Surveys & Tutorials*, *16*(4), 2181–2206.
2. Han, B., Gopalakrishnan, V., Ji, L., & Lee, S. (2015). Network function virtualization: Challenges and opportunities for innovations. *IEEE Communications Magazine*, *53*(2), 90–97.
3. Braun, W., & Menth, M. (2014). Software-defined networking using openflow: Protocols, applications and architectural design choices. *Future Internet*, *6*(2), 302–336.
4. Chowdhury, N. M. M. K., & Boutaba, R. (2010). A survey of network virtualization. *Computer Networks*, *54*(5), 862–876.
5. Galis, A., Clayman, S., Mamatas, L., Loyola, J. R., Manzalini, A., Kuklinski, S., et al. (2013). Softwarization of future networks and services-programmable enabled networks as next gen-

eration software defined networks. In *2013 IEEE SDN for Future Networks and Services (SDN4FNS)* (pp. 1–7). IEEE.
6. Mijumbi, R., Serrat, J., Gorricho, J.-L., Latré, S., Charalambides, M., & Lopez, D. (2016). Management and orchestration challenges in network functions virtualization. *IEEE Communications Magazine*, *54*(1), 98–105.
7. Horing, S., Menard, J. Z., Staehler, R. E., & Yokelson, B. J. (1982). Stored program controlled network: Overview. *The Bell System Technical Journal*, *61*(7), 1579–1588.
8. Tennenhouse, D. L., & Wetherall, D. J. (1996). Towards an active network architecture. *ACM SIGCOMM Computer Communication Review*, *26*(2), 5–17.
9. Khosravi, H., & Anderson, T. (2003). Rfc 3654 requirements for separation of ip control and forwarding.
10. Caesar, M., Caldwell, D., Feamster, N., Rexford, J., Shaikh, A., & van der Merwe, J. (2005). Design and implementation of a routing control platform. In: *Proceedings of the 2nd Conference on Symposium on Networked Systems Design & Implementation-Volume 2*, (pp. 15–28). USENIX Association.
11. Vasseur, J. P., & Le Roux, J. L. (2007). Igp routing protocol extensions for discovery of traffic engineering node capabilities. *IETF RFC5073*.
12. Stewart III J. W. (1998). *BGP4: Inter-domain routing in the Internet*. Addison-Wesley Longman Publishing Co., Inc.
13. Casado, M., Freedman, M. J., Pettit, J., Luo, J., McKeown, N., & Shenker, S. (2007). Ethane: taking control of the enterprise. *ACM SIGCOMM computer communication review*, *37*(4), 1–12.
14. Pfaff, B., Pettit, J., Koponen, T., Jackson, E., Zhou, A., Rajahalme, J., et al. (2015) The design and implementation of open vswitch. In *12th* {USENIX} *Symposium on Networked Systems Design and Implementation* ({NSDI}, *15)*, pp. 117–130.
15. Davie, B., Koponen, T., Pettit, J., Pfaff, B., Casado, M., Gude, N., et al. (2017). A database approach to sdn control plane design. *ACM SIGCOMM Computer Communication Review*, *47*(1), 15–26.
16. Jalili, A., Nazari, H., Namvarasl, S., & Keshtgari, M. (2017). A comprehensive analysis on control plane deployment in sdn: In-band versus out-of-band solutions. In *2017 IEEE 4th International Conference on Knowledge-Based Engineering and Innovation (KBEI)* (pp. 1025–1031). IEEE.
17. Aly, W. H. F. (2019). Controller adaptive load balancing for sdn networks. In *2019 Eleventh International Conference on Ubiquitous and Future Networks (ICUFN)* (pp. 514–519). IEEE (2019).
18. Sufiev, H., & Haddad, Y. (2016). A dynamic load balancing architecture for sdn. In *2016 IEEE International Conference on the Science of Electrical Engineering (ICSEE)* (pp. 1–3). IEEE.
19. Chou, L.-D., Yang, Y.-T., Hong, Y.-M., Hu,J.-K., & Jean, B. (2014). A genetic-based load balancing algorithm in openflow network. In *Advanced Technologies, Embedded and Multimedia for Human-centric Computing*, pp. 411–417. Springer.
20. Zhou, Y., Zhu, M., Xiao, L., Ruan, L., Duan, W., Li, D., Liu, R., & Zhu, M. (2014). A load balancing strategy of sdn controller based on distributed decision. In: *2014 IEEE 13th International Conference on Trust, Security and Privacy in Computing and Communications* (pp. 851–856). IEEE.
21. Wang, C., Bo, H., Chen, S., Li, D., & Liu, B. (2017). A switch migration-based decision-making scheme for balancing load in sdn. *IEEE Access*, *5*, 4537–4544.
22. Wang, K.-Y., Kao, S.-J., & Kao, M.-T. (2018). An efficient load adjustment for balancing multiple controllers in reliable sdn systems. In *2018 IEEE International Conference on Applied System Invention (ICASI)* (pp. 593–596). IEEE.
23. Zhang, J., Xi, K., Luo, M., & Chao, H. J. (2014). Load balancing for multiple traffic matrices using sdn hybrid routing. In *2014 IEEE 15th International Conference on High Performance Switching and Routing (HPSR)*, pp. 44–49. IEEE.
24. He, X., Ren, Z., Shi, C., & Fang, J. (2016). A novel load balancing strategy of software-defined cloud/fog networking in the internet of vehicles. *China Communications*, *13*(2), 140–149.

25. Wang, Y., Tao, X., He, Q., & Kuang, Y. (2016). A dynamic load balancing method of cloud-center based on sdn. *China Communications*, *13*(2), 130–137.
26. Zhong, H., Fang, Y., & Cui, J. (2017). Lbbsrt: An efficient sdn load balancing scheme based on server response time. *Future Generation Computer Systems*, *68*, 183–190.
27. Rangisetti, A. K., & Tamma, B. R., et al. (2017). Qos aware load balance in software defined lte networks. *Computer Communications*, *97*:52–71.
28. Patil, S. (2018). Load balancing approach for finding best path in sdn. In *2018 International Conference on Inventive Research in Computing Applications (ICIRCA)* (pp. 612–616). IEEE.
29. Filali, A., Kobbane, A., Elmachkour, M., & Cherkaoui, S. (2018). Sdn controller assignment and load balancing with minimum quota of processing capacity. In *2018 IEEE International Conference on Communications (ICC)* (pp. 1–6). IEEE.

Heterogeneous Channel Allocation in Mesh Interconnection Network Using Genetic Algorithm

Devasheesh Kumar Sharma and Akash Punhani

Abstract Interconnection networks are widely used in the system on chips. The system of chips is a collection of various intellectual properties that are communicating with each other using a network on chips. This communication is an essential part of any digital system. There have been various interconnection networks that are proposed by various researchers for effective communication between the various intellectual properties. The mesh interconnection network is a tile-based architecture that utilizes the area on chip very effectively; therefore, a large number of variants of mesh interconnection networks have been proposed. In this research paper, we have focused on the mesh interconnection network that will have heterogeneous link capacities to communicate with different intellectual properties. In order for the effective allocation of the bandwidth, the genetic algorithm has been used. From the results, it has been observed that the genetic algorithm is successful in the allocation of the heterogeneous links.

Keywords Mesh · Latency · Genetic algorithm

1 Introduction

In the current era of technology, most of the hardware is designed using the system on chip. Bi-system-on-chip *p* means then the complete system is implemented on the single chip and communication with referral devices for input and output is done through the digital pins. The various intellectual properties are connected with each other with the topology for communication. In earlier systems, there was a dedicated bus to connect all the intellectual properties. The performance of that complete system depends upon the bandwidth of the bus. When the multiple intellectual properties try to communicate with each other with the help of common bus, there is a congestion on

D. K. Sharma · A. Punhani (✉)
Amity University, Noida, Uttar Pradesh, India
e-mail: akashpunhani@gmail.com

D. K. Sharma
e-mail: devasheesh.sharma.1995@gmail.com

S. Tiwari et al. (eds.), *Proceedings of International Conference on Big Data, Machine Learning and their Applications*, Lecture Notes in Networks and Systems 150,
https://doi.org/10.1007/978-981-15-8377-3_29

the bus that acts as a bottleneck for the complete system. However, in order to improve performance, the idea of dedicated bus was introduced to improve the efficiency based on the communication pattern. As the number of intellectual properties on the single chip has increased significantly, this leads to the huge number of dedicated buses and that resembles just like a fully connected network [1–3]. The major drawback of this type of topology on the chip is the area required by the wires on the chip. William J. Dally introduced the tile-based architecture in which the area required by the wires on the VLSI is minimum [4]. This tile-based architecture is commonly referred as mesh topology. In mesh topology, every intellectual property is connected to its dedicated router and the router is connected to the adjacent router that is placed in the manner or in the form of tiles. The mesh architecture has been widely accepted due to the reason of simplicity in design and the simplest routing algorithm required. This led to the study for the exploration of various topologies that can be designed using the mesh interconnection network. The common variants of the mesh interconnection network Torus interconnection network in which the extreme nodes are connected to each other horizontally and vertically [5–7]. The common idea behind the design of all the topologies is that they have homogenous links that means in the designing of the mesh topology the links used have same bandwidth. The fat tree topology that uses the fat links on the nodes at the top has been used to enhance the performance of the tree topology. Idea of using heterogeneous links has been exploited in the horizontal flat mesh [8]. It has been observed that the introduction of fat links in the topology helps in increasing the throughput and the reducing the latency of the network. However, the idea for introducing the fat links in the topology has not explained the reason how this fact links are to be introduced in this paper we have suggested the idea of introducing the heterogeneous links based upon the bandwidth required at a particular node. To get the optimal allocation of the bandwidth, we have used the genetic algorithm to find the solution for designing the topology.

The paper has been divided into five sections. Section 2 discusses the related work. In Section 3, we have proposed the complete algorithm and our problem solution. In Section 4, we have discussed the results, and in Sect. 5, we have concluded the paper.

2 Related Work

Optimization has been used in system on chips at the different areas. It has been used to allocate the task to the processing elements that have the similar computer abilities but different communication requirements as the processing elements are similar so the allocation of the job can be done anywhere in the system; however, if the job with the different communication requirements are allocated the different locations, the performance of the complete system will be affected [9]. In an attempt, the idea of extra link allocation has been addressed by the researchers here and the researchers have stated to use genetic algorithm to explore the extra links in the topology so that the performance of the mesh topology can be increased. However, there is a need for the multiple layers on the topology to design the links that have been introduced in the topology [10]. This has been made possible with the help of the layer architecture that it is used while designing the system on chip in VLSI. Authors have also suggested the use of various techniques on the optimization link placement. In another attempt, the authors have suggested the topological optimization using genetic algorithm for the study of reliability in the network on chips. In [11], authors have suggested the use of genetic algorithm for the energy optimization of two-dimensional network on chips, and 3D topologies' improvement has been suggested using genetic algorithm [12]. In other allocations, deadline, buffer, and energy have been used in the allocation [13]. Similarly, the latency-based mapping in the 3D has been addressed by Wang et al., and here, they have suggested using the multi-objective genetic algorithm that uses the rank-based assignments for the optimization problem [14].

3 Proposed Approach

In the proposed approach, the idea of genetic algorithm has been used to find the optimal allocation of bandwidth for the particular node in the topology. In this idea, we have applied the genetic algorithm for the same.

The various phases of the genetic algorithm application details are provided below:

1. Problem encoding: In this phase, at first the problem is to be converted into the form that can be represented by the chromosome. As we are going to use the heterogeneous allocation of the bandwidth, this has been represented as the weight of the edge. This weight of each node has been used as the part of chromosome to generate the complete genome sequence.
2. Fitness function: Another important part of the genetic algorithm is the fitness function that is to be designed for the genetic algorithm. To do so, we have employed the shortest path algorithm to calculate the cost of the network as we have considered that our bandwidth can be divided into 16 levels that is this allocation is done sequentially. The complete bandwidth is divided by 16 to define each level. The fitness function is described below (1). However, the graph will have weights from 16 and final bandwidth allocation can be generated by Eq. 2 described below:

$$F = \text{Min}\left(\sum_{j=1}^{16} d(1, j)\right) \tag{1}$$

$$B = 16/W \tag{2}$$

3. The initial population of the chromosome has been generated based on the encoding defined, and the size of chromosome is 96 bits. The reason behind the selection of this length is due to the fact that there are 24 edges in the mesh of 4×4 that has been used in the study. Each edge is of 4-bit.

Fig. 1 Describes the mesh topology with edge number

4. The population size is kept as 10, and the total number of generations for the stopping criteria is set to 10.
5. The selection operator selected for this algorithm is tournament selection algorithms where good solutions are compared with each other and the copies are retained.
6. The crossover used is the single-point crossover with probability of 0.90.
7. The mutation is also considered to single-point mutation which is going to flip a single bit of the offspring.

Figure 1 defines the order of the edges that were provided to the mesh topology that helps us in identifying the edges from the chromosome. The initial graph configuration in the form of the chromosome is represented and final optimized configuration for the all the edges has been represented in Table 1.

4 Result Obtained

The trend of the cost function of the genetic algorithm has been represented in Fig. 2.

The performance of the designed topology has been tested with the proposed topology on the end-to-end latency using the omnet++ simulator. The graph for the same has been described in Fig. 3.

5 Conclusion

From Fig. 3, we can observe that the performance of the optimized topology is better in comparison to the normal mesh topology. The genetic algorithm has helped in generating the topology with the heterogeneous links. In future, we suggest to do this mapping considering the complex fitness function which can help us in studying the performance based on the power and heat dissipation by the network on chip.

Table 1 Describes the weights of the edges

Edge number	Normal weight	Optimized weight
1	16	1
2	16	5
3	16	4
4	16	2
5	16	9
6	16	4
7	16	1
8	16	16
9	16	1
10	16	1
11	16	12
12	16	6
13	16	4
14	16	8
15	16	4
16	16	14
17	16	1
18	16	16
19	16	14
20	16	3
21	16	9
22	16	4
23	16	1
24	16	1

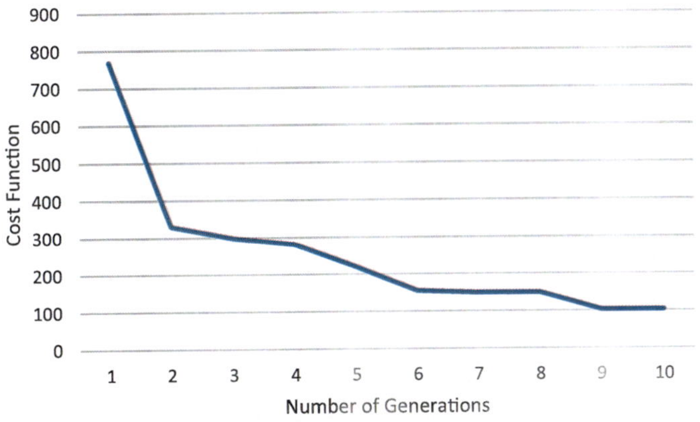

Fig. 2 Describes the cost function value at different iteration of genetic algorithm

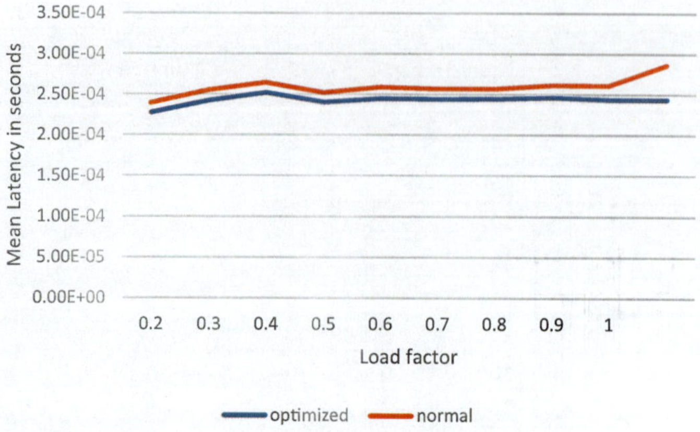

Fig. 3 Describes the comparison of end-to-end latency at different load factors

References

1. Dandamudi, S. P. (2003). *Fundamentals of computer organization and design*. New York: Springer.
2. William, S. (2013). *Computer organisation and architecture* (9th ed.). London: Pearson.
3. Hwang, K. (1993). *Advanced computer architecture with parallel programming*. New York: McGraw-Hill.
4. Dally, W. J., & Towles, B. (2001). Route packets, not wires: On-chip interconnection networks. In *Proceedings of the 38th Conference on Design Automation—DAC'01* (pp. 684–689).
5. Yadav, S., & Krishna, C. R. (2014). CCTorus: A new torus topology for interconnection networks. In *International Conference on Advanced Computational Technologies and Creative Media*.
6. Liu, Y. H., Zhu, M. F., Wang, J., Xiao, L. M., & Gong, T. (2012). Xtorus: An extended torus topology for on-chip massive data communication. In *Proceedings of the 2012 IEEE 26th International Parallel and Distributed Processing Symposium Workshops, IPDPSW 2012* (pp. 2061–2068).
7. Dally, W. J., & Towles, B. P. (2004). *Principles and practices of interconnection networks*. Amsterdam: Elsevier.
8. Punhani, A., Kumar, P., & Nitin (2018). A horizontal fat mesh interconnection network. In *2017 10th International Conference on Contemporary Computing, IC3 2017* (Vol. 2018-January).
9. Tripathy, P. K., Dash, R. K., & Tripathy, C. R. (2012). A genetic algorithm based approach for topological optimization of interconnection networks. *Procedia Technology, 6*, 196–205.
10. Punhani, A., Kumar, P., & Nitin (2017). Optimal extra links placement in mesh interconnection network using improved environmental adaptation method. *Journal of Intelligent & Fuzzy Systems, 32*(32), 3285–3295.
11. Marafie, Z. A. M. R. H. (2016). *Energy optimization for two-dimensional NoCs using genetic algorithms (Ph.D. Dissertation)*. UC Irvine.
12. Elmiligi, H., Gebali, F., & El-Kharashi, M. W. (2014). Power-aware mapping for 3D-NoC designs using genetic algorithms. *Procedia Computer Science, 34*, 538–543.
13. Bruch, J. V., Da Silva, E. A., Zeferino, C. A., & Indrusiak. L. S. (2017). Deadline, energy and buffer-aware task mapping optimization in NoC-based SoCs using genetic algorithms. In *2017 VII Brazilian Symposium on Computing Systems Engineering (SBESC)* (pp. 86–93). IEEE.
14. Wang, J., Li, L., Pan, H., He, S., & Zhang, R. (2011). Latency-aware mapping for 3D NoC using rank-based multi-objective genetic algorithm. In *2011 9th IEEE International Conference on ASIC* (pp. 413–416). IEEE.

Detection of Human Emotion in Text

Pratyush Kumar Singh, Harivans Pratap Singh, Kuldeep Kumar, and Raj Kumar

Abstract The diagnosis of human emotions and the analysis of human emotions are a topic at the moment; the derivation of emotions and feelings from the written text remains a major challenge in this area of research. We do the job of an emotional annotation to find a category of mood states, perhaps the emotional intensity and the words or word expressing emotions in the text. An overview of emotional intelligence research with emphasis on emotional agents has been covering areas like emotional agents, modelling artificial agent's environment, different forms of learning, emotional intelligence in decisions support processes, etc. Machine-learning approaches and knowledge-based approaches are the two main approaches for this task (Agrawal and An in 2012 IEEE/WIC/ACM International Conferences on Web Intelligence and Intelligent Agent Technology (2012), [1]). The performance of these two approaches is quite good but machine-learning approaches always outperform the knowledge-based approaches in all area.

Keywords Machine learning · Human emotion

P. K. Singh · H. P. Singh (✉) · K. Kumar · R. Kumar
ABES Engineering College, Ghaziabad, India
e-mail: harivans.singh@abes.ac.in

P. K. Singh
e-mail: pratyush.16bcs1087@abes.ac.in

K. Kumar
e-mail: kuldeep.16bcs1180@abes.ac.in

R. Kumar
e-mail: raj.16bcs1060@abes.ac.in

S. Tiwari et al. (eds.), *Proceedings of International Conference on Big Data, Machine Learning and their Applications*, Lecture Notes in Networks and Systems 150,
https://doi.org/10.1007/978-981-15-8377-3_30

1 Introduction

Emotions are something which complete human beings and it gives us a sense which can affect our decisions. If we do not know the emotion of a person, we do not know much about the person in real life; just by knowing the emotion of a person, we can predict its behavior very well. Emotions can push the limits of an individual to a completely different level.

Sentiment and emotional analyses are two lines of research that are closely related but have different purposes. Conceptual analysis has the great urge to find a small number of categories such as a good or bad program or a scale of 1–5. It is often associated with NLP or data group mining. While emotion acquisition is often of interest to acquire a larger number of emotions for the purposes of affective computing or text-to-speech [2]. The two areas share in many ways and it is beneficial for them to share information, as they relate to each other in many ways.

1.1 Approaches

- Knowledge-based
- Machine-learning.

Knowledge-based approach started very early and is very resource taking and expensive. While in machine learning, we make machine-learning models based on our use case and then we train our model based on the data we have. This is the process of a supervised learning model. Supervised learning is a little more demanding than other machine-learning techniques like unsupervised learning and semi-supervised learning.

a. **Objective**
The main objective of this research paper is to get a solution of building a system that can detect human emotion in text using machine-learning approaches which is easy to implement and smoother to run. This solution is very versatile that it can be used by anyone without any difficulties. Although this methodology gives very accurate results but it too has some downfalls. We will discuss this later in our report.
b. **Requirement and application**
There are many application of detection of human emotion in text; some of them are as follows:

   1. **Customer experience improvement**
   By knowing the customer's real emotions in the review or comment section will lead a firm to understand the real need of a customer so that a firm can improve or modify its services.
   2. **Integration with chat bots**

During chat with the customer or user, we can actually predict the emotion of the user, so that we can see how much satisfied or disappointed our customer is? From our services.

3. **Improvement in marketing tasks and support conversation**
   Marketing is the biggest industry in this era and by knowing the customer real emotion will automatically benefit the marketing firms.
4. **Social media analysis**
   If we integrate our human emotion detection system in our social media chat box, then, it will be very easy for the government to predict the intention of a human, i.e., whether he or she is planning for any communal riots or risking the other ones life's in danger [3].
5. **Call center performance monitoring**
   We can monitor our client satisfaction and our executive performance also. Whether our executives are generous with the customer while supporting or not [4].

2 Background Details and Related Work

The previous work related to this project is:

We have made machine-learning models using linear regression, support vector regression, polynomial regression, decision tree regression, and random forest regression.

Based on these types of regression models, we have made models and trained them with large data sets to get the output value dependent variable based on the input values of independent variables.

A model to predict the maximum temperature of an area on a particular day based on the date.

A model to predict the salary of an employee based on the amount of experience it has and the number of skills it has.

A model to predict the profit of a company based on the amount of money it invests in R&D, human resource, location of the company, and advertisement.

3 Computational Approaches for Emotion Detection

1. **Keyword-based approaches**
   In this approach, it totally depends upon the keywords which are used to determine the emotion. The words are then split using a parser, then further, the generated tokens are annotated using parts of speech [3].

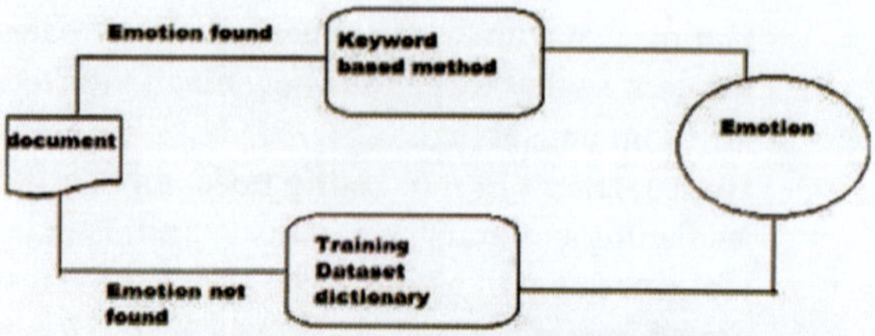

Fig. 1 Basic machine-learning structure

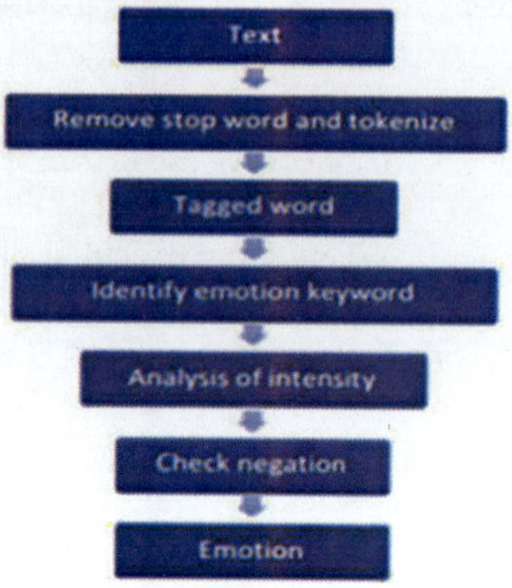

Fig. 2 Keyword-based approach

2. **Learning-based approaches**
   In this approach, trained classifiers are used to separate text with different classes of emotion. It is much faster than the traditional approach and is also easy to understand. It can be easily trained by using large datasets for training our machine-learning model and get accurate results.
3. **Hybrid-based approach**

What this type of approach really does is that it mixes the keyword-based approach and learning-based approach and a new model is formed which is far more accurate and fast than the previous two models [5]. We get the dataset and split it into a training set and testing set. We use around 80% of the data for the training and the rest is used for testing the model and evaluating the errors in it (Fig. 1).

Generally, in every machine-learning-based approaches for detection of human emotion, the architecture is in Fig. 2.

4 Implementation

Algorithm used in emotion detection in text systems.

1. KNN algorithm.
2. Hidden Markov model

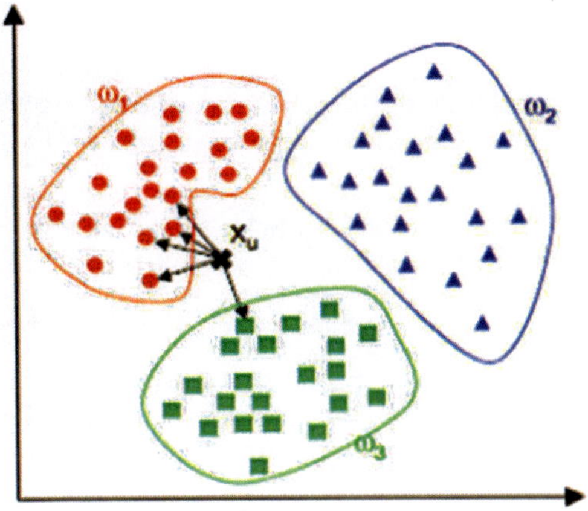

Fig. 3 KNN algorithm

3. Decision tree
4. Random forest.

4.1 KNN Algorithm

1. Nearest neighbor picks nearest on the basis of distance.
2. Distance used is Euclidean distance.
3. Complexity can increase in case of classifying test tuples O(n) (Fig. 3).

4.2 Hidden Markov Model

1. Set of all the states is: $\{s1, s2, s3 \ldots\ldots sn\}$
2. All the processes traverse from one state to another generating a series of states: $s1, s2, \ldots\ldots$
3. Markov chain property: probability of each subsequent state depends only on what was the previous [6].

4.3 Decision Tree

Decision tree is a binary tree which is a decision support tool that uses a tree-like model, and it works in a recursive algorithm.

It splits the dataset into some portion and take and average for the dependent variable for that particular zone (Fig. 4).

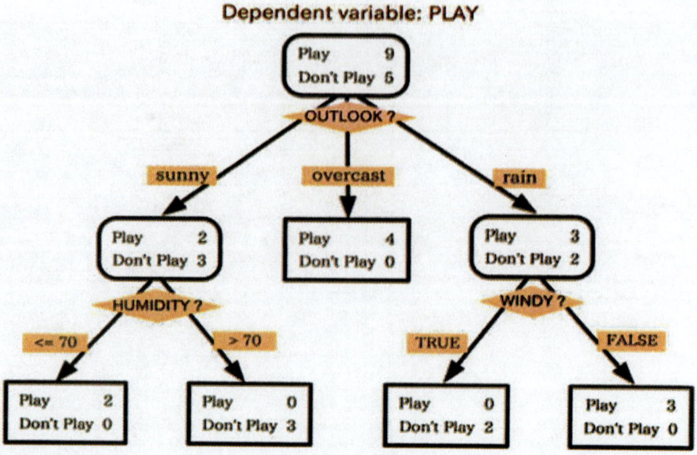

Fig. 4 Decision tree

4.4 Random Forest

In a random forest approach what we do is we make a forest of decision trees and we pass different datasets to each decision tree, after that we take the predicted outcome from each of the decision trees and make an average of the prediction (Fig. 5).

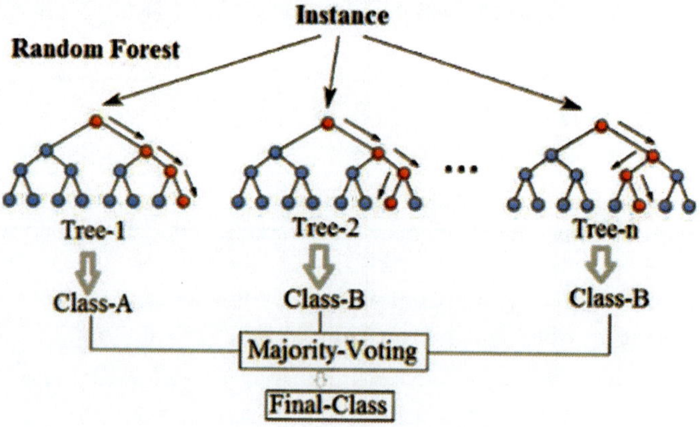

Fig. 5 Random forest classifier [7]

5 Result

Many experiments have been done to check the effectiveness of proposed system and successfully completed all the tests with required result. The table shows the results of our experiments with existing approaches. Each row retrieved from the results by our experiment depicts the emotion detection approach of keyword, machine learning, hybrid based, and proposed system with respect to the input sentence. In the sentences were are providing three common approaches for detecting the emotion, keyword-based approach, machine-learning-based approach and hybrid model-based approach. Here, we take 500+ sentences and we have analyzed them. The results are shown in as below:

Keyword-based approach

Attitude	Positive	Negative	Neutral
Number	79	126	199

Learning-based approach or machine learning

Attitude	Positive	Negative	Neutral
Number	104	134	166

Hybrid-based approach

Attitude	Positive	Negative	Neutral
Number	121	154	129

6 Conclusions

This paper describes methodology for detection of human emotion in text. Proposed real-time emotion detection system helps to predict the emotion of a human written by a human. This technology takes all aspects for predicting the human emotion in text so that a person is able to know the other one's real emotion while chatting or reading the text. It provides an efficient and fast emotion detecting system. This method makes the system cost effective.

Reference

1. Agrawal, A., & An, A. (2012). Unsupervised emotion detection from text using semantic and syntactic relations. In *2012 IEEE/WIC/ACM International Conferences on Web Intelligence and Intelligent Agent Technology* (Vol. 1). IEEE.
2. Shivhare, S. N., & Khethawat, S. (2012). Emotion detection from text." *arXiv preprint* arXiv: 1205.4944.
3. GdBinali, H., Wu, C., & Potdar, V. (2010). Computational approaches for emotion detection in text. In *4th IEEE International Conference on Digital Ecosystems and Technologies*. IEEE.
4. Chew-Yean, Emotion detection and recognition from text. https://devblogs.microsoft.com/cse/2015/11/29/emotion-detection-and-recognition-from-text-using-deep-learning/. November 29, 2015
5. Tao, J. (2004). Context based emotion detection from text input. In *Eighth International Conference on Spoken Language Processing.*
6. Zhou, D., et al. (2016). Emotion distribution learning from texts. In *Proceedings of the 2016 Conference on Empirical Methods in Natural Language Processing.*
7. Carremans. B, Sentiment analysis with text mining. https://towardsdatascience.com/sentiment-analysis-with-text-mining-13dd2b33de27. August 26, 2018

An Improved Hand Gesture Recognition System Based on Optimized MSVM and SIFT Feature Extraction Algorithm

Sonia and Darpan Anand

Abstract In today, the robots and machine, translation tasks are performing the main role in hand gestures. Gesture or detection implementation will help humans in several means. Gesture recognition systems are used in various fields such as DNN, ML, and NNs. The applications of this research are sign language translation, music creation, robot remote controlling, etc. In this research work, proposed HGR methods are using feature extraction and optimized MSVM classification. This process gives the applications of edge detection, interference, filters, SIFT-ALO algorithm, and binary during image preprocessing, in which these approaches add to better extraction and selection. In proposed work, implemented an optimized MSVM method are two classes such as training and testing. Optimized M-SVM is performed on the ASL gesture dataset along with existing SURF and SIFT techniques. This research work is used for the simulation tool (MATLAB) and calculated performance metrics like processing time, error rate, and the accuracy rate with the MSVM value at 99.1 percent as compared to existing feature extraction methods.

Keywords Hand gesture recognition system · Multi-SVM · SIFT-ALOA algorithm

1 Introduction

It is easy and simple data communication between human and human. Once, it is near on humans and machine, it is a very difficult job since machines identified all types of languages human can talk and understand [1]. Thus, for enhancing the characteristics which machines can implement a few interactive methods such as GR. This is a famous research topic in the area of CS and its uses regarded by DL technology. HGR is needed for modelling, control boundaries, no-touch, and in-vehicles. Such DL: Deep learning technologies give drivers to drive through at a similar time connecting with various panels such as cooling and sound [2]. HCI describes the path of how the human communicates to the robot and because the

Sonia · D. Anand (✉)
Chandigarh University, Gharuan, Mohali, India
e-mail: darpan.anand.agra@gmail.com

S. Tiwari et al. (eds.), *Proceedings of International Conference on Big Data, Machine Learning and their Applications*, Lecture Notes in Networks and Systems 150,
https://doi.org/10.1007/978-981-15-8377-3_31

mechanism is not helpful till an HTs the mechanism for a certain job. There are normally two features that would be verified when implementing a man–machine interaction model as described in mechanism usage and evaluation [3]. The postures may be in any format such as pixel and hand image or human-defined pose that need minimum difficulty. Various methods are being implemented by the corporations for gaining essential data for detection hand-made gesture recognition models. Few representative works with main equipments like data glove and color caps to propose difficult data about HG offered by the human [4]. GR: Gesture recognition is normally divided into two classes such as (i) Dynamic and (ii) Static [5]. The SDGs are those that only need the handling of a particular hand gesture image as the input of the classifier, and the main benefit of this method is the minimum cost. DGs need the handling of hand gesture image series more tough GR methods. In the literature survey, it can explore various recognition scenarios based on UL: Unsupervised and SL: Supervised learning methods. They may describe the SVM, CNN, and KNN classifier algorithms. The contribution of this research work: its valued gesture image bases of 16-gestures, certain edge detection, and improved feature extraction methods and the use of MSVM of the classification method. With the research flowchart, it determined that with the basic flow of MSVM, it is possible to attain better outcomes for the ASL database classification technique and to associate the research methods with other prior methods in the survey and other HGR procedures. In research work has implemented an enhanced feature extraction method using the SIFT-ALO algorithm. SIFT algorithm is used to extract the features and the ALO algorithm is used to feature selection in the hand gesture recognition system. After that, in research work has implemented the MSVM algorithm to classify the hand gestures categories. In this proposed work, evaluated the performance metric such as time, accuracy, and error rates (FAR, FRR) and compared with SURF, SIFT, and other optimization algorithms.

2 Literature Review

Benmoussa et al. [6] discussed the HCI expected a nice deal of care during this final period. Prior authors had curved to behavioral communication systems such as gesture human–machine edges. Current works were trying to resolve the issue of HGR: Hand gestures recognition utilizes ML techniques. Few of them were playing to get accurate performance. But, some of them were compelled into account mainly necessaries to relate the workflow of LM: Learning model, normally data unbalance, model selection, and generalize feature selection. In this research section, they implemented an ML technique for real-time classification and recognition of sixteen movements of user-hands utilizing KS that compliments like necessaries. The hand recognition was generated only when there was a moveable HG. This technique was depending on the training of the SVM method on hand depth information from which BoWs (Bag-of-Words) of scale invariant feature transformation and SURF features were extracted. The information was reserved managed and the structure kernel and metrics were chosen utilizing CVP: Cross-validation process. This technique attained 98 percent overall parameters utilizing the field under the ROC: Receiver operating characteristic curve calculates.

2.1 Gesture Using SIFT and SURF Algorithm

HGs train images could be defined by sets of key points created by SURF and SIFT, then the number of key points from the images are various and reduce reasonable ordering. To identify this issue, they utilize the BoWs technique. A BoW, which is a single of the most important techniques CV. Normally, the vital points in the pictures are the VWs: Vision-Words. These sections are characteristics and they are discriminators and even to change and scale variations. These characteristics are fetched utilizing SIFT and SURF technique forming a dataset of VWs then every novel picture is going to be defined as a geometric of VWs that looks in the picture. It is completed utilizing the VQC method (Vector Quantization Clustering), where an individual group defines a VW that correspond to a valuable LP common by the key points in that group. Figure 1 defines a workflow for the model of the technique that utilizes the bag-of-words technique to classified hand signs. The block 1st, the fetched key sections are grouped and create a code-words dictionary, then calculate a histogram utilizing VWs for all the train samples. These histograms are then served

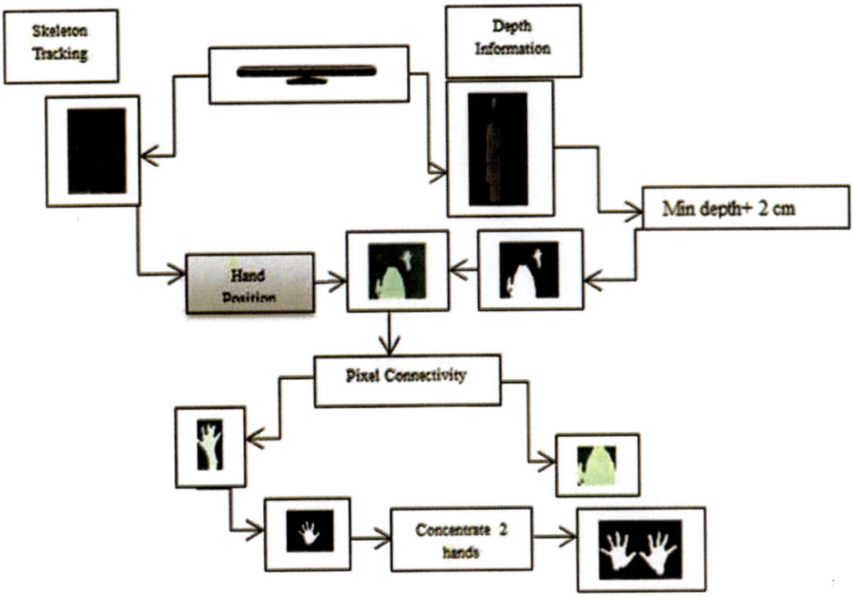

Fig. 1 Three phases of segmentation

to SVM classification. The block 2nd, test pictures are interchanged top-histogram demonstration by including various key sections seemed in each group and classified the consequences utilizing the train support vector machine method [7].

Prakash et al. [8] devised the mechanism for the various latest equipment for social and technological applications such as wireless mobiles, electrical computers. In daily life, they utilize number of the path of communication which contains language, lettering by somebody moveable, but while in case of robot, they quiet held with writing and language so they essential few improvements so that they could interact with robots in somebody moveable also. Hand gesture was a non-vocal route of communication which utilizes hand-motion, several body postures, facial emotions. So, to create a smart robot, they were permitting the machine to take knowledge by recognizing various HGs. HGs were utilized as input in the DIP system.

2.2 *Template Matching*

This technique for HGR and postures used a simulation technique to define the essential number of template of a particular gesture to be full that shall be protected on the dataset for the similarity procedure of the method. If this system would not be able to recognize the posture offered with the templates, an extra template must be stored and trained in the dataset until the HGR system accurately recognizes the postures. This work will sum-up all the time in second under a particular no. the similar no. of template gesture and process is shown in Fig. 2 [9].

Huang et al. [10] defined GR, while had been searching for various years, was still a limitation issue such as complex background, digital camera angles, and brightness situations kind the issue more complex. In this work defines a robust and speedy

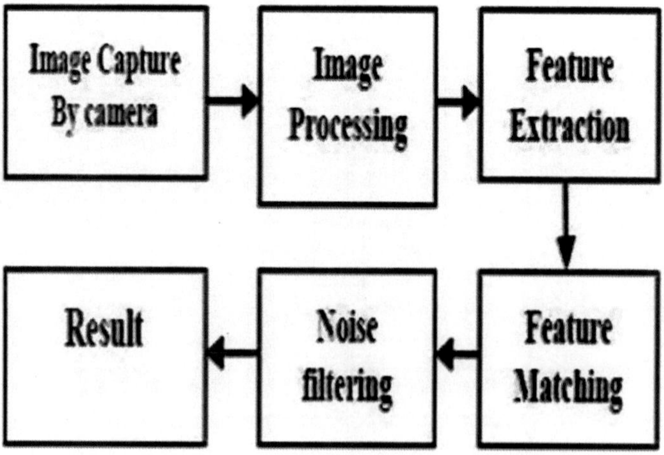

Fig. 2 Flowchart of HGR process

technique for HGR depending on RGB video. Initially, they recognize the skin-based on their color. They fetch the region and edge the hand contour. Lastly, they detect the hand gesture. The outcomes of simulation determine that the research technique was effective to detect postures with increase accuracy rate than the state of the art.

2.3 Hand Region Segmentation and Contour Extraction Method

They eliminate the interference in the hand images. They required extracting regions. They measure each section group as an edge. Between these regions, there is only a single region, which defines HR. Facial and hand regions are the huge twice contours. The issue of searching the hand from the regions develops the difficulty of diving hands from facials. Then, they gather a hundred image samples of facial and hand regions. The VggNet is used for the classification process and training model shown in Fig. 3. In this network, the model is based on DNN established by the visual-geometry cluster and scholars at google-deep-Mind. This model finds the association between the complexity and evaluation of CNNs. By loading the 3 * 3 matrix is used tiny convolution advantages and 2 * 2 max-pooling layer. This model has positively built sixteen or 19 layers DCNNs. This research work has used 16—layers in the vggNet model [11] (Table 1).

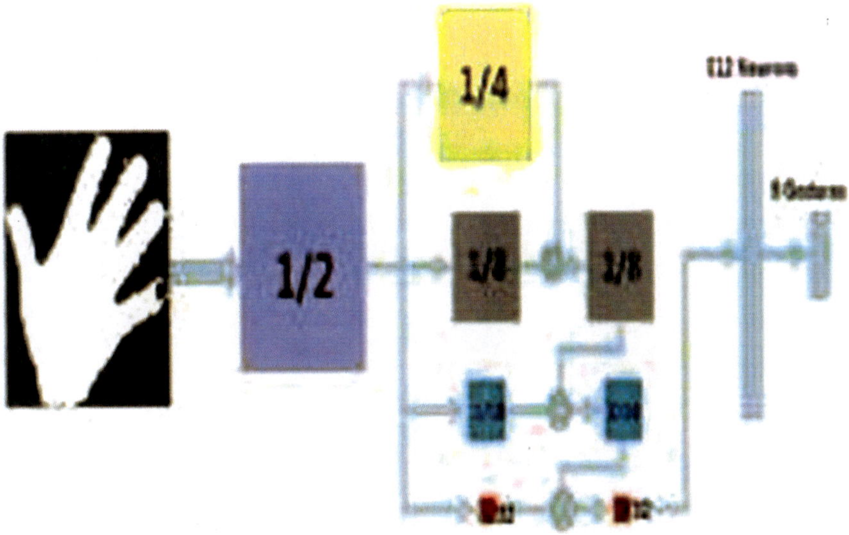

Fig. 3 Training structure in VggNet model

3 Research Challenges and Methodology in HGR System

During the study of the literature survey, we have found out some ways in real-time human communication by posture verification needs future enhancement to recognize hand gesture data accurate and effective design of handshape or size. It looks at difficulties in sign recognition for variable performance by different performers. From the survey, we have found the problem or proposed work in which we are going to continue our work of hand gesture recognition for feature extraction and classify through multi-support vector machine.

3.1 Challenges

The HGR has various challenges described in detailed such as (i) Rotation (ii) Size (iii) Illumination changes, and (iv) Location issues [12].

(i) *Rotation issue*:

The degree of freedom is the main problem. Then, the degree of freedom is modified and gesture initial image might be different and by output, the image can change.

(ii) *Size Issue*:

The human being hand features have different shapes and sizes such as child have small hands and younger have big hands, it could create issues in the system.

(iii) *Location issue*:

Although the defined input image if hands location different such as hand place in corner of the screen, all points which recognizing hand location do not lie on hand image, then it could generate an issue to consider input from people.

(IV) *Illuminate variation issue*:

The brightness variations can affect HG input as it could vary extracted skin contour.

3.2 Methodology

In this section, elaborates the HGRS is separated into four sections like;

- Sign image dataset collection
- Sign image preprocessing
- Segmentation in gesture images
- Feature extraction using SIFT method and
- Optimized MSVM classification algorithm.

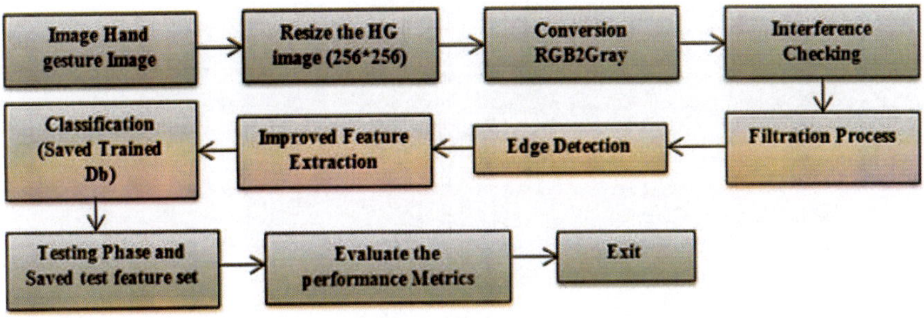

Fig. 4 Diagram of research methodology in work flow

- Performance metrics.

The collection of ASL: American Sign Language gesture dataset from the online repository site. Firstly, upload the sign image from the trained dataset folder. Conversion applied in the resized image that is rgb2grayscale format. It reduces the format of the converted image in 2D. It identifies noises in the grayscale image. It applies the filtration method to remove the interference in the noisy image. After that image converts into a binary pattern and detects the edge in the binary image. SIFT algorithm is used to extract the features in the key-point format. The present work has implemented hybrid optimized MSVM algorithm to classify the gestures and recognition rate. At the last, evaluated and compared the performance metric such as accuracy, processing time, and error rate (Fig. 4).

The proposed work performance is increased as compared with existing methods. The present work has improved the accuracy rate and minimized the processing time (s) as compared with existing methods such as SURF and SIFT algorithms.

4 Experiment Analysis

In this research work of the method was completed in MATLAB 2016a. The ASL Database [13] is used for recognizing the gestures and posture of the hand's images shown in Fig. 7. The database contained 16 English alphabets categories, 80% of train the hand gesture samples, and 20% is used for testing. The research method evaluates a 99% accuracy rate in verifying the test image in the database. The research algorithm workflow is shown in Fig. 4. Such a huge number of HG samples offered different sizes and shapes as can be defined in Fig. 5, which define various alphabetic samples of gesture 'a'. Some other samples of HGs could define in Fig. 6 (Fig. 7).

Figure 8 shows (i) uploaded the hand gesture input train and test image database from the knowledge domain. The properties of uploaded images are pixel size 864 * 1152, bit depth 24, horizontal and vertical resolution 96 dpi. (ii) The uploaded color image coverted into a grayscale image. In conversion, the grayscale format

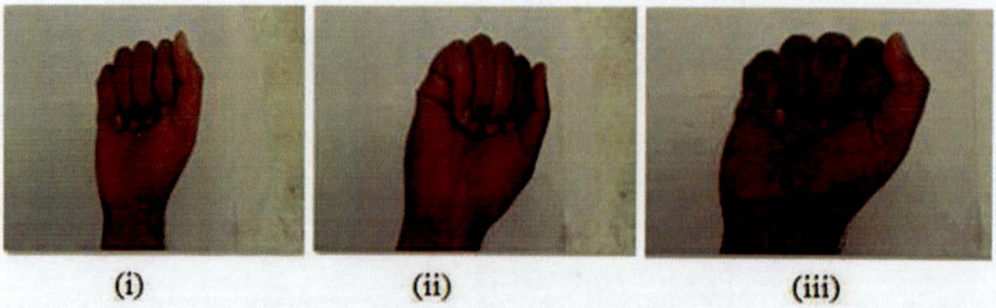

Fig. 5 Hand gesture of 'a' sample

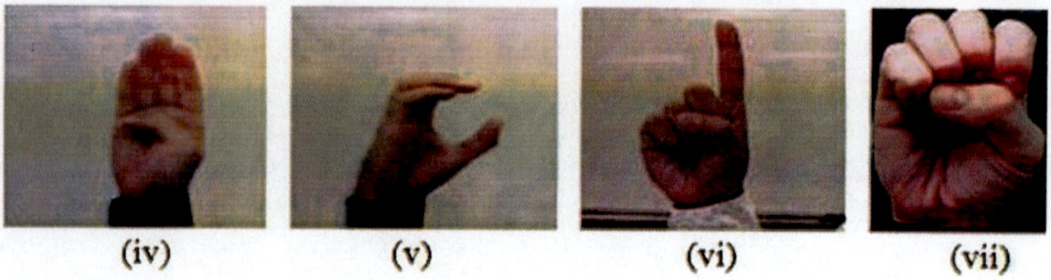

Fig. 6 Other HG samples *L R*: Gesture 'b', 'c', 'd' and 'e'

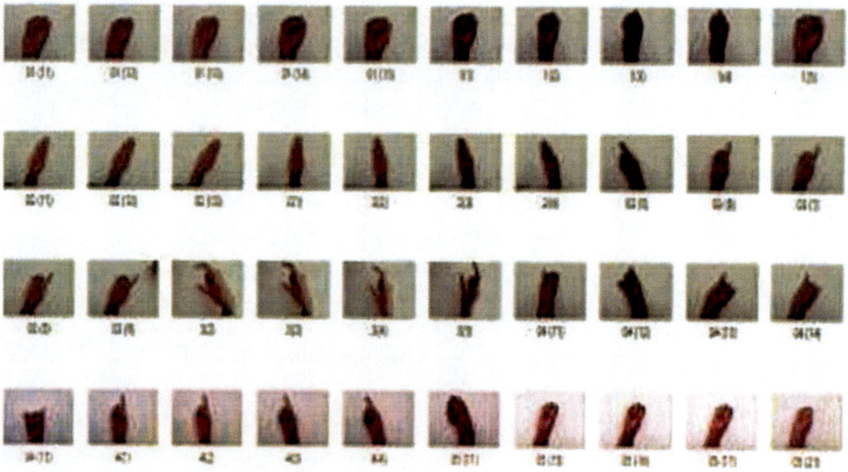

Fig. 7 ASL hand gesture dataset images

image is reducing the dimensionality of the color image. (iii) In this image is identifying the noise level. (iv) If the noise presented in the uploaded image and then applying the filtration method to remove the interference in the image. (iv) After that, it converted a smooth image into a binary format. (vi) Lastly, it detected the edge in the binary image.

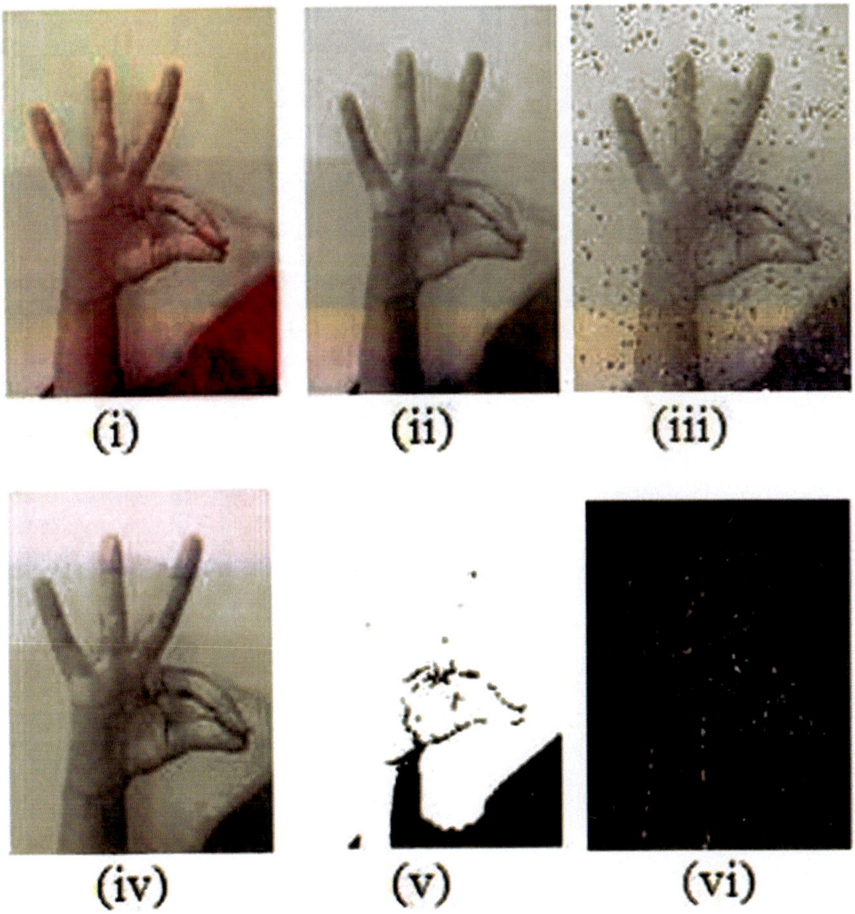

Fig. 8 (i) Input hand gesture image (ii) Conversion HG image (iii) Noisy HG image (iv) Filtered HG image (v) Binary HG images, and (vi) Edge HG image

Figure 9 defines the feature set with all types of local features in numeric format.

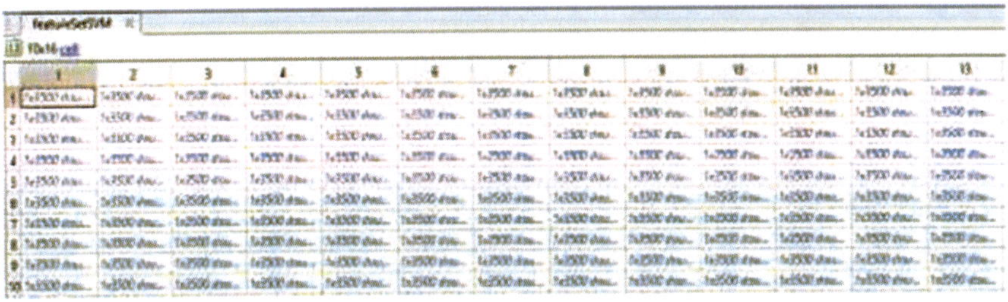

Fig. 9 Extracted features with SIFT algorithm

This section has implemented a SIFT algorithm to extract the features in the form of key points. It extracted the intersect key points from the grayscale image at which nearest features will be included for further mechanism and process. This method is used for a large number of feature extractions in the uploaded image. Figure 10 shows the graph based on the ALO method used for feature selection. This method consists of a search by a random selection of the extracted feature set. It has compressed the extracted feature set. After that optimized MSVM classification method has implemented to recognize the hand gesture in alphabetic format (A-W).

In research work has implemented ALO with MSVM algorithm to improve the HGR system. ALO is metaheuristic algorithm and improved the recognition rate and the best fitness value calculated as compared to other optimization techniques. ALO is a novel phase of the nature-inspired algorithm and less consumed processing time (s).

Table 2 defines the performance parameters with the optimized MSVM method. In the proposed method, parameter performance such as accuracy rate 99%, processing time 0.0369, and MSE value is 0.0322. Table 3 shows the comparative performance analysis with proposed (OMSVM) and existing (SIFT and SURF) methods. In the research algorithm, the accuracy rate is maximum as compared to other feature extraction methods (SIFT and SURF) and the processing time is lesser than other methods.

Figures 11 and 12 shows the comparison between proposed and optimized MSVM method and various feature extraction methods such as SIFT and SURF algorithm with accuracy rate is calculated. In optimized MSVM, accuracy rate is higher than other feature extraction methods. In hand gesture recognition, the rate is maximized with the optimized MSVM algorithm. The second comparison shows the processing time with the proposed and existing feature extraction methods. In the research algorithm, the processing time is less as compared to the existing feature extraction methods.

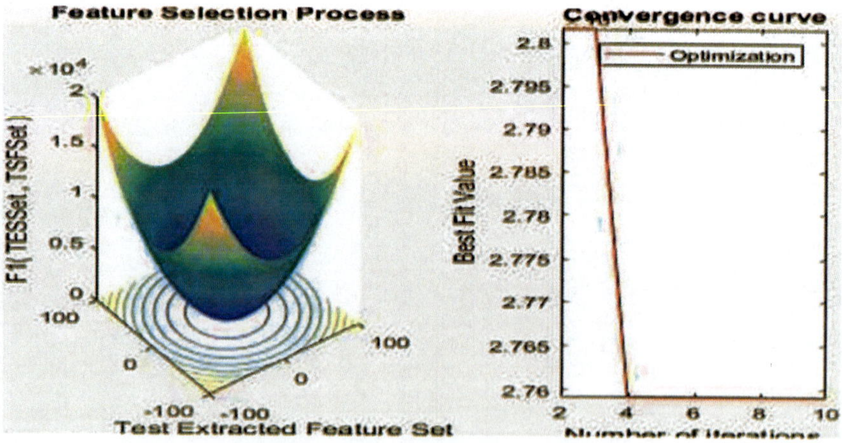

Fig. 10 Ant lion optimization using feature selection

Table 1 Analysis of the various methods and performance metrics

Author name	Method	Parameters	Citation
Marouane et al. (2019)	SIFT	Average time in sec = 0.3 and 0.1 s	[6]
	SURF	Speed = 0.5 and 0.25 s	
	SVM	AUC = 98% and 91%	
Jay Parkash et al. (2019)	Template matching	Price	[8]
	Naïve bayes	Efficient	
	PCA		
Hanwen et al. (2019)	VggNet model	Accuracy = 98.1%	[10]

Table 2 Performance parameters in optimized MSVM algorithm in HGR

S. No.	Parameters name	Proposed values
1	Accuracy rate (%)	99.1
2	FAR	0.022
3	FRR	0.003216
4	MSE	0.0322
5	Processing time	0.036

Table 3 Comparative performance analysis with O-MSVM, SURF, and SIFT methods

Parameter values	SIFT	SURF	OMSVM
Accuracy rate (%)	91	98	99
Processing time (s)	0.118	0.29	0.036

5 Conclusion

The hand gesture recognition system is performing a major role in today's robotics and machine translation tasks. This research makes a message between humans and mechanism is very easy and understandable. The applications of this research are sign language translation, music creation, and robot remote controlling, etc. The main focus of the proposed algorithm is on size, positions of gesture, and recognition rate. After studying various existing techniques such as feature extraction, and classification techniques, we developed pre-processing methods such as image resizing, grayscale conversion, filters, edge detection, and binary process. The proposed hybrid optimized feature extraction and classification technique are developed and simulated. The overall research is based on various performance parameters such as accuracy rate, processing time, and error rate. Optimized M-SVM is performed on the ASL gesture dataset along with existing SURF and SIFT techniques. The performance is calculated after the simulation and the proposed algorithm shows better

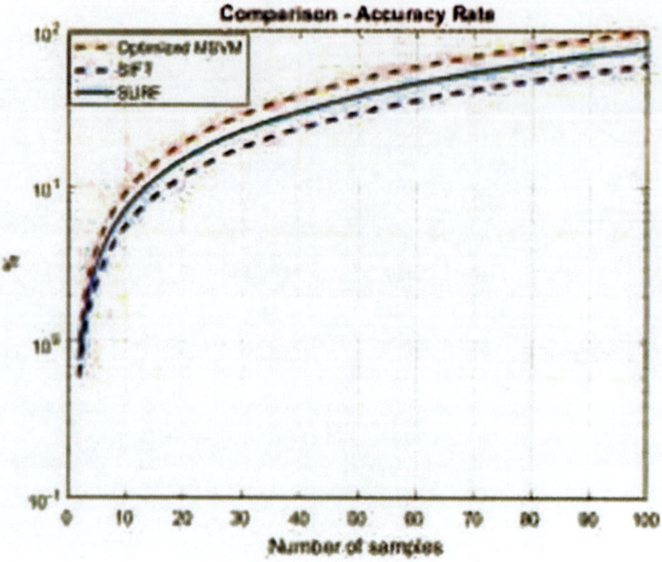

Fig. 11 Comparison based on SIFT, SURF, and optimized MSVM algorithm with accuracy rate (%)

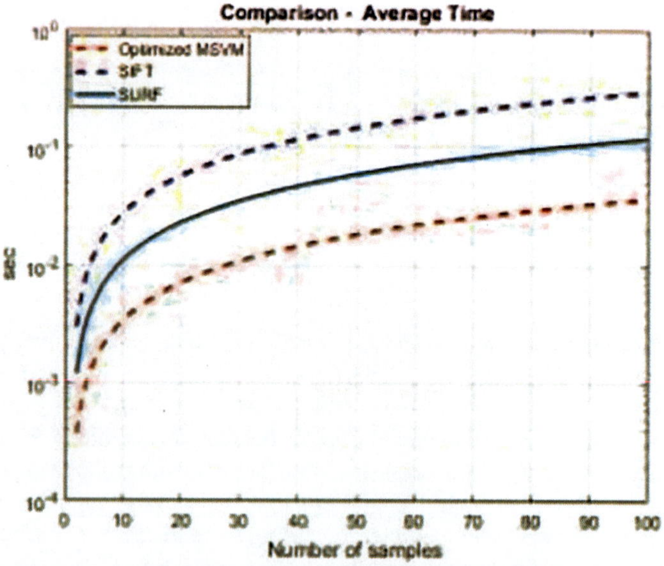

Fig. 12 Comparison based on SIFT, SURF, and optimized MSVM algorithm with processing time in seconds

performance than the previously developed algorithms. In this research MATLAB is used for simulation and achieved an accuracy rate of 99.1%, an error rate of 0.024 in the 0.01 processing time. As per the comparison in Chap. 6, the proposed optimized MSVM shows better performance in all the cases.

In the further work, it can develop a color segmentation method and deep learning method to improve the tracking and capturing of the gesture in any shape or size.

References

1. Ya, D., & Xie, Y. (2018). Design on computer system based on hand gesture recognition. In *2018 IEEE15th International Conference on Networking* (pp. 1–4). Sens Control.
2. Quan, P., & Zhu, M. (2018). Hand gesture recognition system. *International Journal of Computer Applications in Technology, 57*(4), 1–6.
3. Dai, H. (2018). Research on SVM improves algorithm for large data classification. In *IEEE 3rd International Conference on Big Data Analytics* (pp. 181–185).
4. Varun, K. S., Puneeth, I., & Prem Jacob, T. (2019). Hand gesture recognition and implementation for disables using CNN'S. In: *2019 International Conference on Communication and Signal Processing (ICCSP)* (pp. 0592–0595).
5. Pisharady, K., & Saerbeck, M. (2015). Recent methods and databases in vision-based hand gesture recognition: A review. *Computer Vision and Image Understanding, 141*(4356), 152–165.
6. Benmoussa, M., & Mahmoudi, A. (2018). Machine learning for hand gesture recognition using bag-of-words. In: *2018 International Conference on Intelligent Systems and Computer Vision (ISCV)* (pp. 1–7).
7. Zhang, R., Ming, Y., & Sun, J. (2016). Hand gesture recognition with surf-bof based on gray threshold segmentation. In *2016 IEEE 13th International Conference on Signal Processing (ICSP)* (pp. 118–122).
8. Prakash, J., & Gautam, U. K. (2019). Hand gesture recognition. *International Journal of Recent Technology and Engineering (IJRTE), 7*(6).
9. Pavlovic, V. I., Sharma, R., & Huang, T. S. (1997). Visual interpretation of hand gestures for human-computer interaction-review. *IEEE Transactions on Pattern Analysis and Machine Intelligence.*
10. Huang, H., Chong, Y., Nie, C., & Pan, S. (2019). Hand gesture recognition with skin detection and deep learning method. *Journal of Physics: Conference Series, 1213*(2), 022001 (IOP Publishing).
11. Simonyan, K., & Zisserman, A. (2015). Very deep convolutional networks for large-scale image recognition. In *Proceedings of the International Conference Learning Represent.*
12. Shigeo, A. (2005). *Support vector machines for pattern classification.* Kobe University.
13. Barczak, A. L. C., Reyes, N. H., Abastillas, M., Piccio, A., & Susnjak, T. (2011). A new 2D static hand gesture colour image dataset for ASL gestures *15*, 12–20.

An Android Application for Automatic Content Summarization of News Articles Using Multilayer Perceptron

Dasangam Venkat Nikhil, Rupesh Kumar Dewang, Buvaneish Sundar, Ayush Agrawal, Ananta Narayan Shrestha, and Akash Tiwari

Abstract With the advent of digital news, there is an overload of information from various news sites and blogs, on a wide variety of topics. People generally, neither have the time, nor the patience to go through a number of long news articles from various sites to get a gist of what is happening around them. This brings out the need for content summarization. But manual summarization is highly inefficient and practically infeasible. Therefore an automated approach to text summarization must be adopted. This paper is an attempt to implement and build Android Application for Automatic Content Summarization of News Articles Using Multilayer Perceptron, based on extraction. We have employed a feature based machine learning approach multilayer perceptron method, treating this as a classification problem, where for every sentence, we predict the probability of that sentence belonging to the summary. We have collected news articles and generated their extractive summaries using Newspaper module in python. We have considered a list of features that include sentence length and position, sentence similarity with title, presence of numerical values and proper nouns etc., Three different classifiers were used to train, test and compare used dataset, namely Logistic Regression, Artificial Neural Networks (multilayer perceptron) and Gaussian Naive Bayes, with average accuracies of 88.80%, 89.50% and 87.90% respectively. We also developed an android application to act as the interface between user and the summarizer. The user can either get the summary of an article which he/she gives as input, or get the summaries of news articles fetched in real time from various news sites.

Keywords Android application · Content summarization · Multi-layer perceptron · Logistic regression · Naive Bayes · etc

D. V. Nikhil · R. K. Dewang (✉) · B. Sundar · A. Agrawal · A. N. Shrestha · A. Tiwari
Department of Computer Science & Engineering, Motilal Nehru National Institute of Technology, Prayagraj, Allahabad, India
e-mail: rupeshdewang@mnnit.ac.in

D. V. Nikhil
e-mail: dvnikhil68@gmail.com

S. Tiwari et al. (eds.), *Proceedings of International Conference on Big Data, Machine Learning and their Applications*, Lecture Notes in Networks and Systems 150,
https://doi.org/10.1007/978-981-15-8377-3_32

1 Introduction

News are now our daily life event. Morning is not started until we see news paper. But due to busy life style, sometimes users are not having time to see news paper due to its complex structure. In current scenario users are required news which relevant and short in summarize form. Automatic Summarization is the process of shortening a text document with software, in order to create a summary with the major points of the original document [2]. Automatic Content Summarization can be divided into two categories, namely Extractive and Abstractive Summarization. Extractive Summarization is the process by which summary of an article is created using the most important/representative sentences in the article. This is a classic example of a classification problem, in which we classify each sentence as to whether it belongs to the summary or not. In contrast, abstractive summarization methods aim at processing the input article and generating a summary containing new sentences other than that of the input. In other words, they interpret and examine the text using advanced natural language techniques in order to generate a new shorter text that conveys the most critical information from the original text.

Even though summaries created by humans are usually not extractive, most of the summarization research today has focused on extractive summarization. Purely extractive summaries often times give better results compared to automatic abstractive summaries. We have focused only on extractive summarization, in which most important sentences from the article are picked to be part of summary. This is considered as a classification problem, in which for each sentence, based on few determined indicators,the sentence is classified to a 1 or a 0, indicative of whether the sentence is part of the summary or not [2, 3],[?]. An Android application has been developed to be used as a prototype for testing the generated classifier in real time. It has two use cases. The first one is where the user can upload any news article, directly from his phone, and can get the summarized article.The second use case is that the user can also get summarized versions of news articles of the top headlines of the day from various sites in real time.

Our approach is partially based on approaches presented in [3, 5],[?]. The features used in our model were inspired from these paper. But we have attempted to obtain an efficient model by multilayer perceptron and developed an android application which is based multilayer perceptron model. Here, proposed Android application use the available dataset on BBC, CCN, TOI, The Hindu etc. on run time and summarize news for user. In present paper, We have calculated feature values for each feature, after performing pre-processing techniques like stemming, stopwords removal etc., We have also generated our own training data with the help of Newspaper module in python. We have also used three different classifiers namely Logistic Regression, Artificial Neural Networks and Gaussian Naive Bayes, the results of which are given in the subsequent sections.

The remaining paper is organized as follows. Section 2 presents the related work, Section 3, discussion about proposed approaches and model used in paper. Section 4 presents used classification algorithms and results. Section 6 describes proposed Android Application Finally, last section conclusion and future work is presented.

2 Related Work

Text Summarization methods can be classified into extractive and abstractive summarization. In extractive summarization, we select the most important/representative sentences fro the article to be included as part of summary. On the other hand, in abstractive summarization, we convert the text into an intermediate form of representation, based on semantic/structural analysis of text. Then, this intermediate representation is used to generate summaries, using Natural Language Generation techniques.

The first work on this field was done about 50 years ago by Luhn in 1958 [7]. In his approach, he determined the weight/importance of a particular sentence based on word frequency and term filtering. S. Malhotra and A. Dixit have summarization is keyword based extractive summarization [8]. The summarization technique identifies different features like thematic terms, named entity, title terms, numbers etc., The Authors have not use any machine learning technique. In another paper, R. Mihalcea and P. Tarau have proposed the approach makes use of Natural Language Processing and the Text Rank algorithm, which involves creating graphs on the basis of sentence to sentence similarity [9]. On same automatic summrization Sethi et al. [12] proposed lexical chains and using the WordNet thesaurus. They also crated pronoun resolution and lexical chain formation. Authors have not used any machine learning technique. Oi-Mean Foong and Suet Peng Yong have proposed latent semantic analysis model for ATS and draw semantic relation in between terms and sentences. Authors have not focused on quantity of news articles. Cabral et al. has proposed technique for mobile devices which is based on multi-feature language independent summarization application. They also introduced ATS (tool based) techniques to reduce quantity of news articles. Authors have not used any machine learning technique. Evaluating Summarized text is a major issue in this field. There are 2 main approaches, namely intrinsic and extrinsic approaches. The intrinsic techniques focus on the content of the summary, whereas the extrinsic techniques focus on to what extent the summary can replace the original text. Some of the intrinsic evaluation techniques are recall and precision, DUC (Document Understanding Conference), relative utility and ROUGE (Recall Oriented Understudy for Gisting Evaluation) automatic evaluation.

3 Proposed Approach

We have adopted a feature based classification approach, extracting seven features and implementing three classifiers. Figure. 1 shows block diagram of our used model in which the initial step involved is data collection. We have collected data from various sources, as described below. Next, we proceeded to extraction of features mentioned in the subsequent sections. Then after normalization, we decided on implementing algorithms. We used three algorithms, namely Logistic Regression, Artificial Neural Networks and Gaussian Naive Bayes, which are described below. Then the results were compared and evaluated. Finally, android application was created as an interface between the user and the summarizer.

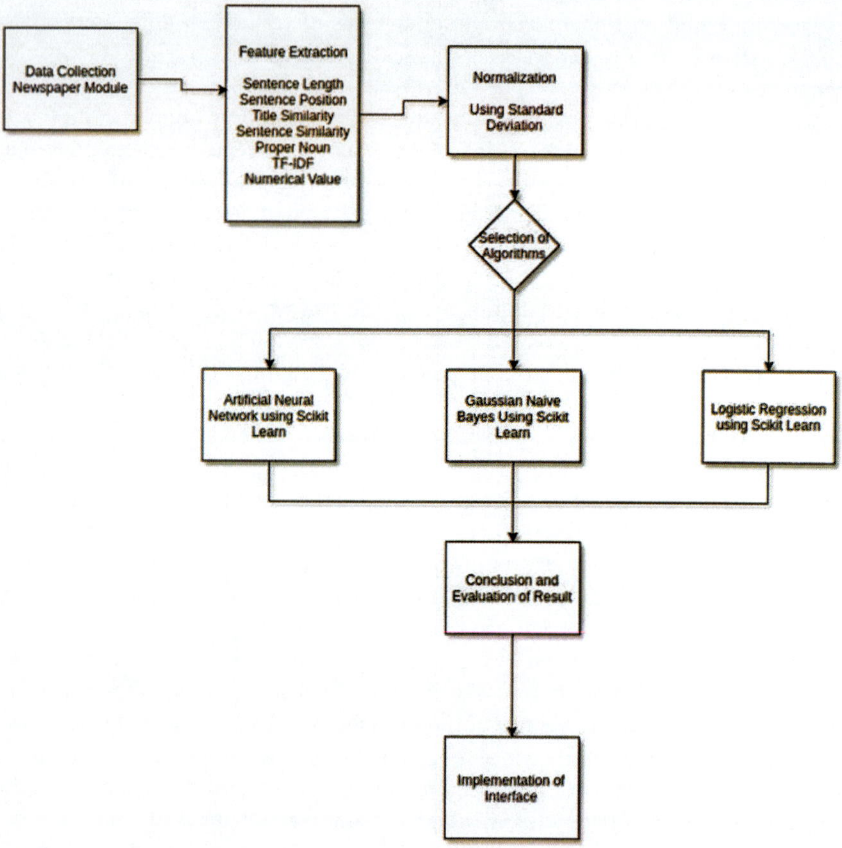

Fig. 1 Flow diagram of the model

3.1 Proposed Model

3.2 Dataset

The training and testing datasets consist of articles along with their summaries, which comprise sentences extracted from the article itself. Due to the unavailability of readily usable articles and their summaries, the training and testing data was created. The articles were scraped from various news sites like BBC, CNN, The Times of India, The Hindu, etc., Then the newspaper module in python to generate extractive summaries of the scraped news articles. These summaries were then used for training and testing purposes. The entire dataset consisted of 1022 articles with a total of 40622 sentences. Out of this, 80% of the data was used for training and 20% was used for testing the trained classifiers.

For the android application, set of 40,622 sentences were used for training the classifier. For producing real time summaries, the News API in python [14] was used to first obtain the articles. News API provides an interface for python programmers to obtain real time news from various news sites across the world. The news can also be filtered using filters like name of the country, section (like entertainment, sports, etc.,) or the time of release of the news. This API is first used to get the top headlines happening in the country, from various news sites along with the urls containing the news articles. Then the newspaper module is used to get the exact news articles, from the urls provided, so that a concise summary can be generated using the classifier developed and then finally provide it to the user.

3.3 Features

We have implemented a total of 7 features that are as follows.

(1) Sentence Length
(2) Sentence Position
(3) Similarity with the title
(4) Sentence to Sentence Similarity
(5) Proper Noun
(6) TF-IDF
(7) Presence of Numerical Value [5]

3.3.1 Sentence Length

The length of a sentence provides a clear indication as to whether a sentence could be a art of summary or not. Generally too long or too short sentences do not tend to be part of summary.

Example: Given the sentence “India is a Secular Country". The length of the sentence is 5. We do not perform any kind of pre-processing before calculating the length of the sentence, as it will lead to inaccurate results.

3.3.2 Sentence Position

The position of a sentence plays a huge role in determining the probability of it being in the summary. The first few sentences of the paragraph generally tend to be more important than the others. For each sentence, we assign a score that depends on the position of the sentence with respect to the article normalized by the number of sentences in the article.

For example: “Grammy nominated Swedish DJ Avicii died Friday, his publicist confirmed to CNN [1]. He was 28. “It is with profound sorrow that we announce the loss of Tim Bergling, also known as Avicii,” a statement to CNN read. “He was found dead in Muscat, Oman this Friday afternoon local time, April 20th. The family is devastated and we ask everyone to please respect their need for privacy in this difficult time. No further statements will be given.” Avicii was one of the most successful touring DJs in the industry, working with artists including Madonna. He had a crossover pop hit in 2013 with “Wake Me Up.” For the above extracts of news article from CNN, we can infer that first sentence of the news article holds the most important aspect of the entire news article. Similarly, we index each sentence from the news article, a possible prospect of being into news summary, by weighing it through the total number of articles. Making the indexing between 0 and 1. In the above example, there are total of 7 sentences. We index first sentence as 1/7, second one as 2/7, and similarly last sentence as 7/7.

3.3.3 Similarity with the Title

Title sentence reflects the main idea of the passage. The more similar a sentence is to the title sentence, the more the chances of it being important, and therefore more the chances of it being in the summary. Both the sentences are represented in the vector space model with a common vocabulary of words. We perform pre-processing techniques like stopwords removal and stemming, inorder to make sure that we don’t treat two similar words as different and also avoid redundancies. Then we calculate the term frequencies of terms in both the sentences. The similarity between the two sentences is calculated by using the cosine similarity function. If there are two vectors say P and Q, then the cosine similarity of P and Q is defined as

$$Sim(P, Q) = \frac{P.Q}{|P| * |Q|} \tag{1}$$

where $P \cdot Q$ is defined as the dot product of vectors P and Q. and $|v|$ represents the magnitude of the vector v. Example: $P = (1, 2, 1)$; $Q = (2, 1, 2)$, then the cosine similarity of vectors P and Q is given as

$$P \cdot Q = 1 * 2 + 2 * 1 + 1 * 2 = 6 \tag{2}$$

$$|P| = \sqrt{1 * 1 + 2 * 2 + 1 * 1} = 2.44 \tag{3}$$

$$|Q| = \sqrt{2 * 2 + 1 * 1 + 2 * 2} = 3.00 \tag{4}$$

$$Sim(P, Q) = \frac{6}{(2.44 * 3.00)} = 0.819 \tag{5}$$

3.3.4 Sentence to Sentence Similarity

The similarity of a sentence to the other sentences in the document is a crucial indicator of the importance of the sentence. The more similar a sentence is to the rest of the sentences, the more it coincides with the main idea of the article, and therefore is more probable to be part of the summary. For every sentence, the similarity of that sentence is calculated with each of the rest of the sentences in the article. Then these values are summed up to be used as a feature. For calculation of this feature, sentences are represented in the vector space model and the similarity is calculated using the vector space model with cosine similarity as described above. In this case also, sentence pre-processing was done before calculating similarity measure inorder to increase the efficiency.

3.3.5 Proper Noun

A proper noun is a noun that in its primary application refers to a unique entity, such as London, Jupiter, Sarah, or Microsoft, as distinguished from a common noun, which usually refers to a class of entities (city, planet, person, corporation), or non-unique instances of a specific class (a city, another planet, these persons, our corporation). The presence of Proper nouns can greatly increase the probability of sentence being included in the summary. For example, if you consider the sentence "London is a fallen city", then "London" is the proper noun, whereas "city" is just a common noun. We have considered both singular and plural proper nouns in the sentence. Firstly, we divide the sentence into words using the word tokenizer provided by nltk [6]. Then we use the part of speech tagger provided by nltk in order to get the part of speech of each word in the sentence. After that, we count the number of singular and plural nouns present in the sentence to be used as a feature. For example, in the sentence considered above, number of proper nouns is 1.

3.3.6 TF-IDF

TF-IDF refers to Term Frequency-Inverse Document Frequency [11]. Term Frequency refers to the frequency of term in a particular article. Inverse Document Frequency of a term refers to the number of documents in the set that contains that particular term. When it comes to vector space models and statistical analysis of text, TF-IDF becomes an indispensable part and Automatic Content Summarization is no exception to that rule. If a sentence contains terms that are highly frequent in the document, then there is a higher probability that the sentence belongs to summary. IDF is included so that the highly frequent terms that are redundant and provide no useful information (like a, an, the etc.,) do not get highly weighted. This ensures that the domain based words get highly rated.

We have also employed stemming and stopwords removal before calculating this feature. For every sentence, we tokenize the sentence into words by using the word tokenizer of nltk. Then, we calculate the tf-idf score of each word in the sentence. These values are then summed up to be used as a feature. The formulas for term frequency and inverse document frequency that we employed can be given as follows. The term frequency is calculated for a term in a single article.

$$tf(t) = freq\_stem\_t \tag{6}$$

$$idf(t) = \log_2 \frac{Total\ No.\ of\ documents}{No.\ of\ documents\ containing\ stem(t)} \tag{7}$$

This plays a crucial role in determining the probability of sentence to be included as part of summary.

3.3.7 Presence of Numerical Value

Python Regular Expression module in news articles, the sentences which feature numerical value are considered more important than the ones which don't have. Therefore, in this paper we have considered into opting a value of 1 as the feature value for the sentence having numerical value while giving value of 0 to the ones not having it. We've parsed through the news article with the help of Python Regular Expression module.

For example: Grammy nominated Swedish DJ Avicii died Friday, his publicist confirmed to CNN. He was 28 [1]. "It is with profound sorrow that we announce the loss of Tim Bergling, also known as Avicii," a statement to CNN read. "He was found dead in Muscat, Oman this Friday afternoon local time, April 20th. The family is devastated and we ask everyone to please respect their need for privacy in this difficult time. No further statements will be given." From the extracts of above news article, second and fourth sentence are given feature value of 1 while others are given 0, considering that these sentences consist of numeric value while others don't possess.

3.4 Normalization

Normalization of data is the process by which we scale all the features inorder to bring their value between a fixed range of either [0,1] or [−1, 1] etc., [13]. This is generally done so as to avoid the result of the algorithm being monopolized by a single or small set of features. In this paper, we have used normalization by standard deviation [8]. In this type of normalization, suppose there is a list of elements, say p. Then every element in p is converted to

$$x = \frac{x - mean(p)}{std(p)} \tag{8}$$

where mean(p) is the average of all the elements in the list p and std(p) is the standard deviation of all the elements in the list p. Let us consider an example, a list p with elements (1, 2, 3, 4). The average of this list p is given as

$$\frac{(1+2+3+4)}{4} = 2.5 \tag{9}$$

The standard deviation of the list p is calculated as follows.

$$std = \sqrt{\frac{\sum_{i=1}^{n}(x_i - x_{avg})^2}{n}} \tag{10}$$

where n is the total number of elements in the list p and x_{avg} is the mean of elements in p. For the given example, standard deviation is given as 1.118.

Therefore,

$$p1 = \frac{(1 - 2.5)}{1.118} = -1.34 \tag{11}$$

$$p2 = \frac{(2 - 2.5)}{1.118} = -0.44 \tag{12}$$

$$p3 = \frac{(3 - 2.5)}{1.118} = +0.44 \tag{13}$$

$$p4 = \frac{(4 - 2.5)}{1.118} = +1.34 \tag{14}$$

Normalization of both the training data and the real time data are done by using the mean and standard deviation values of training data only.

4 Algorithms and Results

We used three classifiers from the Scikit Learn module in python [10], namely Logistic Regression, Artificial Neural Network and Gaussian Naive Bayes, achieving accuracies of 88.80%, 89.50% and 87.90% respectively which has shown in Table 3.

4.1 Logistic Regression

Logistic regression is a parametrized supervised classification algorithm, which works based on the logistic function [15]. The hypothesis is a logistic function of linear combination of features, whose parameters are obtained by Gradient Descent algorithm.

4.1.1 Logistic Regression Equation

The logistic function can be given as below.

$$Logit(p) = \frac{1}{(1 + (e^{-p}))} \tag{15}$$

For Automatic Content Summarization of News Articles, We have a total of two classes in our problem. In this model, the probability that the given sentence belongs to both the classes is calculated. For whichever class, we obtain the maximum probability, the given sentence is mapped to that particular class. Table 1 shows hyper parameter values. For the problem in the context (Automatic Content Summarization of News Articles), for the implementation, we have used the logistic regression model in the Scikit learn module of python. First, we obtain a classifier of the model, and use the fit method in order to train the data based on the given training set. Then, in order to obtain the accuracy of the trained classifier, we use the score method on the testing data set.

The tolerance refers to the condition that can be placed to stop further iterations in the algorithm. The inverse of regularization strength is used to control over-fitting. Smaller the value, larger is the regularization.The solver used is liblinear and the maximum no. of iterations of the gradient descent algorithm is 100. We have taken these parameters in Logistic Regression Classifier and We have obtained an average final accuracy of 88.80 %.

Table 1 Logistic regression hyper parameter values

Parameter	Value
Tolerence	0.0001
Regularization Strength C	1.0
Max-iter (No. of Iterations)	100
Solver	Liblinear

4.2 *Artificial Neural Network*

Artificial Neural Network is a computation system that is inspired by biological neural networks [4]. It consists of input, output and hidden layers. Each layer consists of neurons that are fully connected to the next layer. Each neuron acts as a sigmoid neuron and implements logistic function, that is described below. Sigmoid neuron, alike its counterpart perceptrons has input for different weights $(w1, w2)$ and bias b, but holds significant distinction to have output values ranging from 0 to 1 than holding to 0 or 1. The output is given by the following equation.

$$\sigma\ddot{I}(w_x + b), \tag{16}$$

Here sigmoid function is given by following equation,

$$\sigma\ddot{I}\sigma\ddot{I}(z) = \frac{1}{1 + e^{-z}} \tag{17}$$

Having inputs as $x1, x2, \ldots$weights as $w1, w2, \ldots$and bias as b, the output of sigmoid neuron is given by the following equation:

$$\frac{1}{1 + exp(-\sum_j w_j x_j - b)} \tag{18}$$

In the algorithm, the probability for the input feature vector being mapped to each of the class values is calculated, and class which is having the maximum probability is the class to which the input given is mapped. In this problem, we only have two classes. For the problem in the context (Automatic Content Summarization of News Articles), we have used the Multi Layer Perceptron Classifier (MLP Classifier) model in the Scikit learn module of python. First, we obtain a classifier of the model, and use the fit method in order to train the data based on the given training set. Then, in order to obtain the accuracy of the trained classifier, we use the score method on the testing data set.

Number of hidden layers, Learning rate, Layer sizes(number of neurons in each layer), Mini Batch size, Epochs (number of iterations) are known as Hyper Param-

Table 2 ANN hyper parameters

	Hyper parameters	Value
1	Number of hidden layers	2
2	Hidden layer size of each hidden layer	200
3	Tolerance factor	0.0001
4	Constant learning rate	0.01
5	Maximum number of iterations	300
6	Batch size	500

eters. By varying the hyper parameters, the accuracy varies. In this paper, our input layer contains all the input features. So the input layer has a total of 7 features. The output layer has 2 neurons corresponding to two classes. For a given a test case, these output neurons each have a value between 0 and 1. Feature vector will be assigned a class that has the maximum probability. The hyperparameter list and their values are as given below. We have chosen the solver to be 'sgd' which stands for **Stochastic Gradient Descent**.

There are six parameters which were varied (Table 2).

The tolerance factor refers to the stopping condition of the gradient descent algorithm. The learning rate of gradient descent is chosen as a constant value. The other parameters varied are the maximum no. of iterations per instance of running the gradient descent, no. of hidden layers, no. of neurons per hidden layer and batch size for implementing stochastic gradient descent. After executing the algorithm with the below mentioned hyper-parameters, we have obtained an average final accuracy of 89.50%.

4.3 Naive Bayes Algorithm

Naive Bayes algorithm is a machine learning algorithm based on Bayes probability theorem used widely for text classification [?]. It helps in cases where there are high dimensional training data sets. This algorithm is called "Naive" because it makes the assumption that the occurrence of a certain feature is independent of the occurrence of other features. The mathematical formulation behind this theorem is

$$P(H \mid E) = \frac{P(E \mid H) * P(H)}{P(E)} \tag{19}$$

For Automatic Content Summarization of News Articles, We have a total of 2 classes in our problem. In this model, the probability that the given sentence belongs to both the classes is calculated. For whichever class, we obtain the maximum probability, the given sentence is mapped to that particular class. For implementation, we used the SKlearn module's Gaussian Naive Bayes classifier. Firstly, we used the train and test split method to split the data into training and testing sets in the ratio of 0.8:0.2. Then we use the fit method of the model to train the classifier against the training data. Then we can either use the predict method to get the classification for any new input, or use the score method to test the accuracy of the classifier,by using the testing data. We have implemented Gaussian Naive Bayes with the following features and obtained an average accuracy of 87.90%.

5 Android Application

We have created a Android Application to act as an interface between the user and the summarization system. We have created a client server based system. Initially, in the application, the user is provided with an activity containing two use cases. The model of the application is given in Fig. 2.

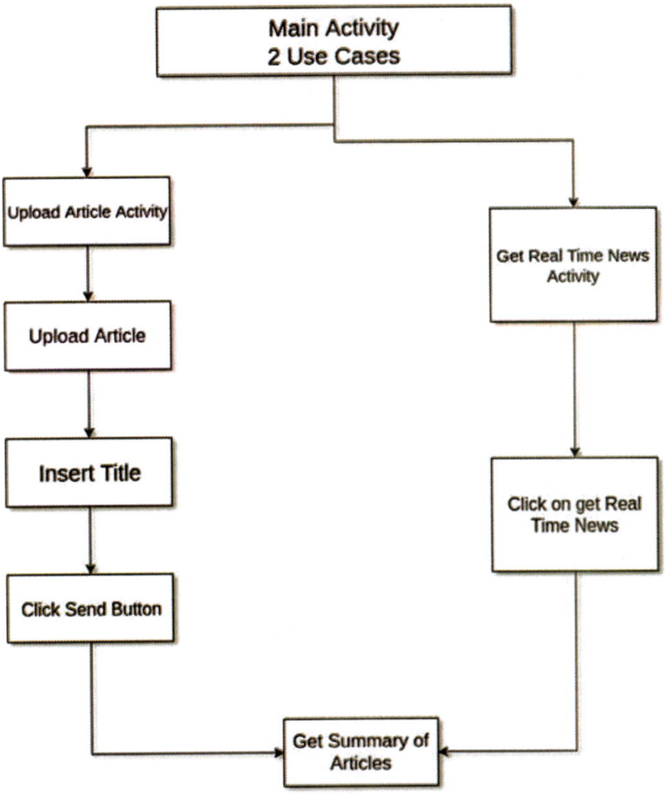

Fig. 2 The working of the application is illustrated

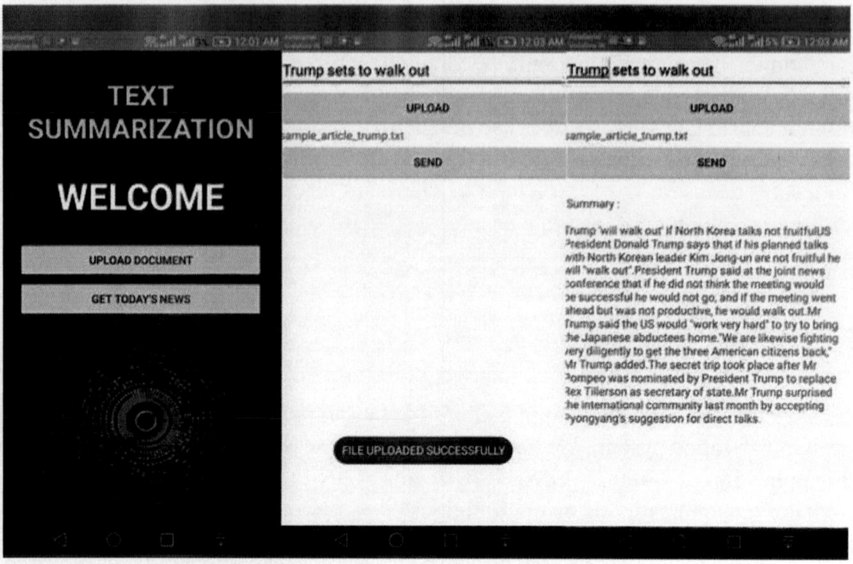

Fig. 3 App-View of Home Screen, Sample upload screen and Summary screen

In Fig. 3 shows the screen shots of app. Firstly, the user can get summary of any article by clicking on upload document button, which leads to the following activity.

Here the user should enter the title of the article that is to be summarized. Then the user can upload the file containing the article, after which the status of upload is displayed. Then when the user clicks send, the article is sent through a socket connection to a server, which runs the classifier to summarize the article, and returns the summary as the result, which is shown in Fig. 3.

From the first activity, the user can also select the option of getting real time news. Here, a request is sent to the server, which results in the server extracting the top headlines and URLs of news stories currently happening in the country from various news sites, using the News API in python. Then the server scrapes information from all the URLs of news sites, summarizes the articles and returns the summaries as result. The output screen is shown in Fig. 4 (Table 3).

6 Conclusion and Future Work

We have obtained an accuracy of 88.80% using Logistic Regression Classifier, an accuracy of 89.50% using Artificial Neural Networks and an accuracy of 87.90% using Gaussian Naive Bayes Classifier. This is indicating that the features used like Title Similarity, Sentence Similarity and TF-IDF are able to model the problem statement well, even for news articles collected from varied sources. On comparing

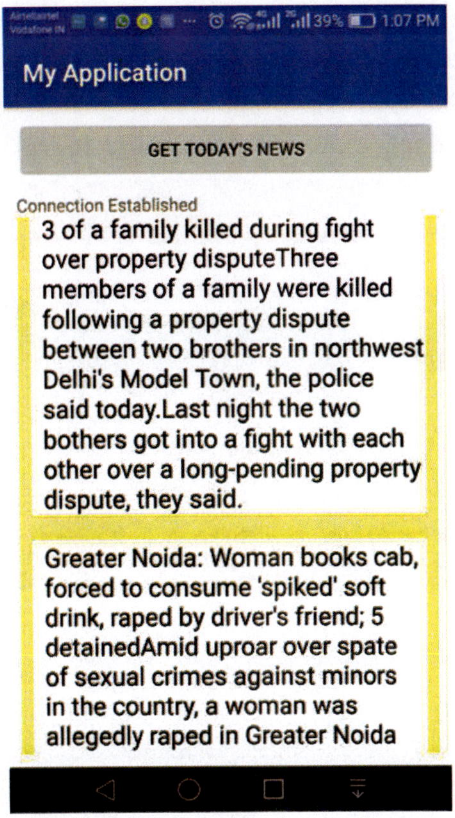

Fig. 4 Real time news summary

Table 3 Result summary

Algorithm	Accuracy (%)
Logistic Regression	88.80
Artificial Neural Network	89.50
Naive Bayes Algorithm	87.90

the three algorithms that we used, Artificial Neural Network performs slightly better than the other two. However, we've used Naive Bayes algorithm in our android application, to produce concise summaries. In future work, we can Working on Hindi News and developed algorithm for abstraction based text summarization.

References

1. .
2. Allahyari, M., Pouriyeh, S., Assefi, M., Safaei, S., Trippe, E. D., Gutierrez, J. B., & Kochut, K. (2017). *Text summarization techniques: A brief survey*. arXiv preprint arXiv:1707.02268.
3. Cabral, L., Lima, R., Lins, R., Neto, M., Ferreira, R., Simske, S., & Riss, M. (2015). *Automatic summarization of news articles in mobile devices* (Vol. 10, pp. 8–13).
4. Hassoun, M. H., et al. (1995). *Fundamentals of artificial neural networks*. MIT Press.
5. Kumar, Y. J., Goh, O., Halizah, B., Ngo, H., & Puspalata, C. (2016). A review on automatic text summarization approaches. *J. Comput. Sci*, *12*(4), 178–190.
6. Loper, E., & Bird, S. (2002). Nltk: The natural language toolkit. In *Proceedings of the ACL Workshop on Effective Tools and Methodologies for Teaching Natural Language Processing and Computational Linguistics*. Philadelphia: Association for Computational Linguistics.
7. Luhn, H. P. (1958). The automatic creation of literature abstracts. *IBM Journal of Research and Development, 2*(2), 159–165.
8. Malhotra, S., & Dixit, A. (2013). An effective approach for news article summarization. *International Journal of Computer Applications, 76*(16).
9. Mihalcea, R., & Tarau, P. (2004). Textrank: Bringing order into text. In *Proceedings of the 2004 Conference on empirical methods in natural language processing*.
10. Pedregosa, F., Varoquaux, G., Gramfort, A., Michel, V., Thirion, B., Grisel, O., et al. (2011). Scikit-learn: Machine learning in python. *Journal of Machine Learning Research, 12*, 2825–2830.
11. Ramos, J. *Using tf-idf to determine word relevance in document queries*.
12. Sethi, P., Sonawane, S., Khanwalker, S., & Keskar, R. B. (2017). Automatic text summarization of news articles. In *2017 International Conference on Big Data, IoT and Data Science (BID)* (pp. 23–29).
13. Sherman, B. T., Lempicki, R. A., et al. (2009). Systematic and integrative analysis of large gene lists using David bioinformatics resources. *Nature Protocols, 4*(1), 44.
14. Van Rossum, G., & Drake Jr, F. L. (1995). *Python reference manual*. Centrum voor Wiskunde en Informatica Amsterdam.
15. Wright, R. E. (1995). *Logistic regression*.

Chronic Kidney Disease Prediction Using Artificial Neural Network

Gaurav Dubey, Yashdeep Srivastava, Aman Verma, and Shriyansh Rai

Abstract Disease prediction/detection is a crucial and challenging problem, because it helps in early diagnosis of disease by supporting the pathologists and doctors in making their decision. In the recent studies and research, "The Artificial Neural Network" giving a practical approach to solve many day-to-day to life or in many sectors such as in medical sector in which it can be used to predict a particular disease based on some data given. In this research paper, we are here describing a chronic kidney disease prediction system by using ANN technique. Back propagation algorithms are used for the training of ANN over the given dataset having several attributes containing information of patients. This neural network gives us an output which distinguish the patients having chronic kidney disease or not. As after doing so much of research regarding the above issue of prediction of kidney disease, artificial neural network shows a more accurate results than other machine learning algorithms.

Keywords Chronic kidney disease (CKD) · Artificial neural network (ANN) · Glomerular filtration rate (GFR) · Back propagation algorithm (BPA) · Data preprocessing · Correlation-based feature subset selection (CFS)

1 Introduction

Healthcare organizations are evolving day by day toward technology enhancement in the medical sector using the artificial neural network [1] techniques to improve delivery of care of patients at a reduced cost. In many developed countries, the use of

G. Dubey (✉) · Y. Srivastava · A. Verma · S. Rai
ABES Engineering College, Ghaziabad, UP 201009, India
e-mail: gaurav.dubey@abes.ac.in; gdubey1977@gmail.com

Y. Srivastava
e-mail: yashdeep.16bcs2048@abes.ac.in

A. Verma
e-mail: aman.16bcs1063@abes.ac.in

S. Rai
e-mail: shriyansh.16bcs2031@abes.ac.in

S. Tiwari et al. (eds.), *Proceedings of International Conference on Big Data, Machine Learning and their Applications*, Lecture Notes in Networks and Systems 150,
https://doi.org/10.1007/978-981-15-8377-3_33

advance healthcare systems using machine learning and artificial neural network is becoming popular. In developing a neural network-based system, we will face some new complexities related to improve the structure or model to manage healthcare system. A research showed that almost 10% of the population [2] of the world is suffering from a chronic disease and its day by day increasing among the people. While the prediction of these chronic diseases are based on some old methodologies in various countries. The systematic procedure and an automated technologies are used in very few countries. Chronic kidney disease [2] is a type of disease which is related to kidney and its functioning. CKD affects the kidney in a way that kidney fails to purify the blood in a proper manner. The disease is called as "chronic" because it damage the kidneys in a slowly manner for over a long period of time. CKD can also cause other health problems [3]. Firstly, data preprocessing is applied to the dataset to remove the less important features related to the pathological investigation involving more resources, money, time, and uncertainties. After that, a reduced set of attributes is obtained by applying correlation-based feature subset selection [4] using data preprocessing over the dataset.

2 Background Details & Related Work

Here, we are highlighting some major researches regarding the CKD prediction that are done by various techniques of machine learning as follows:-This is a theory of developing an optimal fuzzy K-NN [2] for the prediction of CKD with a mindset of improving the accuracy and minimizing the errors. The technique used in developing this model is OF-KNN technique, in which an optimization algorithm such as Bat is used to tune the fuzzy, and then the OF is used to measure similarity in the KNN [5]. This technique produces better results and in an acceptable time but the accuracy is less compared to ANN. Another study which proposes an adaptive neuron fuzzy inference system [6] for predicting the timeframe of CKD which is based on the clinical data methods. They used the clinical data of patients of around ten years. The threshold value of GFR was used as 15 cc/kg/min 1.73 m^2 and this value known as the counter value of renal failure. Various attributes are selected for the input of ANN as described in the dataset section of this paper.

A recent study of producing a reduced set of attributes [4] for producing accurate results on predicting CKD using neural network because it reduces the complexity of the neural network and less load on the network. The reduced set of attributes are produced by applying CFS in which a feature subset is considered good which are highly correlated with the class. Another approach of applying machine learning in the kidney disease prediction which involves the comparison of the two machine learning techniques such as support vector machine [1] and artificial neural network (ANN) began with the technique of data mining. The data mining method is used over the dataset to gain the hidden information from the dataset which is required for the prediction of the disease. The two above approaches are compared on the basis of classification of kidney diseases and prediction of kidney disease for the given data of patients. Support vector algorithm is a supervised algorithm that uses the method of nonlinear mapping which is used to renovate the unique training information into some higher dimension. Whereas the other algorithm uses back propagation technique which process the neural network iteratively by processing a data of set of training tuples. For each input, weights are modified iteratively. After comparing both the algorithms, the artificial neural network possesses more accurate results than the support vector machine.

3 Proposed Approach

See Fig. 1.

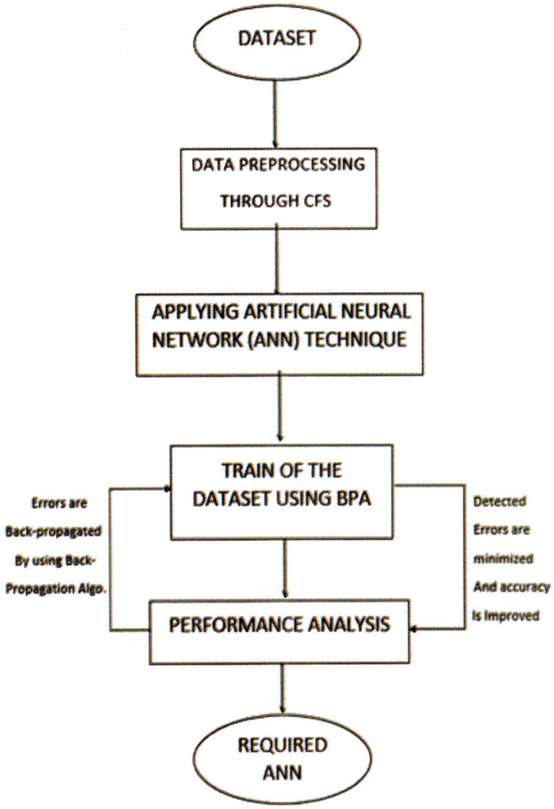

Fig. 1 Proposed methodology

3.1 Dataset & Data Preprocessing

We have taken a dataset consisting of 25 features as columns and a total of 400 patients as rows. Here are the features:

> Age, Blood pressure, Specific gravity, Albumin, Sugar, RBC, Pus cells, Pus cell clumps, Bacteria, Blood glucose random, Blood urea, Serum creatinine, Sodium, Potassium, Hemoglobin, Packed cell volume, White blood cell count, Red blood cell count, Hypertension, Diabetes mellitus, Coronary artery disease, Appetite, Pedal edema, Anemia, Classification(ckd or notckd).

Data preprocessing involves transformation of raw dataset into a more valuable dataset. In this research paper, we use correlation-based algorithm [4] which has a method known as correlation-based feature subset (CFS) which gives us a reduced set of attributes of the given dataset. We have to give proper inputs to the neural network. It really depends upon the neural network and what type of format the input data is in. Since artificial neural networks take the numeric vectors as an input, we just need to convert the data into a numeric form. For the most general kind of data, tabular data (e.g., CSV, Excel spreadsheet, SQL), the simplest way to process the data would be to load the data into a pandas DataFrame, then convert it to a NumPy array. The resulting NumPy array can be used as input for neural networks in any of the major machine learning frameworks (Keras, TensorFlow, PyTorch, etc.).

3.2 Applying Artificial Neural Network

Artificial neural networks are the parallel computing devices that is built in accordance with a motive to make computerized model of human brain. The main reason to make ANN is to develop a system which is faster and more accurate than the existing traditional systems. Artificial neural network is a type of network in which each node is connected to the every node in the neural network through which information passes from one node to another until it reaches output layer then the error is calculated and minimized to make model more productive (Fig. 2).

In this diagram, we can clearly see the connections used in the ANN. The training algorithm which we are using is back propagation algorithm (BPA).

Pseudo code for back propagation algorithm:

1. *Initializing the weights and biases used in ANN.*
2. *while terminate condition is not satisfied{// Starting while condition*
3. *For each and every training tuple X in dataset D{*
4. *// This network propagates the input data forward:*
5. *For each and every input layer unit k{*
6. $O_k = I_k$; *// Output of an input layer unit is equal to its actual input value.*

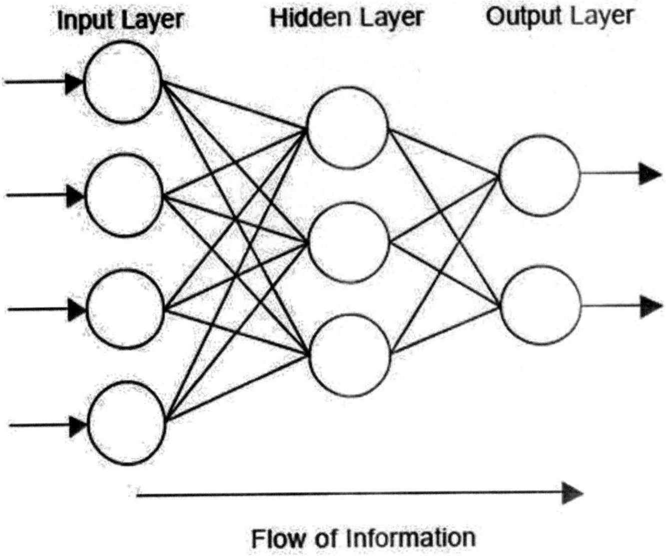

Fig. 2 Architecture of the ANN

7. *For each and every hidden or output layer unit k:*
8. $I_k = \sum i\, W_{ik} O_i$; *// net input of the nodes of unit k w.r.t previous layer.*
9. *Using sigmoidal activation function,* $O_j = 1/(1+ e^{-1})k$; *//Output of the unit k*
10. *// Errors are reducing using back propagation technique.*
11. *Updating the respective weights and biases.*

3.3 Training the Neural Network and Analyzing the Performance

This ANN is used for the prediction of chronic kidney disease (CKD) over the given data set. After building the neural network (ANN), we have to train the ANN over the given data set. Main objective of training the dataset is to minimize the occurring errors or losses in the neural network and to improve the accuracy. We split the dataset into training size 80% and 20% testing size. In this neural network, we have used 100 epochs to minimize the error and to improve the accuracy. An epoch is defined as an instance of using all the training data once for updating the weights. By applying this technique, we get an accuracy of 90.28% on predicting the CKD of the patients in the given dataset.Training the network refers to refresh or update the weights of the nodes. The update in the weights continues until the output of the model that is

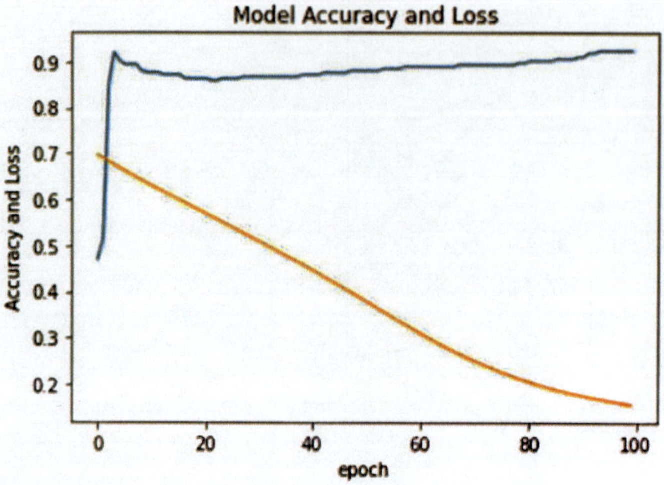

Fig. 3 Output of the ANN

actual output will be very near to the desired output and the error in each output step is minimized by the back propagation algorithm which we are using in this neural network.

4 Experimental Setup and Results

The trained artificial neural network gives an accuracy of 90.28% and a loss of 15.57% in the process of predicting chronic kidney disease for several patients. We have used keras and tensor-flow library of neural network to implement the project (Fig. 3).

The artificial neural network we use here is trained over 100 epochs to improve the accuracy. That is our main goal to improve accuracy and provide a neural network approach to solve the problem of prediction of chronic kidney disease for several patients according to the data set we have used and achieve a reasonable accuracy with a very minimum loss.

5 Conclusions

This research is proposed in a manner to provide a proper approach for the prediction of CKD by the phenomena of artificial neural network which gives us results with high accuracy and minimum loss. So, after doing proper research as shown above in the paper we can conclude that artificial neural network can give more accurate and fast results which can be used to diagnose the disease at the right time.

References

1. Vijayarani, S., & Dhayanand, S. (2016). Kidney disease prediction using SVM and ANN algorithms. *International Journal of Computing and Business Research (IJCBR), 6*(2), 2229–6166
2. Subhashini, R., & Jeyakumar, M. K. (2017). OF-KNN technique: An approach for chronic kidney disease prediction. *International Journal of Pure and Applied Mathematics, 116*(24), 331–348.
3. Varma, P. P. (2015). Prevalence of chronic kidney disease in India—Where are we heading. *Indian Journal of Nephrology, 25*(3), 133–135.
4. Misir, R., Mitra, M., & Samanta, R. K. (2017). A reduced set of features for chronic kidney disease prediction. *Journal of Pathology Informatics, 8*(24).
5. Tekale, S., & Shingavi, P. (2018). Prediction of chronic kidney disease using machine learning algorithm. *International Journal of Advanced Research in Computer and Communication Engineering, 10*, 2319–5940.
6. Norouzi, J., & Yadollahpour, A. (2016). Predicting renal failure progression in chronic kidney disease using integrated intelligent fuzzy expert system. *Computational and Mathematical Methods in Medicine, 2016*(3), 137–166.

Vehicle Detection and Speed Tracking Based on Object Detection

Gaurav Dubey, Ayush Tiwari, Akash Singh, and Aryan Upadhyay

Abstract With the increase in the traffic and vehicles on road, there is a need for smart and intelligent traffic monitoring system and surveillance. A smart and intelligent traffic monitoring and surveillance system is basic requirement for the development of smart cities in India. This can be achieved by the detection of moving vehicles along with the speed tracking for detection of speed limit violations. This research paper shows some effective approach for the vehicle detection and estimation of their speed with the help of machine learning concept. The given methods are used to detect and track the speed of the vehicles passing through the traffic surveillance area.

Keywords Machine learning · Object detection · Speed tracking · YOLOv3 · Darknet · OIDv6

1 Introduction

In India, plan of government is to develop many cities having smart features in the future. In all the smart cities, there is the requirement of smart services like high-level traffic management, traffic surveillance, etc. And for delivering these smart services, several information and communication technologies will be used. In smart traffic monitoring, we have to develop a system that can automatically detect that

G. Dubey (✉) · A. Tiwari · A. Singh · A. Upadhyay
ABES Engineering College, Ghaziabad, UP 201009, India
e-mail: gaurav.dubey@abes.ac.in; gdubey1977@gmail.com

A. Tiwari
e-mail: ayush.16bcs2054@abes.ac.in

A. Singh
e-mail: akash.16bcs2045@abes.ac.in

A. Upadhyay
e-mail: aryan.16bcs2072@abes.ac.in

S. Tiwari et al. (eds.), *Proceedings of International Conference on Big Data, Machine Learning and their Applications*, Lecture Notes in Networks and Systems 150,
https://doi.org/10.1007/978-981-15-8377-3_34

which vehicle is violating the traffic rules. Smart traffic management needs a automatic traffic surveillance system. This paper proposes the methods of vehicle detection, classification, and speed tracking by the concept of machine learning. Traffic surveillance is the significant application of video-based monitoring system. So, in the vision-based intelligent transportation system, many years of researches have been done. The researchers got motivation for developing more efficient solution for computer vision-based applications due to the economic and easily availability of hardware. Computer vision based on images processing has become a successful technology for real-time supervision, monitoring, and control that is useful in major areas varying from civil applications to industries using deep learning methods [1, 2]. If we consider the importance of intelligent transport system, the vision-based supervision has a major contribution, as it can make easier for real-time monitoring, vehicle tracking, and identification as well as face detection also [3]. Instead of this, other parametric identifications such as vehicle speed, vehicle density, and vehicle count are used for effective traffic monitoring and control.

2 Background Details and Related Work

2.1 Speed Tracking: Physical Methods

Physical method (e.g., laser and radar) and image processing method are the main method by which we estimate the speed of any object [4]. These two methods which are used to detect the speed of object have both its advantages and disadvantages, respectively (Fig. 1).

The physical method works on the principle that the ejector ejects the wave or light signal of radar or laser and the returned or blocked signal is received by the receiver. Since the signal frequency will change with the change in the speed of object, we can calculate the speed of object by comparing the received and ejecting signal frequency [5, 6].

Apart from some merits, there are some demerits of this method also:

(1) It can measure the speed of only one object at one time [4, 7],
(2) The device in itself must be stable,
(3) It can only measure limited distance.

2.2 Image Processing Method

There are too many solutions for the image processing method, the correlation among the camera and real world can be obtained by using camera calibration method, and for estimating the speed of vehicle with respect to change of location, we can use image matching method. And for calculating the speed of camera itself, we can

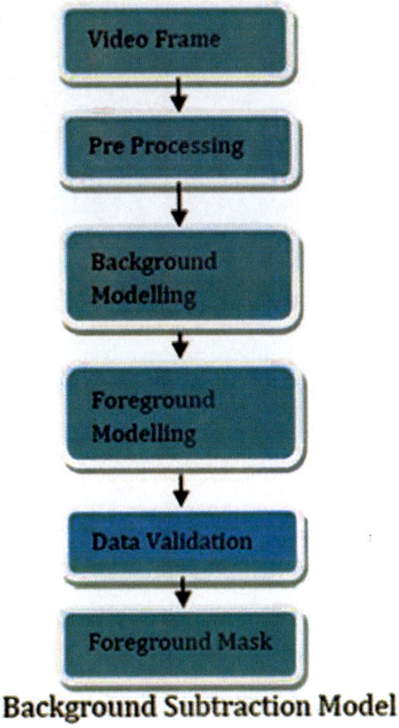

Fig. 1 Background subtraction model

use the information of fixed point of camera [7, 8]. All these methods are used for calculating the speed of the vehicle. One of the major drawbacks is that in most of these methods camera should be stable because in most of the method the frame differencing is used to detect the object, and in this method the background should not be changed [5].

3 Proposed Method

(i) **Object detection**: We have chosen YOLOv2 as the basic network for achieving the object detection in real time. YOLOv2 working is same as YOLOv1 in which it acts as end-to-end CNN and object detection problem is taken as a regression problem and making it possible to run in real-time speed. In YOLOv2, 98 anchor boxes are used in with preselected size to represent object and the height and width are near about equal to the height and width of the object.

(ii) **Vehicle target detection**: Figure 2 is showing the vehicle target detection process. To constitute the VOC car datasets, firstly we have to screen out samples from datasets of VOC 2007 and VOC 2012 which contain car target. And then in YOLOv3 training network, put the training samples, and then the

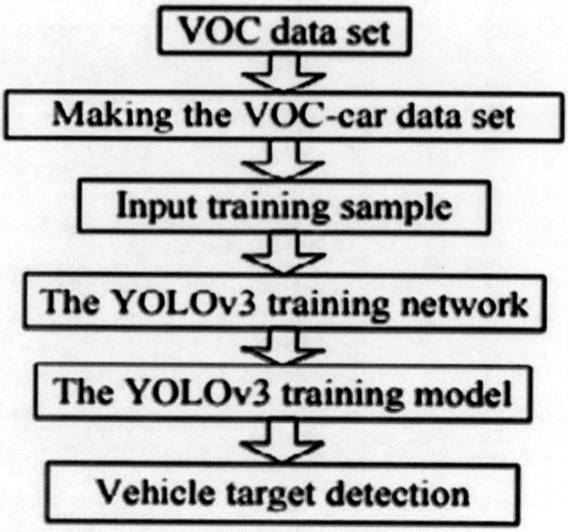

Fig. 2 Car target detection flow chart

training is carried out until the network convergence. Finally, in the YOLOv3 model, the weight.h5 file was loaded to test samples (Fig. 3).

(iii) **YOLO v3 introduction**: The YOLO algorithm is given by Redmon in 2016, and the YOLO algorithm is a CNN that can anticipate location of multiple boxes and categories at same time. In reality, its accuracy decreased although it has end-to-end target detection and use the advantage of fast speed. However, to improve the accuracy and maintaining the speed at same time, we have an improvement of YOLO 9000 algorithm on the original YOLO algorithm. In

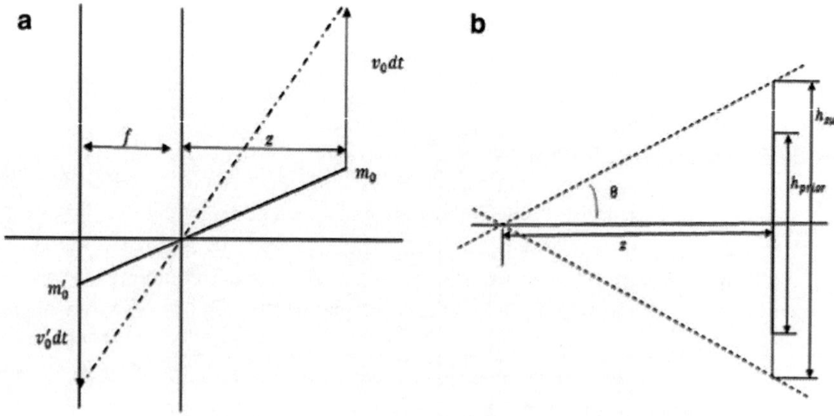

Fig. 3 **a** Camera vector diagram. **b** The vector diagram for showing focusing principle

this way, the datasets of COCO and ImageNet can be simultaneously trained by YOLO9000 algorithm, and 9000 objects can be detected by the trained model.

(iv) **Calculating true speed**: For calculating the real speed of object, the main thing is that we need to figure out the relation between the real world and the camera optics. The overview of this relation is assuming a point m_0 has real speed v_0 (here, we are assuming that the camera is stable and the direction of camera plane and speed is parallel), and the projection is on image m_0' that has speed v_0'. Their journey in unit time is $v_0 \mathrm{d}t$ and $v_0' \mathrm{d}t$. Equation representing the concept of camera projection:-

$$\frac{v_0 * dt}{z} = \frac{v_0' * dt}{f} \tag{1}$$

In the above formula, f represents the camera's focal length and z represents the object and camera distance [5]. Connection between the speed of image v_0' and optical flow speed v_{opt} is given below:

$$v_{\text{opt}} = v_0' * k_0 \tag{2}$$

Here, k_0 represents a unique value. From the above Eqs. (1) and (2), we got:

$$v_0 : v_{\text{opt}} = z : (f * k_0) = z : k \tag{3}$$

where $k = f * k_0$ specific value of f is different with different camera.

In that way, we can figure out relation between v_0 and v_{opt}. Figure 5 shows vector diagram of camera.

Relationship between z, θ, h_{sum} and h_{prior}. We can calculate the true speed v_0 by using the other parameters, since we know that z_2, h_2, v_1, v_{opt1}, z_1 are known parameter, r_{yolo} is the output of the YOLO network and h_{prior} is the prior value and v_{opt} is the output of flownet. One thing to be noted down is that v_0 and v_{opt} are of vector type so we cannot only calculate the speed of it but also the direction of it. For our convenience, we are assuming that camera is still but we can calculate optical flow even if camera is moving.

4 Experimental Setup and Results

The below graph is showing the experimental result in the ROC curve in which the vertical line represents the detection rate and horizontal line represents the number of false positive per image (FPPI). In the curve, the terms our 5, our 10, our 30 represent the classifier and the number of subspaces in the classifier is 5, 10, 20, and 30, respectively (Fig. 4).

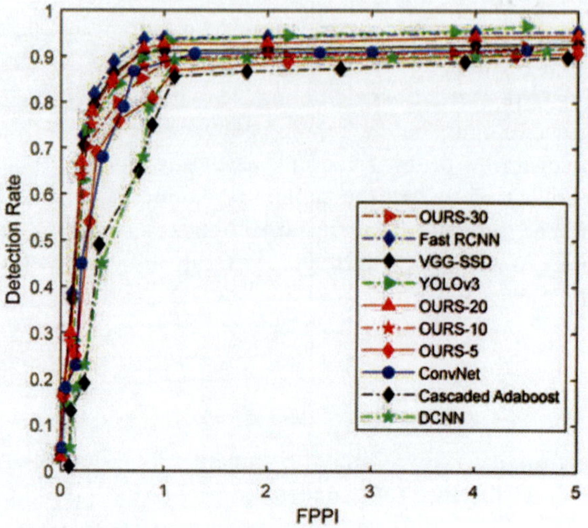

Fig. 4 Detection rate versus FPPI

Figure 5 mentioned above represents the average loss iteration as the number of iteration increases the average loss decreases.

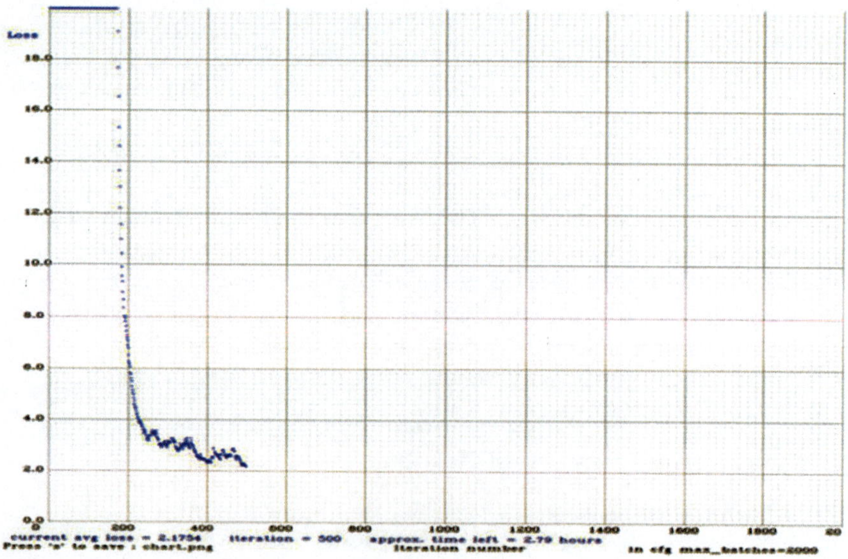

Fig. 5 Average loss per iteration

5 Conclusions

In this research paper, we have shown some approach by which we can detect and track the moving vehicles and can calculate their speed. The detection of vehicle is based on machine learning concept, which further uses the knowledge of transfer learning to detect and classify the vehicle. For tracking the speed, it uses the concept of optical flow calculation.

References

1. LeCun, Y., Bengio, Y., & Hinton, G. (2015). Deep learning. *Nature, 521*(7553), 436.
2. Ren, S., He, K., Girshick, R., & Sun, J. (2015). Faster R-CNN: Towards real-time object detection with region proposal networks. In *Advances in neural information processing Systems* (pp. 91–99). Montré.
3. Viola, P., & Jones, M. J. (2004). Robust real-time face detection. *International Journal of Computer Vision, 57*(2), 137–154.
4. Felzenszwalb, P. F., Girshick, R. B., & McAllester, D. (2010). Cascade object detection with deformable part models. In *2010 IEEE Conference on Computer Vision and Pattern Recognition (CVPR)* (pp. 2241–2248). IEEE.
5. Liu, W., Anguelov, D., Erhan, D., Szegedy, C., Reed, S., Fu, C. Y., & Berg, A. C. (2016) SSD: Single shot multibox detector. In *European Conference on Computer Vision* (pp. 21–37). Amsterdam.
6. Felzenszwalb, P. F., Girshick, R. B., McAllester, D., & Ramanan, D. (2010). Object detection with discriminatively trained part-based models. *IEEE Transactions on Pattern Analysis and Machine Intelligence, 32*(9), 1627–1645.
7. Girshick, R., Donahue, J., Darrell, T., & Malik, J. (2014). Rich feature hierarchies for accurate object detection and semantic segmentation. In *Proceedings of the IEEE Conference on Computer Vision and Pattern Recognition* (pp. 580–587).
8. Pan, S. J., & Yang, Q. (2010). A survey on transfer learning. *IEEE Transactions on Knowledge and Data Engineering, 22*(10), 1345–1359.